Mineralogy of Quartz and Silica Minerals

Mineralogy of Quartz and Silica Minerals

Special Issue Editor

Jens Götze

MDPI • Basel • Beijing • Wuhan • Barcelona • Belgrade

MDPI

Special Issue Editor
Jens Götze
Institut für Mineralogie, TU Bergakademie Freiberg
Germany

Editorial Office
MDPI
St. Alban-Anlage 66
4052 Basel, Switzerland

This is a reprint of articles from the Special Issue published online in the open access journal *Minerals* (ISSN 2075-163X) from 2017 to 2018 (available at: https://www.mdpi.com/journal/minerals/special_issues/Silica_Minerals)

For citation purposes, cite each article independently as indicated on the article page online and as indicated below:

LastName, A.A.; LastName, B.B.; LastName, C.C. Article Title. *Journal Name* **Year**, *Article Number, Page Range.*

ISBN 978-3-03897-348-5 (Pbk)
ISBN 978-3-03897-349-2 (PDF)

Cover image courtesy of Jens Götze.

Contents

About the Special Issue Editor

Jens Götze is professor of Technical Mineralogy at the Technical University Bergakademie Freiberg (Germany). He graduated in Mineralogy and Geochemistry in 1985 and received his Doctorate in Mineralogy from the Bergakademie Freiberg in 1989. Since 2005, he is Professor at the Institute of Mineralogy of the Technical University in Freiberg. His activities in research and education cover the field of applied mineralogy, in particular non-metallic raw materials, crystal growth, and analytics. His research is focused on the typomorphic properties and industrial use of quartz and silica minerals and on cathodoluminescence microscopy and spectroscopy. From 2000 to 2001, he was the president of the international Society of Luminescence Microscopy and Spectroscopy (SLMS). He was guest lecturer at universities in Austria, Brazil, Norway, Italy, Poland, Spain, and Russia and was invited as a plenary speaker to various international conferences. Since 2015, he works in the Springer Advisory Board in Earth Sciences.

Preface to "Mineralogy of Quartz and Silica Minerals"

The various modifications of silica, especially quartz, play a central role in the composition of geological materials. In addition, quartz is widely used as raw material in numerous industrial fields. Therefore, the knowledge of the specific properties of SiO_2 rocks and minerals is indispensable for the understanding and reconstruction of geological processes, as well as for specific technical applications.

This Special Issue contains 13 contributions dealing with the formation, mineralogy, and geochemistry of quartz and other silica minerals. These papers discuss aspects of the formation of quartz deposits and problems of processing, the analysis of high-purity quartz, as well as the the specifics of SiO_2 modifications and varieties (e.g., opal, chalcedony, agate, quartz, amethyst). Following the fundamental idea of the Special Issue series of Minerals, this book combines results from theoretical, analytical, and industrial studies, with the aim of exchanging information and discussing recent developments in the research on SiO_2 materials.

Jens Götze
Special Issue Editor

minerals

MDPI

Editorial

Editorial for Special Issue "Mineralogy of Quartz and Silica Minerals"

Jens Götze

Institute of Mineralogy, TU Bergakademie Freiberg, 09599 Freiberg, Germany;
jens.goetze@mineral.tu-freiberg.de

Received: 12 October 2018; Accepted: 17 October 2018; Published: 19 October 2018

Quartz and other silica minerals make up 12.6 wt % of the Earth's crust and belong to the most frequent rock-forming constituents. Despite the simple chemical formula SiO_2, at least 14 crystalline and amorphous silica modifications with varying crystallographic order exist which play an important role in geological as well as industrial processes [1]. In respect to the occurrence in nature and the amount of technical material used, quartz (trigonal alpha-quartz) is the most important silica phase.

The chemical and physical properties of quartz and the other silica phases are determined by their real structure. The type and frequency of lattice defects are influenced by the thermodynamic conditions during mineralization or secondary processes of alteration. Accordingly, the real structure is a fingerprint of the genetic conditions of formation. The knowledge of the interrelations between genesis and specific properties can, therefore, be used both for the reconstruction of geological processes and for specific technical applications [2].

Although quartz and silica research has a long history, the questions concerning the chemical and physical properties are far from being answered completely. However, modern analytical methods brought a lot of new mineralogical and geochemical data concerning the origin of quartz and the other silica phases. In particular, those methods which allow investigations with high resolution, low detection limits or spatially resolved analyses (e.g., electron microscopy, trace element analysis, electron paramagnetic resonance spectroscopy, infrared spectroscopy or cathodoluminescence) are useful for the extension of the state of the art. The contributions of this special issue of *Minerals* clearly demonstrate that complex investigations by a combination of different advanced methods will have the greatest potential for the successful completion of upcoming geological or industrial problems.

The papers by Götze et al. [3], Lin et al. [4], Pei et al. [5] and Guatame-Garcia and Buxton [6] impressively demonstrate how advanced analytical methods are being used for the characterization of mineral properties and how this knowledge can be used for processing. The material that was investigated in these studies includes high-purity quartz from metamorphic host rocks, hydrothermal vein quartz, as well as diatomite. The results of the investigations emphasize that a thorough mineralogical and geochemical characterization of different kinds of SiO_2 raw materials is indispensable for the successful use of high-quality materials in the industry.

Another topic covers several complex investigations concerning the reconstruction of the formation conditions of different types of SiO_2 mineralization. In their contribution about silica colloid ordering in a dynamic sedimentary environment, Liesegang and Milke [7] show how ordered arrays of amorphous silica spheres form in deeply weathered sediments. The formation of such ordered particle arrays not only takes place in inorganic photonic structures in the geosphere, however is also important for nanotechnology and biological systems. Other formation environments for quartz and silica minerals are discussed by Müller et al. [8], Voudouris et al. [9] and Trümper et al. [10]. Spectacular quartz crystals of various colours and habits were reported from a hydrothermal breccia of Berglia-Glassberget, Norway [8] and also in volcanic rocks in different occurrences of Greece [9]. Both papers try to reconstruct the specific conditions leading to the formation of the quartz crystals based on thorough mineralogical and geochemical analyses (trace elements, fluid inclusions, oxygen isotopes).

Trümper et al. [10] studied fossil wood from five late Paleozoic settings using field observations, taphonomic determinations as well as mineralogical analyses to reconstruct the silicification process. The results indicate that silicification is sometimes a monophase, however it is often a multiphase process under varying physico-chemical conditions.

Other studies in the present book show how the knowledge about the processes of SiO_2 mineralization can also help to decipher the origin of gold deposits and provide information about the mineralization conditions. Taksavasu el el. [11] found implications for the formation of bonanza veins in low-sulfidation epithermal deposits from the textural characteristics of non-crystalline silica in sinters and quartz veins. Wertich et al. [12] developed a multi-stage model of gold-bearing hydrothermal quartz veins at the Mokrsko gold deposit (Czech Republic) based on cathodoluminescence and trace element data. Based on the results of such studies, the prognosis of potential gold deposits could be significantly improved.

The importance of methodological studies for the further development of analytical methods is illustrated by two contributions providing new data concerning thermogravimetry-mass-spectrometry [13] and cathodoluminescence microscopy and spectroscopy [14], respectively. Richter-Feig et al. [13] studied volatile components in micro- and macro-crystalline quartz of agates (chalcedony) and used these data for the reconstruction of the kind and composition of mineralizing fluids of these spectacular forms of silica. In the study by Sittner and Götze [14], defects and micro-textures of quartz in different metamorphic rocks from the Kaoko belt (Namibia) representing metamorphic zones from greenschist to granulite facies were analyzed by cathodoluminescence. The results illustrate that the cathodoluminescence properties of quartz can also be used to get information about the conditions of mineral formation.

The book content is completed by a review article of Kayama and co-authors [15] about Lunar and Martian silica phases. Although silica polymorphs, such as quartz, tridymite, cristobalite, coesite, stishovite, seifertite, high-pressure silica glass, moganite or opal, are relatively rare in extraterrestrial materials, the occurrence of these phases can provide valuable information about different pressure and temperature conditions. Thus, igneous processes (e.g., crystallization temperature and cooling rate), shock metamorphism (e.g., shock pressure and temperature) or hydrothermal fluid activity can be reconstructed based on the presence and properties of specific silica phases, implying the importance of SiO_2 minerals in planetary science.

All of these examples show that the knowledge of the relationships between the genetic conditions of the formation of silica phases and the development of specific properties is an important factor in geological research as well as in many technical applications. Therefore, it is my hope that this special issue is a valuable and substantive resource for anyone who is interested in studies of quartz and silica minerals and that it will serve as a basis for further research.

Conflicts of Interest: The author declares no conflict of interest.

References

1. Götze, J. Chemistry, textures and physical properties of quartz—Geological interpretation and technical applications. *Mineral. Mag.* **2009**, *73*, 645–671. [CrossRef]
2. Götze, J.; Möckel, R. *Quartz: Deposits, Mineralogy and Analytics*; Springer Geology: Heidelberg, Germany; New York, NY, USA; Dordrecht, The Netherlands; London, UK, 2012; p. 360. ISBN 978-3-642-22160-6.
3. Götze, J.; Pan, Y.; Müller, A.; Kotova, E.; Cerin, D. Trace Element Compositions and Defect Structures of High-Purity Quartz from the Southern Ural Region, Russia. *Minerals* **2017**, *7*, 189. [CrossRef]
4. Lin, M.; Pei, Z.; Lei, S. Mineralogy and Processing of hydrothermal vein quartz from Hengche, Hubei Province (China). *Minerals* **2017**, *7*, 161. [CrossRef]
5. Pei, Z.; Lin, M.; Liu, Y.; Lei, S. Dissolution Behaviors of Trace Muscovite during Pressure Leaching of Hydrothermal Vein Quartz Using H_2SO_4 and NH_4Cl as Leaching Agents. *Minerals* **2018**, *8*, 60. [CrossRef]
6. Guatame-Garcia, A.; Buxton, M. The Use of Infrared Spectroscopy to Determine the Quality of Carbonate-Rich Diatomite Ores. *Minerals* **2018**, *8*, 120. [CrossRef]

7. Liesegang, M.; Milke, R. Silica colloid ordering in a dynamic sedimentary environment. *Minerals* **2018**, *8*, 12. [CrossRef]

8. Müller, A.; Ganerød, M.; Spjelkavik, S.O.S.S.; Selbekk, R. The hydrothermal breccia of Berglia-Glassberget, Nord Trøndelag, Norway: Snapshot of a Triassic earthquake. *Minerals* **2018**, *8*, 175. [CrossRef]

9. Voudouris, P.; Melfos, V.; Mavrogonatos, C.; Tarantola, A.; Götze, J.; Alfieris, D.; Maneta, V.; Psimis, I. Amethyst occurrences in Tertiary volcanic rocks of Greece: Mineralogical, fluid inclusion and oxygen isotope constraints on their genesis. *Minerals* **2018**, *8*, 324. [CrossRef]

10. Trümper, S.; Rößler, R.; Götze, J. Deciphering silicification pathways of fossil forests: Case studies from the late Paleozoic of Central Europe. *Minerals* **2018**, *8*, 432. [CrossRef]

11. Taksavasu, T.; Monecke, T.; Reynolds, T.J. Textural characteristics of non-crystalline silica in sinters and quartz veins: Implications for the formation of bonanza veins in low-sulfidation epithermal deposits. *Minerals* **2018**, *8*, 331. [CrossRef]

12. Wertich, V.; Leichmann, J.; Dosbaba, M.; Götze, J. Multi-stage evolution of gold-bearing hydrothermal quartz veins at the Mokrsko gold deposit (Czech Republic) based on cathodoluminescence, spectroscopic, and trace elements analyses. *Minerals* **2018**, *8*, 335. [CrossRef]

13. Richter-Feig, J.; Möckel, R.; Götze, J.; Heide, G. Investigation of Fluids in Macrocrystalline and Microcrystalline Quartz in Agate Using Thermogravimetry-Mass-Spectrometry. *Minerals* **2018**, *8*, 72. [CrossRef]

14. Sittner, J.; Götze, J. Cathodoluminescence (CL) characteristics of quartz from different metamorphic rocks within the Kaoko belt (Namibia). *Minerals* **2018**, *8*, 190. [CrossRef]

15. Kayama, M.; Nagaoka, H.; Niihara, T. Lunar and Martian Silica. *Minerals* **2018**, *8*, 267. [CrossRef]

minerals

MDPI

Article

Trace Element Compositions and Defect Structures of High-Purity Quartz from the Southern Ural Region, Russia

Jens Götze [1,*], Yuanming Pan [2], Axel Müller [3,4], Elena L. Kotova [5] and Daniele Cerin [2]

[1] Institute of Mineralogy, TU Bergakademie Freiberg, Brennhausgasse 14, 09596 Freiberg, Germany
[2] Department of Geological Sciences, University of Saskatchewan, Saskatoon, SK S7N5E2, Canada; yuanming.pan@usask.ca (Y.P.); dac176@mail.usask.ca (D.C.)
[3] Naturhistorisk Museum, Universitet i Oslo, P.O. Box 1172, Blindern, 0318 Oslo, Norway; a.b.muller@nhm.uio.no
[4] Natural History Museum, Cromwell Road, London SW7 5BD, UK
[5] Mining State University St. Petersburg, 21st Line, St. Petersburg 199106, Russia; museum_spmi@spmi.ru
* Correspondence: jens.goetze@mineral.tu-freiberg.de; Tel.: +49-3731-392638

Received: 14 September 2017; Accepted: 5 October 2017; Published: 11 October 2017

Abstract: Quartz samples of different origin from 10 localities in the Southern Ural region, Russia have been investigated to characterize their trace element compositions and defect structures. The analytical combination of cathodoluminescence (CL) microscopy and spectroscopy, electron paramagnetic resonance (EPR) spectroscopy, and trace-element analysis by inductively coupled plasma mass spectrometry (ICP-MS) revealed that almost all investigated quartz samples showed very low concentrations of trace elements (cumulative concentrations of <50 ppm with <30 ppm Al and <10 ppm Ti) and low abundances of paramagnetic defects, defining them economically as "high-purity" quartz (HPQ) suitable for high-tech applications. EPR and CL data confirmed the low abundances of substitutional Ti and Fe, and showed Al to be the only significant trace element structurally bound in the investigated quartz samples. CL microscopy revealed a heterogeneous distribution of luminescence centres (i.e., luminescence active trace elements such as Al) as well as features of deformation and recrystallization. It is suggested that healing of defects due to deformation-related recrystallization and reorganization processes of the quartz lattice during retrograde metamorphism resulted in low concentrations of CL activator and other trace elements or vacancies, and thus are the main driving processes for the formation of HPQ deposits in the investigated area.

Keywords: quartz; cathodoluminescence; electron paramagnetic resonance; trace elements

1. Introduction

Quartz and other silica minerals are some of the most important rock-forming minerals of the Earth's crust, and are important industrial raw materials. Owing to their abundance and physical and chemical properties, natural silica raw materials have a wide range of industrial and technological applications [1,2].

In particular, high-purity quartz (HPQ), with less than 50 ppm of contaminating trace elements [3,4], is of high economic value, resulting in prices up to 20 times higher than those of low-quality ("common") silica raw materials [5,6]. High-purity quartz is of strategic importance for the high-tech industry, because it is a critical material for the manufacture of crucibles used for single crystal growth of silicon metal (needed for solar panel and micro-chip production), high-temperature lamp tubing, telecommunications, optics, and semiconductor materials. Because of the increasing demand for HPQ there are increasing exploration activities underway to search for potential deposits worldwide.

The specific quality requirements of the quartz material are challenging with respect to analytics due to the very low concentrations (from 0.1 to 50 ppm) of impurity trace elements [2]. Natural silica materials, in particular quartz, are characterized by specific properties—including lattice defects, abundance of lattice-"foreign" trace elements, degree of recrystallization, etc.—which are the result of the regional geological history and the related specific conditions of formation. Therefore, the knowledge of the interrelations between genetic conditions and quartz properties can be used both for the reconstruction of geological processes and for the prediction of deposit location and quality as well as for specific industrial applications [7]. In particular, information about the number and types of defects is important for the processing of the raw materials and the potential technical applications.

The present study presents results of a comprehensive mineralogical and geochemical study on potential HPQ deposits of different genetic types from 10 sites in the Southern Ural region, Russia. The investigation aims to obtain detailed information about the type and abundance of lattice defects and contaminating trace elements of these quartz materials in order to determine the critical processes and conditions responsible for the formation of HPQ deposits. This aim is achieved by a combination of multiple high-sensitivity analytical techniques ranging from cathodoluminescence (CL) microscopy and spectroscopy to electron paramagnetic resonance (EPR) spectroscopy and trace-element analysis by inductively coupled plasma mass spectrometry (ICP-MS) and laser ablation ICP-MS.

2. Materials and Methods

2.1. Geological Background and Sample Material

Potential high-purity quartz from 10 different localities of the Southern Ural region northwest of Chelyabinsk (Russia) was investigated (Figure 1, Table 1). The sample material includes quartz from a pegmatite, hydrothermal quartz veins, tectonically deformed and partially recrystallized hydrothermal quartz, and quartz from two quartzite occurrences.

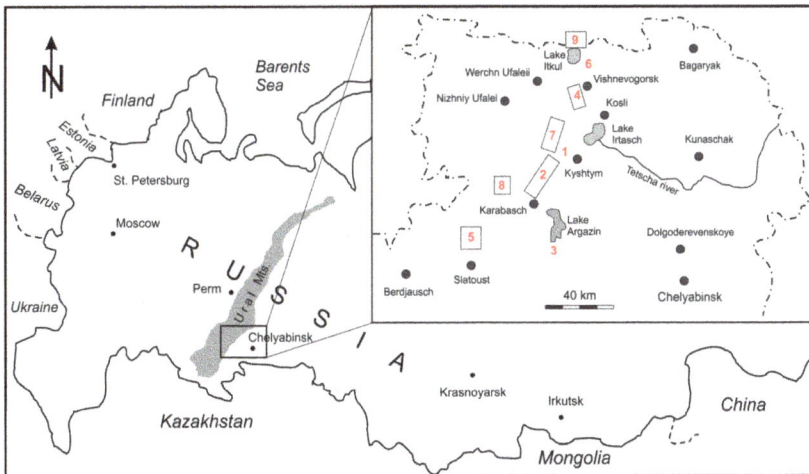

Figure 1. Topographic sketch showing the investigated quartz occurrences of the Southern Ural region northwest of Chelyabinsk (Russia); 1—Berkutinskaya (Berkut), 2—Kyshtym (sample Ky-175), 3—Argazinskoe (Arg), 4—Vjasovka (Vja), 5—Itkulskoe (Itkul), 6—Bolotnaya (Bol), 7—Kuznechikhinsk (Ku-414, Ku-2136), 8—Yurma ridge (Yur), and 9—Taganai ridge (MT-09). Numbers relate to the locations and quartz types in Table 1.

Table 1. Investigated quartz samples from the Southern Ural region (Russia).

Location	Type	Sample
1 Berkutinskaya	Pegmatite	Berkut
2 Kyshtym	Hydrothermal vein	Ky-175
3 Argazinskoe	Hydrothermal vein	Arg
4 Vjazovka	Hydrothermal vein	Vja
5 Itkulskoe	Hydrothermal vein	Itkul
6 Bolotnaya	Hydrothermal vein	Bol
7 Kuznechikhinsk	Hydrothermal vein, metamorphic overprint	Ku-414
7 Kuznechikhinsk	Hydrothermal vein, metamorphic overprint	Ku-2136
8 Yurma ridge	Quartzite	Yur
9 Taganai ridge	Quartzite	MT-09

All quartz bodies occur within the Ufalei metamorphic (gneiss–migmatite) complex, consisting of two tectonometamorphic units which lay on top of each other. Highly metamorphosed Upper Proterozoic rocks of the Ufalei suite belong to the lower unit and form the core of an anticlinorium. The hydrothermal vein deposits from Kyshtym (sample Ky-175), Kuznechikhinsk (10 km southwest of Kosli, samples Ku-414, Ku-2136), and Argazinskoe (on the southwest coast of Lake Argazin, sample Arg), and the quartzite massifs of Yurma (north of Karabasch on the Yurma ridge) as well as the Taganai ridge (north of Slatoust, sample MT-09) are situated within this unit (Table 1). The pegmatite body of the Berkutinskaya deposit is situated within the Berkut ridge near Kyschtym. The upper unit of the Ufalei metamorphic complex comprises Ordovician and Lower Devonian sequences represented by terrigenous schistose meta-sediments surrounding the anticlinorium core. This unit hosts the hydrothermal vein deposits Vjazovka near Vyschnevogorsk, Bolotnaya and Itkulskoe (north coast of Lake Itkul) (Figure 1, Table 1).

All quartz bodies were formed during long-lasting and multi-stage metamorphism in the Ural region [8]. Two main stages of metamorphism can be distinguished in the investigated area: (1) the Late Cambrian stage, which correlates with the formation of the Ufalei anticlinorium and is subdivided into two sub-stages—an early phase of progressive metamorphism (sillimanite-almandine subfacies of the amphibolite facies), followed by retrograde metamorphism; and (2) the Middle Paleozoic stage, which is characterized by metamorphic transformations of the gneiss core and schist frame (staurolite–quartz subfacies of the amphibolite facies). Silica mobilization and formation of hydrothermal quartz veins and bodies are supposedly related to the retrograde stages of both metamorphic events [9].

2.2. Analytical Methods

Polished thin sections were prepared for microscopic and cathodoluminescence (CL) investigations from all samples listed in Table 1. Polarizing microscopy was carried out using a Zeiss Axio Imager A1m (ZEISS Microscopy, Jena, Germany) to document the grain-size and microstructure of the different quartz types. Micrographs were recorded using a digital camera MRc5 and the software Axiovision (ZEISS Microscopy, Jena, Germany).

CL microscopy and spectroscopy were performed on carbon-coated thin sections using a hot-cathode CL microscope HC1-LM (LUMIC, Bochum, Germany) [10]. The system was operated at 14 kV and 0.2 mA (current density ~10 $\mu A/mm^2$) with a defocused electron beam. Luminescence images were captured during CL operations using a peltier cooled digital video-camera (OLYMPUS DP72, OLYMPUS Deutschland GmbH, Hamburg, Germany). CL spectra in the wavelength range of 370–920 nm were recorded with an Acton Research SP-2356 digital triple-grating spectrograph with a Princeton Spec-10 charge-coupled device (CCD) detector (OLYMPUS Deutschland GmbH, Hamburg, Germany) that was attached to the CL microscope by a silica-glass fibre guide. CL spectra were measured under standardized conditions (wavelength calibration by a Hg-halogen lamp, spot width 30 μm, measuring time 5 s). Irradiation experiments were performed to document the behaviour of the quartz samples under electron irradiation. Samples were irradiated 5 min under constant conditions (14 kV, 0.2 mA) and spectra were measured initially and after every 1 min.

The paramagnetic centres of quartz-powder samples were investigated by EPR spectroscopy using a Bruker EMX spectrometer (Bruker Corporation, Billerica, MA, USA) operated with the X-band microwave frequencies at both room temperature and liquid-nitrogen temperature. Experimental conditions for room-temperature EPR included a microwave frequency of ~9.63 GHz, modulation frequency of 100 kHz, modulation amplitude of 0.1 mT, and microwave powers from 0.02 mW to 20 mW. The spectral resolutions were ~0.146 mT for wide scans 50–6500 mT and 0.024 mT for narrow scans 300–350 mT. All samples after room-temperature EPR measurements were irradiated at room temperature in a ^{60}Co cell for a dose of ~10 kGy. Low-temperature (85 K) EPR measurements were made immediately after gamma-ray irradiation, with similar experimental conditions used for the room-temperature analyses except for a microwave frequency of ~9.39 GHz.

The chemical composition of bulk quartz samples (dissolved powders) was first analysed using solution ICP-MS. The samples (400–500 mg) for ICP-MS analysis were milled to a grain size of <30 μm using a pre-cleaned agate mortar. The powdered sample was digested in a glassy carbon vessel with 5 mL concentrated HF and 3 mL concentrated HNO_3 at 50 °C (35 min). Rhenium solution (1 mL of 100 μg·L^{-1} concentration) was added as an internal standard for the ICP-MS measurements. The analysis was performed using a Perkin Elmer Sciex Elan 5000 quadrupole instrument (Percin Elmer Inc., Baesweiler, Germany) with a cross-flow nebulizer and a rhyton spray chamber. The precision and accuracy of the ICP-MS measurements were evaluated by analysis of the glass sand reference material UNS-SpS. The relative standard deviations for most analytes were below 10%. The ICP-MS results showed procedural limits of detection ranging from 0.22 μg·L^{-1} to 3.1 μg·L^{-1} for Na, Mg, Al, K, Ca and Ba. Elements such as Li, Mn and Sr had procedural limits of detection ranging from 0.02 to 0.04 μg·L^{-1}, whereas these limits range from 1 to 7 μg·L^{-1} for the other elements investigated [11].

In addition, 200-μm polished thick sections of the samples were prepared for laser ablation inductively-coupled plasma mass spectrometry (LA-ICP-MS) to determine trace elements of individual quartz crystals in situ. Concentrations of Li, Be, B, Na, Al, P, K, Ca, Ti, Mn, Fe, Ge, Rb, Sr, Ga, and Sb were analysed with a double-focusing sector field mass spectrometer ELEMENT XR coupled with a NewWave 193-nm excimer laser probe (Thermo Scientific, Waltham, MA, USA) [12]. The laser had a pulse rate of 20 Hz, a speed of 15 μm·s^{-1}, a spot size of 50 μm and energy fluence of 5–7 mJ·cm^{-2} on the sample surface. Raster ablation was applied on an area of approximately 150 μm × 300 μm. The approximate depth of ablation was about 50 μm. The carrier gas for transport of the ablated material to the ICP-MS was He mixed with Ar. External calibration was performed using three silicate glass reference materials produced by the National Institute of Standards and Technology, USA (NIST SRM 610, 612 and 614). In addition, the NIST SRM 1830 soda-lime float glass (0.1% *m/m* Al_2O_3), the certified reference material BAM No.1 amorphous SiO_2 glass from the Federal Institute for Material Research and Testing in Germany, and the Qz-Tu synthetic pure quartz monocrystal provided by Andreas Kronz from the Geowissenschaftliches Zentrum Göttingen (GZG), Germany, were used. Each measurement comprised 15 scans of each isotope, with a measurement time varying from a 0.15 s/scan for K in high resolution to a 0.024 s/scan of, for example, Li in low resolution. A linear regression model, including several measurements of the different reference materials, was used to define the calibration curve for each element. For the calculation of P concentrations, the procedure of Müller et al. [13] was applied. Ten sequential measurements on the Qz-Tu synthetic pure quartz monocrystal were used to estimate the limits of detection (LOD—3σ of 10 measurements; see Table S3). The analytical error ranges within 10% of the absolute concentration of the element.

3. Results

3.1. Cathodoluminescence (CL)

CL imaging revealed heterogeneities, intra-crystal micro-structures and micro-inclusions in the quartz samples. The detected micro-inclusions include carbonate (orange CL) in quartz from Bolotnaya, and feldspar (microcline—bright blue CL, albite—bluish-violet CL) in sample Ku-2136, as well as zircon/monazite (bright radiation halos) and mica (non-luminescent) in sample MT-09 (Figure 2).

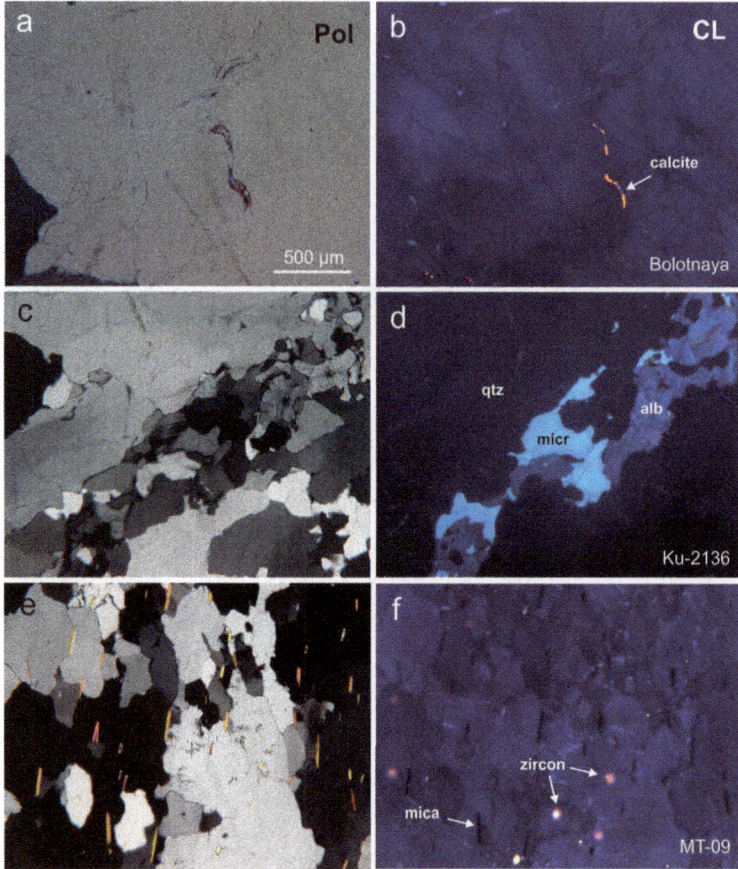

Figure 2. Micrograph pairs in transmitted light (crossed polars—Pol) and cathodoluminescence (CL) showing micro-inclusions of minerals in the quartz samples: (**a,b**) calcite in the hydrothermal quartz from Bolotnaya (Bolot); (**c,d**) microcline (micr) and albite (alb) in metamorphically overprinted hydrothermal quartz from Kuznechikhinsk (Ku-2136; (**e,f**) mica (non-luminescent) and zircon with radiation haloes in the quartzite from the Taganai ridge (MT-09).

The micro-inclusions can be related to the composition of the host rocks (e.g., hydrothermal carbonate veins) or the educt material of the metamorphic rocks. Heterogeneities were detected in quartz grains of almost all samples, which appeared homogeneous under polarized light. Moreover, features of alteration and recrystallization/reorganization as well as trails of fluid migration could be revealed by CL (Figure 3).

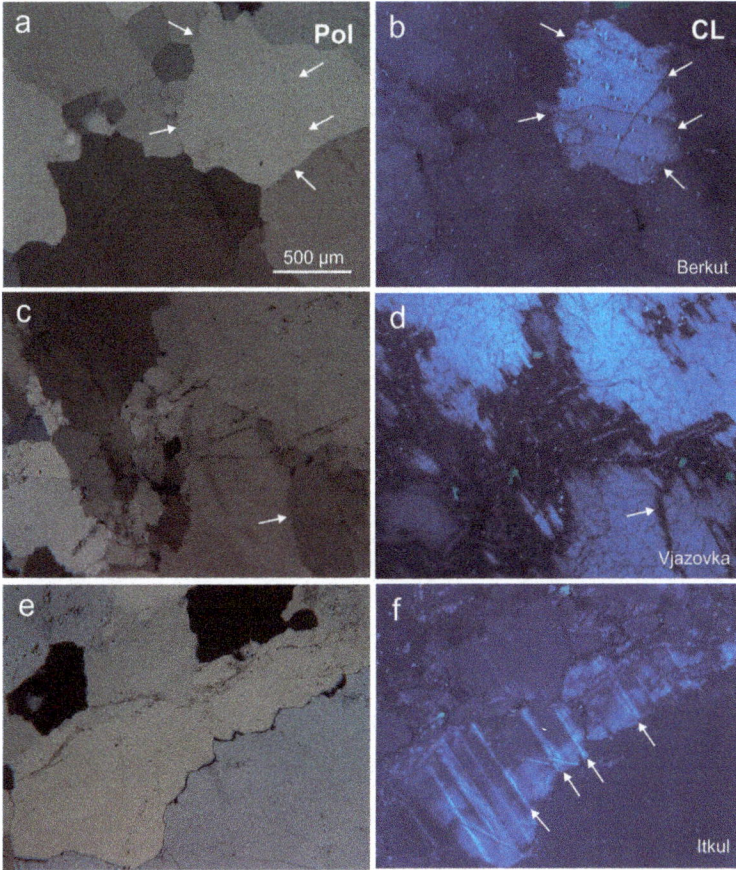

Figure 3. Examples of micro-structural heterogeneities detected by cathodoluminescence (CL) in the investigated quartz samples mostly invisible in transmitted light (crossed polars—Pol): (**a**,**b**) brightly luminescing sub-grain areas in the pegmatite quartz from Berkutinskaya (Berkut); (**c**,**d**) trails of reduced CL intensity (arrow) due to migration of fluids in the hydrothermal quartz from Vjazovka (Vja); (**e**,**f**) dislocation planes in the hydrothermal quartz from Itkulskoe (Itkul).

Pegmatite quartz shows a more or less homogeneous bluish-green CL with a characteristic CL emission band at ca. 500 nm (Figure 4a). Sub-grain areas with strong CL (Figure 3b) have the highest intensities of this 500 nm band indicating the highest amounts of the luminescence related defect(s). The intensity of the 500 nm emission strongly decreases under electron irradiation. The resulting CL spectrum after 5 min of electron bombardment consists of an emission band at 450 nm and a weak band at 650 nm (Figure 4a).

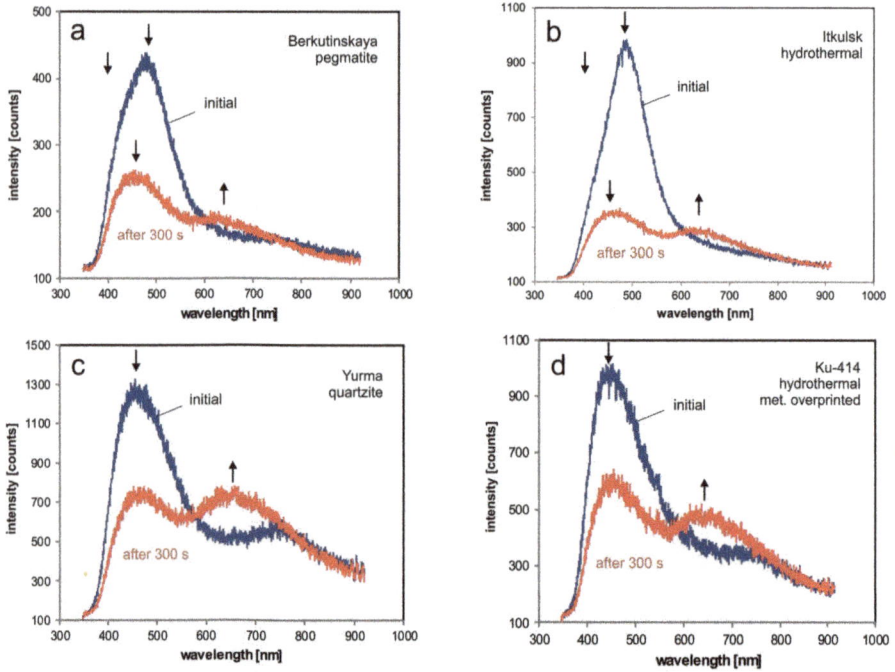

Figure 4. Representative CL emission spectra of quartz of different origin (blue spectrum = initial, red spectrum = after 5 min of electron irradiation): (**a**) pegmatite quartz; (**b**) hydrothermal vein quartz; (**c**) quartzite; (**d**) tectonically deformed and partially re-crystallized hydrothermal quartz.

Quartz from hydrothermal veins exhibits a typical short-lived blue CL. The initial spectra are mainly composed of a strong band at 500 nm and a second emission band at 390 nm, which is only visible as a shoulder (see arrow in Figure 4b) because of the spectral transmissibility of the used equipment (absorption of the spectral UV region due to glass optics). The CL spectrum after 5 min of electron irradiation is dominated by emission bands at 450 nm and 650 nm, respectively. A conspicuous feature of CL imaging in most hydrothermal quartz samples is a heterogeneous pattern (Figure 3d). Brightly luminescent areas alternate with areas of low CL intensity. Interactions of migrating fluids with the host quartz left their traces in trails of reduced CL intensities (Figure 3d). Moreover, features of deformation are visible during initial electron radiation, but disappear during electron bombardment (Figure 3f).

With increasing deformation degree of hydrothermal quartz the luminescence intensity decreases and the visible CL colour becomes more homogeneous. The typical luminescence emission bands for undeformed hydrothermal quartz at 390 nm and 500 nm are missing. The spectra of deformed hydrothermal quartz are dominated by two bands at 450 nm and 650 nm, respectively (Figure 4d).

Quartzite samples appear commonly heterogeneous under CL and may contain micro-inclusions of minerals probably originating from the primary source rocks (sample MT-09; Figure 2e,f). The quartzite from Yurma does not contain any visible mineral inclusions and represents high-purity material. Quartz from the quartzite samples is characterized by a deep blue CL showing emission bands at 450 nm and 650 nm (Figure 4c).

An orientation-dependent behaviour of the CL during electron irradiation was observed in the hydrothermal quartz from Kyshtym (Ky-175, Figure 5). Sub-grains which are cut perpendicular to the c-axis in thin section (dark in polarized light) show a change of the initial blue CL colour into red-violet (increase of the 650 nm band), whereas sub-grains with other orientations show only a decrease of the initial blue CL (decreasing composite blue emission band).

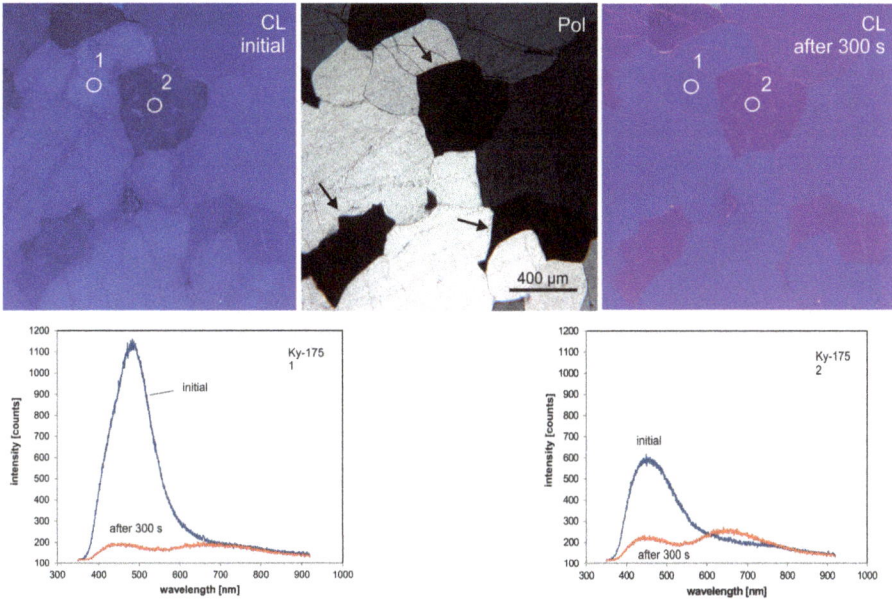

Figure 5. Micrographs in cathodoluminescence (CL) and transmitted light (crossed polars—Pol) of the hydrothermal quartz from Kyshtym (Ky-175); the images and related spectra show the initial CL and the CL after 5 min of electron irradiation. Note the different CL behaviour of quartz sub-grains with varying crystallographic orientation; sub-grains cut perpendicular to the crystallographic c-axis (see arrows) have a lower initial blue CL intensity and develop a reddish CL (650-nm emission band) due to the electron bombardment.

3.2. Electron Paramegnetic Resonance (EPR)

Figures 6 and 7 present the EPR spectra of all quartz samples measured at room temperature. The EPR spectra are essentially featureless, except for the presence of trace and variable amounts of the rhombic Fe^{3+} signal at the effective g value of ~4.38 (Figure 6) [14,15].

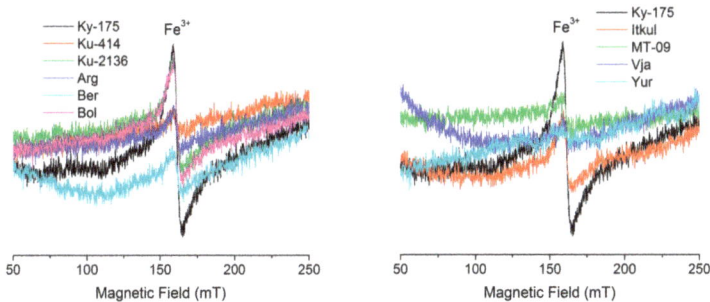

Figure 6. Electron paramagnetic resonance (EPR) spectra of investigated quartz samples taken at room temperature showing the weak rhombic Fe^{3+} signal at $g_{eff} = 4.38$; the intensities are given in arbitrary units.

Obviously, the intensity of this signal is low in all samples (i.e., low signal-to-noise ratios) and is variable between samples. Even without a proper standard, it is apparent that the Fe^{3+} signal is exceedingly small in the samples Berkutinskaya (pegmatite), Argazinskoe, Vjasovka (hydrothermal veins), Ku-414 (metamorphically overprinted hydrothermal quartz), MT-09, and Yurma (quartzite). The Fe^{3+} signal is somewhat more elevated in the samples Ky-175, Itkulskoe, Bolotnaja (hydrothermal vein), and Ku-2136 (metamorphically overprinted hydrothermal quartz).

The spectra in Figure 7 show the resonance signals at the central magnetic field region. Again, all spectra have low signal-to-noise ratios. Indeed, none of the radiation-induced defects at the effective g values of ~2.00 are present [16–19]. Even the common E'_1 is exceedingly rare. It should be mentioned that the silica tubes used as sample containers show a very weak E'_1 signal, which contributes to the weak E'_1 signal in the measured spectra (Figure 7).

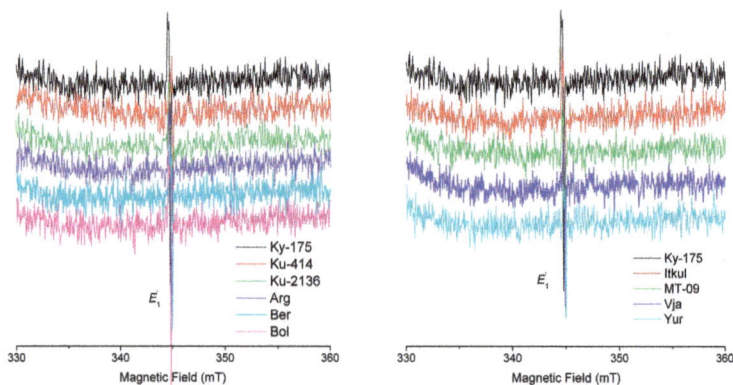

Figure 7. EPR spectra of investigated quartz samples taken at room temperature showing a slight E'_1 but a general absence of other radiation-induced defects at g_{eff} = 2.0.

The EPR spectra of gamma-ray-irradiated samples, measured at 85 K, all show the presence of varying amounts of the well-known $[AlO_4]^0$ centre with the characteristic ^{27}Al hyperfine structure [20] (Figure 8). The paramagnetic $[AlO_4]^0$ centre in quartz has been shown to form from neutral $[AlO_4/M^+]^0$ (M = H, Li, Na, K) precursors (i.e., the monovalent charge compensators migrated away during room-temperature irradiation) [20,21]. Figure 8 compares the individual spectra using the sample Ky-175 as a common reference. The intensities of measured $[AlO_4]^0$ signals of samples Argazinskoje, Bolotnaya and Yurma are notably higher than those of the other samples.

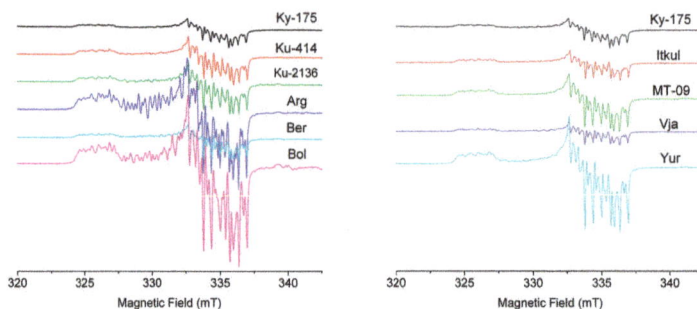

Figure 8. EPR spectra of investigated quartz samples taken at 85 K showing varying quantities of the $[AlO_4]^0$ centre.

EPR spectra measured at 85 K show again the presence of the Fe^{3+} and the E'_1 centre but a general absence of the Ti- and Ge-associated defects with diagnostic ^{47}Ti, ^{49}Ti and ^{73}Ge hyperfine structures [16,19], confirming the results observed at room temperature.

3.3. Trace Elements

"Bulk" solution ICP-MS: Trace-element concentrations of all investigated quartz samples determined by solution ICP-MS are summarized in Tables S1 and S2. The data illustrate that most quartz samples have low concentrations of most trace elements compared with average concentrations in natural quartz [5,7]. Al concentrations are generally below 100 ppm, and those of Fe and Ti are below 10 ppm. An exception is the quartzite from the Taganai ridge (sample MT-09) which shows elevated concentrations of Al, K, Mg, Ti, Fe, Mn, Zr, U, Th, and Hf. Based on microscopic investigations the elevated concentrations of Al, K, Mg, Ti, Fe and Mn are caused by micro-inclusions of mica and those of Zr, U, Th and Hf by zircon (Figure 2e,f). There are similar effects of micro-inclusions of feldspars (albite, microcline—Figure 2d) in quartz from Kuznechikhinsk (samples Ku-2136—K, Na, Al and Fe) as well as carbonate (Figure 2b) in the sample from Bolotnaja (sample Bolot—Ca, Mg and Fe).

Sodium, Sr, K, Mg, and Rb show positive correlations in the different samples (Figure 9). The abundance of these alkali and alkali earth elements can predominantly be related to the presence of fluid inclusions, although mineral micro-inclusions of feldspar and mica (e.g., samples MT-09, Ku-2136) can also influence these correlations. However, in most of the investigated quartz no mineral inclusions were detected.

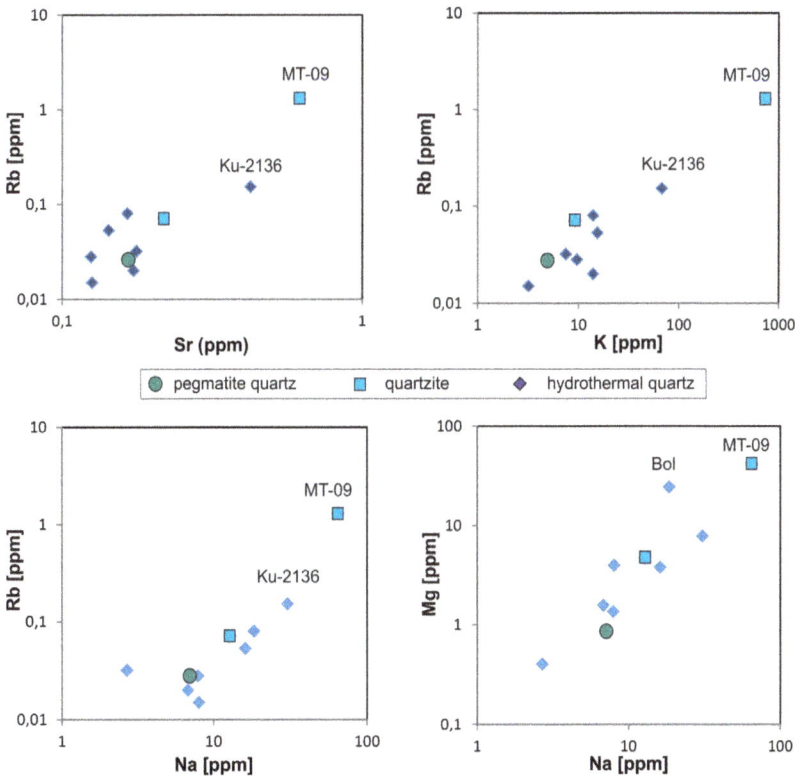

Figure 9. Element ratios of selected alkali and alkali earth elements in the investigated quartz samples (the labelled samples are those with detected mineral inclusions).

The rare earth element (REE) concentrations are listed in Table S2 and illustrated in Figure 10. In some samples (Berkutinskaya, Bolotnaya, Ku-414, and Ku-2136) the concentrations of certain rare earth elements are below the detection limit of the applied ICP-MS analysis and, thus, their chondrite-normalized REE distribution patterns are not shown in Figure 10. The REE patterns of the two quartzite samples are different to the patterns of the hydrothermal quartz. In particular, the quartzite sample MT-09 has high absolute REE concentrations with a typical crustal signature of enriched light REE (LREE) and depleted heavy REE (HREE) (Figure 10a). The REE enrichment is most likely be related to micro-inclusions found in this sample. The quartzite from Yurma has much lower REE contents but shows a similar LREE pattern. However, the HREE are slightly enriched.

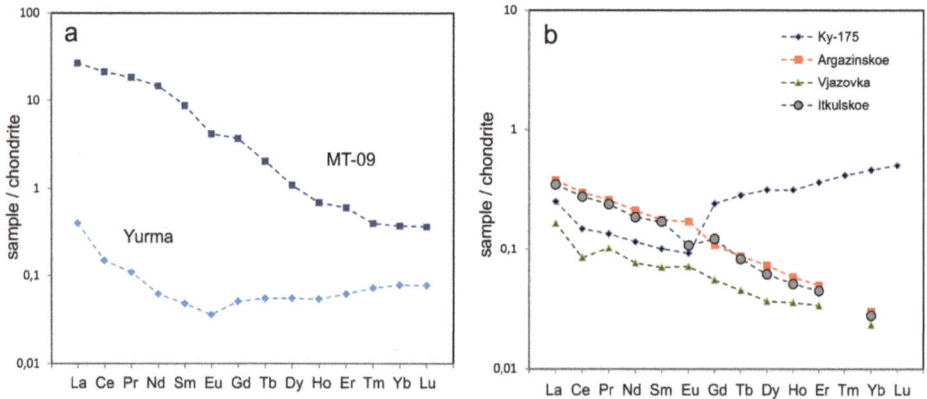

Figure 10. Chondrite-normalized rare earth element (REE) distribution patterns of quartzite (**a**) and hydrothermal vein quartz (**b**) (normalization according to data of Anders and Grevesse [22].

Three of the hydrothermal quartz samples (Argazinskoe, Vjasovska, Itkulskoje) are similar both in absolute REE values and in their chondrite-normalized REE patterns, which correspond to the common crustal distribution (Figure 10b). Only a slight negative Ce anomaly (Vjazovka) and negative (Itkulskoe) or positive (Argazinskoe, Vjazovka) Eu anomalies are detectable. Quartz Ky-175 shows comparable LREE patterns but steeply increasing HREE (Figure 10b), which is most likely caused by zircon or xenotime micro-inclusions.

Summarizing, the "bulk" solution ICP-MS provides concentrations of lattice-bound trace elements plus concentrations of elements bound in mineral and fluid micro-inclusions, which could not be removed during sample preparation.

In situ LA-ICP-MS: Element concentrations determined by in situ LA-ICP-MS are generally lower than those determined with solution ICP-MS, except Ge, Ti and Li (Tables S1 and S3). This is mainly due to the fact that during laser ablation the analyses of visible (>0.5 μm) micro-inclusions can be avoided by choosing clear, inclusion free ablation areas. Thus, concentrations measured by LA-ICP-MS reflect almost the values of lattice-bound trace elements.

Aluminium concentrations analysed by LA-ICP-MS are consistently low (<15 ppm) for all samples except sample Vjazovka (from 21 to 32 ppm). Titanium has more variable concentrations (compared with its general abundance in quartz, e.g., [5]), ranging from <1 ppm to 37 ppm. Lithium concentrations are very low in all samples (<2.3 ppm). Boron and Ge were also found in low concentrations, except in the sample Argazinskoe, which has up to 5.3 ppm B and 1.6 ppm Ge. The Berkutinskaya sample (pegmatite quartz) has high P concentrations, of up to 7.6 ppm.

4. Discussion

4.1. Trace Element Incorporation Into Quartz

The applied combination of CL, EPR and ICP-MS methods permits the identification of types of structural defects and provide insights into their incorporation and/or transformation mechanisms during quartz crystal genesis. In addition, the results provide first indications concerning the very low trace-element concentrations and show the potential of the sampled quartz deposits for high-purity applications in the industry. These data are essential for the prediction of theoretical limits of the processing procedure, assuming that impurities from mineral and fluid inclusions could be minimized or completely removed during refinements.

Figure 11 shows the concentrations of Al and Ti measured by laser ablation ICP-MS (LAICP-MS) and the field defining the quartz economically as high-purity quartz (HPQ) in accordance with Harben [3] and Müller et al. [4]. It can be concluded that all quartz samples except MT-09 plot into the HPQ field. The best chemical quality is of quartz of the hydrothermal veins at Kyshtym (Ky-175) and Itkulskoe.

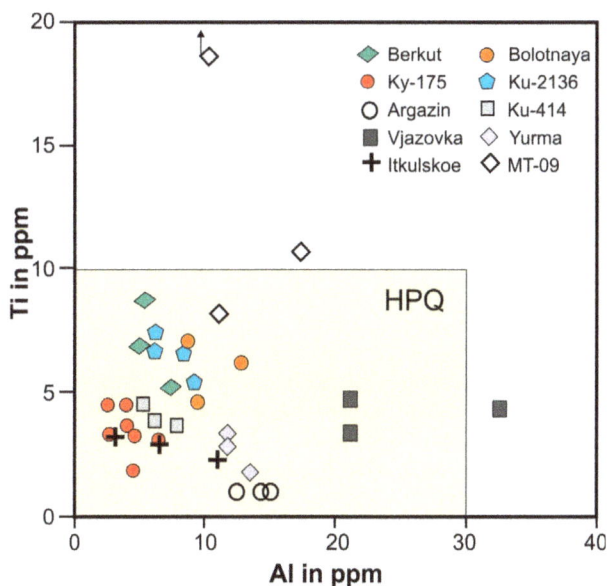

Figure 11. Ti and Al concentrations from laser ablation inductively-coupled plasma mass spectrometry (LA-ICP-MS) analyses of the quartz samples plotted into the field of high-purity quartz (HPQ) according to Harben [3] and Müller et al. [4].

Some elements (Ge, Ti, Li) show more or less constant concentrations when comparing the data from solution ICP-MS analyses of bulk quartz and LA-ICP-MS measurements of selected inclusion-free areas. For these elements it can be concluded that they are preferentially incorporated in the quartz structure. Because of the very low concentrations of these elements no EPR signals were detectable for potential paramagnetic defects.

In contrast to the above mentioned elements, Al concentrations measured by LA-ICP-MS are much lower than those analysed in the bulk quartz by ICP-MS. Although the EPR data show that the paramagnetic $[AlO_4]^0$ centre is the most frequent in the investigated quartz; not all of the Al measured by solution ICP-MS seems to be structurally bound. The paramagnetic $[AlO_4]^0$ centre and its diamagnetic precursors $[AlO_4/M^+]^0$ play an important role as imperfections in quartz. The monovalent ions H^+, Li^+, Na^+ commonly charge-compensate the Al defects. However, Na (together with other alkali and alkali earth elements) seems to be mainly related to fluid inclusions (see Figure 9) and the Li content is low in all samples. Data from literature show that hydrothermal quartz mineralization and pegmatite quartz often contain elevated concentrations of Li [23], and Li is a common charge-balancing cation for Al [24]. The data of the present study, however, indicate that the charge compensator of Al^{3+} substituting Si^{4+} is predominantly H^+. This is in accordance with the results of Miyoshi et al. [25] who detected a preferred incorporation of H^+ as the charge-balancing cation of Al defects in hydrothermal quartz. Müller et al. [26] have already shown that igneous quartz preferentially incorporates H^+ as a charge-balancing cation in lieu of Li, K and Na. Similar findings have been made more recently for pegmatite quartz [27,28].

Although the comparison of EPR and trace-element data reveals a general trend of an increasing number of paramagnetic Fe centres with increasing contents of chemically measured Fe concentrations (with the exception of sample MT-09), there are nevertheless variations in the absolute values. The differences could be explained by two different facts. First, Fe-bearing micro-inclusions are the reason for elevated Fe concentrations in the bulk quartz samples. This is especially true for sample MT-09, where a couple of mineral inclusions (e.g., mica) have been detected by microscopy. Second, Fe is known to occur as both Fe^{2+} and Fe^{3+} in quartz [15,29]. However, Fe^{2+} is not detectable by EPR at X-band frequencies [14]. Therefore, 1:1 correlations of the Fe^{3+} EPR signal with the total Fe contents from chemical analyses are not expected.

4.2. Cathodoluminescence and Structural Defects

Characteristic luminescence colours of quartz with different geological history, and associated spectral CL measurements together with the results from the EPR spectroscopy provided information about the defect structures and incorporated trace elements, primary growth conditions, and processes of secondary overprint. These features not only influence the properties and quality of the potential quartz raw material, but also allow insights into the genetic history of the quartz occurrences.

According to Ramseyer and Mullis [30] and Götze et al. [31] the activation of the visible greenish-blue CL of the pegmatite quartz from Berkutinskaya is associated with cation compensated trace-element centres in the quartz structure. The intensity of the luminescence emission falls off rapidly after electron bombardment within 30–60 s, which can be related to ionization-enhanced diffusion of luminescence centres as was shown by Ramseyer and Mullis [30] with electro-diffusion experiments. However, the absolute concentration of Al in this sample is low (<10 ppm). This is confirmed by the low abundance of paramagnetic trace-element defects of Al, Ti and Ge. Also, the contents of Li and other cations are far below the common abundance of such elements in pegmatite quartz [24,31]. The heterogeneous CL textures of grains indicate a heterogeneous distribution of trace elements responsible for the CL (Figure 3b). Only areas with higher intensity (brighter CL) of the transient CL point to elevated contents of responsible trace elements (e.g., Al, Li). Moreover, a possible role of H^+ for the 500-nm luminescence signal must be taken into consideration.

All hydrothermal quartz samples show a characteristic short-lived blue CL, which is elicited by two main emission bands at ~390 nm and 500 nm. According to Ramseyer and Mullis [30] and Perny et al. [32] these luminescence bands are activated by cation-compensated $[AlO_4/M^+]$-centres in the quartz structure. The strong decrease of these emission bands during electron irradiation is due to the interaction of the cation-balanced Al centres with the electron beam and indicates elevated trace element contents in areas with high CL intensity [30,33] (Figure 4b). Such heterogeneous pattern is depicted in the quartz from Vjazovka (compare Figure 3d), which has an elevated concentration of Al compared to the other samples but shows no remarkable EPR signal for the $[AlO_4]^0$ centre.

The relatively low contents of Al and charge-balancing cations in some of the investigated hydrothermal quartz suggest that either the $[AlO_4/M^+]$ defects are effective CL activators (i.e., even low abundance of $[AlO_4/M^+]$ can activate the characteristic CL), or that other activators exist for this specific luminescence. Gorton et al. [34] found the typical short-lived blue CL in synthetic quartz with extremely low impurity concentrations and concluded that probably other activators than [AlO_4/M^+]-defects might be additionally responsible for the blue CL. However, the EPR measurements in the present study did not provide any indication for other luminescence-active paramagnetic defects. EPR measurements revealed an almost complete absence of intrinsic lattice defects associated with oxygen or silicon vacancies (e.g., the E' centre, O_2^{3-} centre), and even the abundance of paramagnetic trace-element defects of Al, Ti, Ge is in general low.

Areas of dull CL in hydrothermal quartz can be the result of post-crystallization overprint. For instance, the interaction of migrating fluids with the host quartz left their traces in trails of reduced CL intensity (Figure 3d). Van den Kerkhof and Hein [35] explained this phenomenon with the loss of trace elements during recrystallization of the interacting area with the fluids. Decreased luminescence intensities were also detected along grain boundaries. This can probably be related to the opening of fluid inclusions and the migration of inherited fluids (especially along grain boundaries).

Indications of secondary overprint and deformation are also visible in features of apparent dislocation planes (Figure 3f). Supplementary electron back scattered diffraction (EBSD) measurements provided evidence of Dauphine twinning, which is preferentially initiated by mechanical deformation. In addition, these analyses revealed low-angle tilting of sub-grains due to deformation (Figure 12).

Figure 12. SEM forescattered image showing low-angle tilting of sub-grains due to deformation in the hydrothermal quartz from Kyshtym (Ky-175).

The quartz samples from Kuznechikhinsk (Ku-414, Ku-2136) illustrate that deformation of hydrothermal quartz resulted in a general decrease of the luminescence intensity and a homogenization of the CL pattern. This fact can probably be related to healing of defects and reduction of trace-element impurities due to recrystallization and reorganization processes of the quartz lattice. Measured trace-element contents are low (e.g., Al < 10 ppm, Ti < 2 ppm, Li ~1 ppm) and even the detected intensities of paramagnetic defects of Al and Fe as well as vacancy-related defects are very low.

The visible CL colour of both metamorphically overprinted hydrothermal quartz and quartzite is dark blue, and the spectra are dominated by two bands at 450 and 650 nm, respectively (Figure 4c,d). The typical luminescence emission bands at 390 nm and 500 nm that were detected in the hydrothermal quartz samples are missing. The emission observed at ~450 nm is associated with O-deficiency centres (ODC) in quartz [36,37]. It can be related to the recombination of the self-trapped exciton (STE), which involves an irradiation-induced oxygen Frenkel pair consisting of an oxygen vacancy and a peroxy linkage [38].

The 650-nm emission is attributed to non-bridging oxygen-hole centres (NBOHC), which are formed from different precursor defects [39,40]. A number of different precursors of this NBOHC have been proposed, such as H- or Na-impurities, peroxy linkages (O-rich samples), or strained Si-O bonds [38]. The time dependent spectra reveal an increase of the 650 nm band under the electron beam, which points to a conversion of precursor centres into the NBOHC. Considering the mechanical deformation of the samples, strained Si–O bonds may result in bond breaking during electron bombardment and thus, in the formation of NBOHC and the related increase of the 650-nm emission band.

Another conspicuous CL feature is the orientation-dependent CL behaviour that was observed in the hydrothermal quartz from Kyshtym (sample Ky-175; Figure 5). The initial blue CL colour of quartz sub-grains cut perpendicular to the crystallographic c-axis turned into a red-violet CL colour during electron irradiation (due to increasing 650-nm emission intensity). Surprisingly, sub-grains with other orientations do not show the same effect and only the intensity of the initial blue CL decreases (strong decline of the 450-nm emission band).

It cannot be ruled out that the crystallographic orientation of the crystals influences the interaction with the electron beam due to the relatively open structural channels along the c-axis of the quartz lattice. An alignment of the c-axis along the microscope axis would than provide a more intensive interaction with the electron beam, i.e., an increased conversion of precursor centres of the NBOHC associated with a stronger increase of the 650-nm emission. Another explanation is probably the influence of polarization effects of the luminescence light. Walderhaug and Rykkje [41] compared c-axis orientations of quartz grains and related CL colours and found clear indications that the observed colour variations are a function of crystallographic orientation. They concluded that this might be due to selective absorption of light along different crystallographic directions, a phenomenon well known from spectroscopy. Sippel [42] reported similar effects of the quartz CL when observing the luminescence under a nicol prism. The quartz CL showed strong polarization effects with variations of the CL colour from red to blue through a rotating polarizer. He found that the blue oscillators are aligned with the c-axis, whereas the red emission is unpolarized. Such a behaviour might explain the observed colour variations in the investigated quartz sample Ky-175. The effect may be interpreted by anisotropic luminescing centres that have oriented the oscillation in a strong internal crystal field.

4.3. Genetic Implications for the Formation of High-Purity Quartz

The investigated quartz localities in the northern Ural were all formed during multiple metamorphism and related deformation. According to Kelman [9], silica mobilization and formation of quartz veins and bodies can especially be related to the retrograde stage of the two metamorphic events. Although quartz from the different locations may originate from varying processes, all types show in general low concentrations of trace elements and low abundance of other point defects. Elevated trace-element contents in several samples can unambiguously be related to micro-inclusions (e.g., samples MT-09, Ku-2136).

The low abundance of structural point defects such as E'_1 centres in the quartz samples investigated requires a combination of the following two conditions: (1) low intrinsic structural imperfection (i.e., low oxygen and silicon vacancies); and (2) weak natural irradiation. This fact implies that there was sufficient time for crystallization under more or less equilibrium conditions during the formation of these quartz occurrences. On one hand, the metamorphic processes caused the mobilization of silica-rich fluids (hydrothermal veins) and melts (pegmatite), and the precipitation as high-purity quartz. On the other hand, recrystallization, healing of defects or disintegration of fluid inclusions under metamorphic conditions resulted in a natural "purification" of the pre-existing quartz crystals and a lowering of trace-element contents. Larson et al. [43] and Müller et al. [5] found that Li concentrations in quartz can be reduced during recrystallization, and Al contents may both decrease or increase depending on the specific environment. Germanium and Ti seem to be more or less unchanged.

The formation of such metamorphic mobilisates appears to be an appropriate process for the formation of HPQ deposits. Compared with data from the literature, the analysed absolute trace-element concentrations as well as chondrite-normalized REE distribution patterns are similar to those of metamorphic and hydrothermal quartz from other regions [44]. Even the pegmatite quartz from Berkutinskoe (Berkut) follows this trend, with atypical very low concentrations of characteristic elements for pegmatite quartz such as Al, Ge or Li.

The REE patterns mostly show a crustal signature indicating a source of the silica from mobilization processes of crustal material. Only the hydrothermal quartz from Kyshtym (Ky-175) has elevated concentrations of HREE. This unusual pattern is most likely caused by zircon or xenotime micro-inclusions.

In conclusion, the results of the present study show a strong link between metamorphic processes and formation of quartz bodies of high purity. If the quartz bodies are large enough and the abundance of pre- and syn-genetic micro-inclusions is low, they may represent potential HPQ deposits.

5. Conclusions

The investigation of quartz samples of different genetic types (hydrothermal, pegmatite, and quartzite) from 10 localities in the Southern Ural region (Russia) using an analytical combination of CL microscopy and spectroscopy, EPR spectroscopy, and bulk as well as spatially resolved trace-element analysis, proved to characterize the type and abundance of trace elements and structural defects in quartz, and reconstruct processes responsible for their formation. In situ LA-ICP-MS analysis showed cumulative trace-element concentrations of <50 ppm with <30 ppm Al and <10 ppm Ti for almost all samples, defining the quartz economically as high-purity quartz. EPR data confirmed the low abundances of substitutional Ti and Fe and showed Al to be the only significant trace element structurally bound in the investigated quartz samples. Elevated concentrations of selected trace elements that were analysed with "bulk" solution ICP-MS could be related to mineral (Ti, Al, Fe, Mn, Mg, K, Zr, U, Th, Hf) and fluid (Na, K, Rb, Ca, Mg, Sr) micro-inclusions.

CL imaging reveals a heterogeneous distribution of luminescence centres, i.e., varying amounts of activator trace elements in the quartz grains. Hydrothermal and pegmatite quartz exhibit a distinct short-lived greenish-blue CL with main CL emission bands at ~390 and 500 nm, which can be attributed to $[AlO_4/M^+]$ (M = H, Li, Na, K) defects. Low contents of Li, Na and K indicate that H^+ is the main charge-balancing cation of Al^{3+}-related defects. Accordingly, areas with bright greenish-blue CL could be related to higher abundances of the $[AlO_4/M^+]$ defects. Moreover, an orientation-dependent behaviour of the CL during electron irradiation was observed that can be related to anisotropy effects of the crystal lattice, which results in an orientation-dependent interaction with the electron beam and the selective absorption of emitted light along different crystallographic directions.

Certain features of alteration, recrystallization/reorganization as well as trails of fluid migration revealed by CL in the quartz samples can be related to metamorphism-related deformation. All quartz bodies were formed during retrograde stages of complex metamorphic processes, resulting in silica mobilization and formation of quartz veins and lenses. Deformation of hydrothermal quartz resulted in a general decrease of the luminescence intensity and a homogenization of the CL pattern. This feature can probably be related to healing of defects and reduction of trace-element impurities due to recrystallization and reorganization processes of the quartz lattice. The geochemical and spectroscopic data demonstrate that the formation of such metamorphic mobilisates is an appropriate process for the formation of high-purity quartz (HPQ) deposits. Also, the quartz occurrences formed prior to the metamorphic events (pegmatite, quartzites) were suggestively purified during regional metamorphism and deformation, resulting in a high-purity quartz province in the Southern Urals.

Supplementary Materials: The following are available online at www.mdpi.com/2075-163X/7/10/189/s1, Table S1: Trace-element concentrations of investigated quartz samples analysed by solution ICP-MS (results in ppm), Table S2: REE concentrations (results in ppm) and chondrite-normalized interelemental ratios of investigated quartz samples analysed by solution ICP-MS, Table S3: Results of spatially resolved trace-element analyses by LA-ICP-MS (in ppm).

Acknowledgments: We gratefully acknowledge analytical efforts by G. Bombach (Freiberg) and B. Flem (Trondheim) during trace-element analyses as well as G. Nolze (Berlin) for EBSD measurements. M. Magnus and G. Geyer (Freiberg) are thanked for their help with sample preparation.

Author Contributions: Jens Götze and Elena Kotova conceived and designed the experiments; Elena Kotova provided the sample material; Jens Götze, Yuanming Pan, Axel Müller, Elena Kotova, and Daniele Cerin performed the experiments; Jens Götze, Yuanming Pan and Axel Müller analysed the data; Jens Götze wrote the paper.

Conflicts of Interest: The authors declare no conflict of interest.

References

1. Heaney, P.J.; Prewitt, C.T.; Gibbs, G.V. Silica—Physical Behaviour, Geochemistry and Materials Application. In *Reviews in Mineralogy*; Mineralogical Society of America: Washington, DC, USA, 1994; Volume 29, p. 606, ISBN 0-939950-35-9.

2. Götze, J.; Möckel, R. *Quartz: Deposits, Mineralogy and Analytics*; Springer: Berlin/Heidelberg, Germany, 2012; p. 360, ISBN 978-3-642-22161-3.

3. Harben, P.W. *The Industrial Mineral Hand Book—A Guide to Markets, Specifications and Prices*, 4th ed.; Industrial Mineral Information: London, UK, 2002; p. 412, ISBN 978-1900663519.

4. Müller, A.; Ihlen, P.M.; Wanvik, J.E.; Flem, B. High-purity quartz mineralisation in kyanite quartzites, Norway. *Miner. Depos.* **2007**, *42*, 523–535. [CrossRef]

5. Müller, A.; Wanvik, J.E.; Ihlen, P.M. Petrological and chemical characterization of high-purity quartz deposits with examples from Norway. In *Quartz: Deposits, Mineralogy and Analytics*, 1st ed.; Götze, J., Möckel, R., Eds.; Springer: Berlin/Heidelberg, Germany, 2012; pp. 71–118.

6. Müller, A.; Ihlen, P.M.; Snook, B.; Larsen, R.; Flem, B.; Bingen, B.; Williamson, B.J. The chemistry of quartz in granitic pegmatites of southern Norway: Petrogenetic and economic implications. *Econ. Geol.* **2015**, *110*, 137–157. [CrossRef]

7. Götze, J. Chemistry, textures and physical properties of quartz—Geological interpretation and technical application. *Mineral. Mag.* **2009**, *73*, 645–671. [CrossRef]

8. Melnikov, E.P. Main controlled factors of localization of the granulated quartz in the Southern part of the Ufaleisky anticlinorium. In *Vein Quartz of the Eastern Hang of the Urals*; Melnikov, E.P., Melnikova, N.I., Eds.; Nedra: Moscow, Russia, 1970; Issue No. 80, pp. 41–50. (In Russian)

9. Kelman, G.A. *Migmatite Complexes of the Mobile Belt*; Nedra: Moscow, Russia, 1974; p. 191. (In Russian)

10. Neuser, R.D.; Bruhn, F.; Götze, J.; Habermann, D.; Richter, D.K. Kathodolumineszenz: Methodik und Anwendung. *Zent. Geol. Paläontologie Teil I* **1995**, *H 1*, 287–306.

11. Monecke, T.; Bombach, G.; Klemm, W.; Kempe, U.; Götze, J.; Wolf, D. Determination of trace elements in the quartz reference material UNS-SpS and in natural quartz samples by ICP-MS. *Geostand. Newsl.* **2000**, *24*, 73–81. [CrossRef]

12. Flem, B.; Müller, A. In situ analysis of trace elements in quartz using Laser ablation inductively coupled plasma mass spectrometry. In *Quartz: Deposits, Mineralogy and Analytics*, 1st ed.; Götze, J., Möckel, R., Eds.; Springer: Berlin/Heidelberg, Germany, 2012; pp. 219–236.

13. Müller, A.; Wiedenbeck, M.; Flem, B.; Schiellerup, H. Refinement of phosphorus determination in quartz by LA-ICP-MS through defining new reference material values. *Geostand. Geoanal. Res.* **2008**, *32*, 361–376. [CrossRef]

14. Weil, J.A. EPR of iron centers in silicon dioxide. *Appl. Magn. Reson.* **1964**, *6*, 1–16. [CrossRef]

15. Sivaramaiah, G.; Lin, J.; Pan, Y. Electron paramagnetic resonance spectroscopy of Fe^{3+} ions in amethyst: Thermodynamic potentials and magnetic susceptibility. *Phys. Chem. Miner.* **2011**, *38*, 159–167. [CrossRef]

16. Weil, J.A. A review of electron spin resonance and its applications to the study of paramagnetic defects in crystalline quartz. *Phys. Chem. Miner.* **1984**, *10*, 149–165. [CrossRef]
17. Nilges, M.J.; Pan, Y.; Mashkovtsev, R. Radiation-damage-induced defects in quartz. I. Single-crystal W-band EPR study of hole centers in an electron-irradiated quartz. *Phys. Chem. Miner.* **2008**, *35*, 221–235. [CrossRef]
18. Pan, Y.; Nilges, M.J.; Mashkovtsev, R.I. Radiation-indiced defects in quartz. II. Single-crystal W-band EPR study of a natural citrine quartz. *Phys. Chem. Miner.* **2008**, *35*, 387–397. [CrossRef]
19. Mashkovtsev, R.I.; Pan, Y. Nature of paramagnetic defects in α-quartz: Progresses in the First Decade of the 21st Century. In *New Developments in Quartz Research: Varieties, Crystal Chemistry and Uses in Technology*; Novak, B., Marek, P., Eds.; Nova Science Publishers: Hauppauge, NY, USA, 2013; pp. 65–104.
20. Walsby, C.J.; Lees, N.S.; Claridge, R.F.C.; Weil, J.A. The magnetic properties of oxygen-hole aluminum centres in crystalline SiO_2. VI: A stable AlO_4/Li centre. *Can. J. Phys.* **2003**, *81*, 583–598. [CrossRef]
21. Botis, S.M.; Pan, Y. First-principles calculations on the $[AlO_4/M^+]^0$ (M = H, Li, Na, K) defects in quartz and crystal-chemical controls on the uptake of Al. *Mineral. Mag.* **2009**, *73*, 537–550. [CrossRef]
22. Anders, E.; Grevesse, N. Abundances of the elements: Meteoritic and solar. *Geochim. Cosmochim. Acta* **1989**, *53*, 197–214. [CrossRef]
23. Blankenburg, H.-J.; Götze, J.; Schulz, H. *Quarzrohstoffe*; Deutscher Verlag für Grundstoffindustrie: Leipzig-Stuttgart, Germany, 1994; p. 296.
24. Götze, J.; Plötze, M.; Graupner, T.; Hallbauer, D.K.; Bray, C. Trace element incorporation into quartz: A combined study by ICP-MS, electron spin resonance, cathodoluminescence, capillary ion analysis and gas chromatography. *Geochim. Cosmochim. Acta* **2004**, *68*, 3741–3759. [CrossRef]
25. Miyoshi, N.; Yamaguchi, Y.; Makino, K. Successive zoning of Al and H in hydrothermal vein quartz. *Am. Mineral.* **2005**, *90*, 310–315. [CrossRef]
26. Müller, A.; Koch-Müller, M. Hydrogen speciation and trace element contents of igneous, hydrothermal and metamorphic quartz from Norway. *Mineral. Mag.* **2009**, *73*, 569–583. [CrossRef]
27. Baron, M.A.; Stalder, R.; Konzett, J.; Hauzenberger, C.A. OH-point defects in quartz in B- and Li-bearing systems and their application to pegmatites. *Phys. Chem. Miner.* **2015**, *42*, 53–62. [CrossRef]
28. Frigo, C.; Stalder, R.; Hauzenberger, C.A. OH defects in quartz in granitic systems doped with spodumene, tourmaline and/or apatite: Experimental investigations at 5-20 kbar. *Phys. Chem. Miner.* **2016**, *43*, 717–723. [CrossRef]
29. Di Benedetto, F.; Innocenti, M.; Tesi, S.; Romanelli, M.; D'Acapito, F.; Fornaciai, G.; Montegrossi, G.; Pardi, L.A. A Fe K-edge XAS study of amethyst. *Phys. Chem. Miner.* **2010**, *37*, 283–289. [CrossRef]
30. Ramseyer, K.; Mullis, J. Factors influencing short-lived blue cathodoluminescence of alpha-quartz. *Am. Mineral.* **1990**, *75*, 791–800.
31. Götze, J.; Plötze, M.; Trautmann, T. Structure and luminescence characteristics of quartz from pegmatites. *Am. Mineral.* **2005**, *90*, 13–21. [CrossRef]
32. Perny, B.; Eberhardt, P.; Ramseyer, K.; Mullis, J.; Pankrath, R. Microdistribution of Al, Li, and Na in α-quartz: Possible causes and correlation with short-lived cathodoluminescence. *Am. Mineral.* **1992**, *77*, 534–544.
33. Götze, J.; Plötze, M.; Habermann, D. Cathodoluminescence (CL) of quartz: Origin, spectral characteristics and practical applications. *Mineral. Petrol.* **2001**, *71*, 225–250. [CrossRef]
34. Gorton, N.T.; Walker, G.; Burley, S.D. Experimental analysis of the composite blue CL emission in quartz. *J. Lumin.* **1996**, *72–74*, 669–671.
35. Van den Kerkhof, A.M.; Hein, U.F. Fluid inclusion petrography. *Lithos* **2001**, *55*, 27–47. [CrossRef]
36. Skuja, L. Optically active oxygen-deficiency-related centers in amorphous silicon dioxid. *J. Non-Cryst. Solids* **1998**, *239*, 16–48. [CrossRef]
37. Pacchioni, G.; Ierano, G. Optical absorption and nonradiative decay mechanism of E′ centre in silica. *Phys. Rev. Lett.* **1998**, *81*, 377–380. [CrossRef]
38. Stevens-Kalceff, M.A.; Phillips, M.R. Cathodoluminescence microcharacterization of the defect structure of quartz. *Phys. Rev. B* **1995**, *52*, 3122–3134. [CrossRef]
39. Siegel, G.H.; Marrone, M.J. Photoluminescence in as-drawn and irradiated silica optical fibers: An assessment of the role of non-bridging oxygen defect centres. *J. Non-Cryst. Solids* **1981**, *45*, 235–247. [CrossRef]
40. Stevens-Kalceff, M.A. Cathodoluminescence microcharacterization of point defects in α-quartz. *Mineral. Mag.* **2009**, *73*, 585–606. [CrossRef]

41. Walderhaug, O.; Rykkje, J. Some Examples of the Effect of Crystallographic Orientation on the Cathodoluminescence Colors of Quartz. *J. Sediment. Res.* **2000**, *70*, 545–548. [CrossRef]
42. Sippel, R.F. Luminescence petrography of the Apollo 12 rocks and comperative features in terrestrial rocks and meteorites. In *Lunar and Planetary Science Conference Proceedings*; The M.I.T. Press: Cambridge, MA, USA; Volume 1, pp. 247–263.
43. Larsen, R.B.; Henderson, I.; Ihlen, P.M.; Jacamon, F. Distribution and petrogenetic behavior of trace elements in granitic quartz from South Norway. *Contrib. Mineral. Petrol.* **2004**, *147*, 615–628. [CrossRef]
44. Monecke, T.; Kempe, U.; Götze, J. Genetic significance of the trace element content in metamorphic and hydrothermal quartz: A reconnaissance study. *Earth Planet. Sci. Lett.* **2002**, *202*, 709–724. [CrossRef]

Article

Mineralogy and Processing of Hydrothermal Vein Quartz from Hengche, Hubei Province (China)

Min Lin *, Zhenyu Pei and Shaomin Lei

School of Resources and Environmental Engineering, Wuhan University of Technology, Wuhan 430070, China; zhypei@163.com (Z.P.); shmlei@163.com (S.L.)
* Correspondence: 208726@whut.edu.cn; Tel.: +86-027-8788-5647

Received: 19 July 2017; Accepted: 31 August 2017; Published: 2 September 2017

Abstract: Quartz occurs in many geological materials, and is used in numerous industrial fields as a raw material. Mineralogy and the processing of hydrothermal quartz were studied by optical microscope, electron probe microanalysis, scanning electron microscope, inductively coupled plasma-optical emission spectrometry, and inductively coupled plasma mass spectrometer. A combination of the geological occurrence of the quartz deposit, mineralogical studies, and the processing technologies of the hydrothermal quartz was accomplished. The results show that impurities within the quartz mainly include muscovite, hematite, apatite, and secondary fluid inclusions. The main chemical impurities are Al (353 $\mu g \cdot g^{-1}$), K (118 $\mu g \cdot g^{-1}$), Fe (61.2 $\mu g \cdot g^{-1}$), P (15.5 $\mu g \cdot g^{-1}$), Na (13.4 $\mu g \cdot g^{-1}$), Mg (11.8 $\mu g \cdot g^{-1}$), Ti (8.31 $\mu g \cdot g^{-1}$), and B (10.8 $\mu g \cdot g^{-1}$). Based on these results, a combined process consisting of calcination and fluoride-free pressure acid leaching was established to effectively decompose and dissolve the quartz, and remove gangue minerals and fluid inclusions. The calcination process not only removed volatile components; it also destroyed the crystal structure of gangue minerals and enhanced their release probabilities. The calcination process has a positive influence on the removal of impurity elements by the fluoride-free pressure acid leaching process. A total of 85.2 wt % and 84.0 wt % of impurity elements was removed using the leaching systems of HCl-NH$_4$Cl and H$_2$SO$_4$-NH$_4$Cl, respectively.

Keywords: hydrothermal quartz; geological setting; mineralogy; processing technology

1. Introduction

SiO$_2$ minerals and rocks play an important role in geological processes and industrial applications [1]. Quartz is one of the most important silica minerals, and it is abundant in the Earth's crust in igneous, metamorphic, and sedimentary rocks [1]. SiO$_2$ minerals and rocks have been formed by primary and secondary magmatic, hydrothermal, or sedimentary processes or during diagenesis and metamorphosis [2].

Hydrothermal quartz is an important substitute of crystal quartz that has been studied in detail because it is rich in SiO$_2$ [3,4]. The hydrothermal quartz commonly contains a significant amount of associated minerals and fluid inclusions, which leads to difficulties in quartz processing [5]. Some primary and fine-grained mineral inclusions including mica, feldspar, and hematite are difficult to be removed because they are closely included within quartz grains. Unlike metal ores, the quartz can't be freely ground to cause the decomposition of quartz and fine-grained gangue minerals, because there is a special demand for particle size in quartz sand production [6]. Further, some micron-size fluid inclusions are widely distributed in hydrothermal quartz. Although decrepitation methods at low and medium temperatures (below 700 °C) can enhance the removal of volatile materials in quartz fluid inclusions, the micron-size fluid inclusions rarely are removed, especially in high-temperature quartz [7,8].

Quartz processing technologies have been studied in detail, but systemic research on geological occurrence and mineralogy, combined with optimized processing technologies, is warranted [9,10]. The research on the geological formation and mineralogy of hydrothermal quartz provides an important basis for quartz processing, and helps to characterize the quartz ore and its value.

Processing technologies of high-grade quartz mainly include pre-processing, physical processing (magnetic separation and flotation), chemical treatment (acid leaching and hot chlorination), and thermal treatment (calcination) [11–14]. Furthermore, physical processing and chemical leaching are common processing technologies used in industry. Magnetic separation is mainly used for removing iron-bearing magnetic minerals, while flotation is mainly used for removing aluminum-bearing silicate minerals from quartz ore [15]. However, a common weakness of both processes is that they can only remove specific kinds of impurity minerals. This weakness unavoidably leads to a long process flow during the effective purification of quartz. Although the acid leaching process can effectively remove different impurities at the same time, its leaching solution commonly contains HF, NaF, or CaF. Fluoric reagents, especially hydrofluoric acid, can cause environment pollution [16,17]. Leaching techniques are more effective than flotation for removing intergrown impurities of quartz and gangue minerals [18,19], but the selectivity of acid leaching with hydrofluoric acid is typically low, and this process could sharply reduce particle size of quartz sand as it removed impurities as a result [20,21].

The calcination process, as a thermal treatment technology, is commonly used to reduce reaction time and agent consumption by increasing the exposure probability of gangue minerals and destroying their crystal structures [10,11]. In addition, pressure acid leaching with mixed agents consisting of acids and inorganic salts is deemed to be an effective method to dissolve and remove muscovite without using any fluorides [22]. The metallic ions in the inorganic salts are unacceptable in quartz purification because they would unavoidably drag in metallic impurities during processing, but the use of NH_4Cl avoids the problem. Therefore, leaching systems including $HCl-NH_4Cl$ and $H_2SO_4-NH_4Cl$ could be introduced into quartz purification alongside the calcination process to effectively remove impurities from quartz.

This study presents an analysis of the regional geological setting and mineralogy of the hydrothermal quartz from Hengche Town in the Qichun County of the Hubei Province of China, and further develops a suitable process flow for the purification of hydrothermal quartz. In particular, the research is focused on analyzing occurrences and distributions of impurity elements in the quartz. Based on those results of this mineralogy, a combination process consisting of calcination and fluoride-free acid leaching was developed to remove impurities in the hydrothermal quartz. More importantly, the effects of the calcination and pressure acid leaching on quartz sand were evaluated.

2. Materials and Methods

2.1. Materials and Geological Situation

Quartz ore samples studied are from a hydrothermal vein deposit in Hengche Town in the Qichun County of the Hubei Province in China. The hydrothermal vein deposits are located in the south of a fold belt along the Qinling and Dabie mountains, and close to the Yangtze block. Figure 1 shows a sketch of the geology of the southeast of Hubei Province, China. The Qichun County belongs to the fold belt along the Qinling and Dabie mountains (I), which are bordered by the Yangtze block along the Zengjiaba–Qingfeng–Xiangguang (ZQX) fault zone [23,24]. The reentry belt of high pressure and ultrahigh pressure metamorphism from Tongbai to Dabie (I23) [23,24] was formed by the subduction and collision of the area from late Paleozoic to Triassic, and ended in late Triassic and early Jurassic age. In general, the quartz deposits in the area were formed in the combined belt of Shangdan, influenced by the orogenesis of the Qinling Mountains, and activated by the Yangtze block [23].

Subduction and collision from Paleozoic to Jurassic age caused ore-forming conditions, and the high pressure and ultrahigh pressure rarely contribute to the deposit. As a result, the agglutinated fine-grained quartz likely contains secondary mineral inclusions in assemblages. In late Triassic to Mesozoic age, the northward subduction and reentry of the Yangtze block or Shangdan Ocean resulted in an orogenic belt south of the Qinling Mountains. This indicates that the geological environment provides a good framework for the deposit formation. Since the quartz deposit is close to the block in the south of the Qinling Mountains, some wall rocks rich in Si, Fe, P, S, and Ca unavoidably have an influence on assemblages of quartz and enlarged sides.

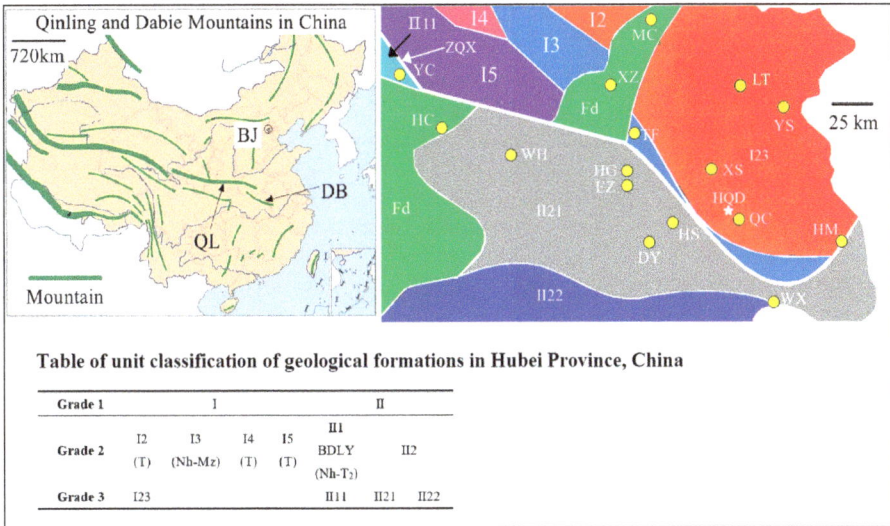

Table of unit classification of geological formations in Hubei Province, China

Grade 1	I				II		
Grade 2	I2 (T)	I3 (Nb-Mz)	I4 (T)	I5 (T)	II1 BDLY (Nb-T₂)	II2	
Grade 3	I23				II11	II21	II22

Figure 1. Locations of the Qinling and Dabie Mountains, and a sketch map of geological formations in the southeast of the Hubei Province in China (Revised according to Mao et al., 2014) [23]: QL—Qinling Mountain, DB—Dabie Mountain, BJ—Beijing, HQD—the hydrothermal quartz deposit, MC—Macheng, YC—Yingcheng, XZ—Xinzhou, LT—Luotian, HC—Hanchuan, TF—Tuanfeng, YS—Yingshan, WH—Wuhan, HG—Huanggang, EZ—Ezhou, XS—Xishui, HS—Huangshi, QC—Qichun, HM—Huangmei, DY—Daye, WX—Wuxue; Fd—Fault depressions, I—Qinling and Dabie fold belt, II—Yangtze block, I2-Combining belt of Shangdan, I3—Block in the southern of Qinling Mountains, I4—Arc basin system from Maota to Suizhou, I5—Combining belt from Liangzhu to Suinan, I23—Reentry belt of high pressure and ultrahigh pressure metamorphism from Tongbai to Dabie, II11—Passive margin in the north rim of Yangzi block, II21—Forelandbasin in the south rim of Yangzi block, II22—Passive margin in the south rim of Yangzi block.

When compared with the few ultrahigh purity quartz deposits (c(Al) < 20 µg·g^{-1}) in Qichun County [24], the Hengche quartz is from a typical, low-grade and small-scale deposit. In the Luliang age, a hydrothermal solution rich in SiO_2 was split from granite magma. Along the bedding, the vein quartz was formed when the hydrothermal solution invaded into biotite-plagioclase gneiss, granitic gneiss, and amphibolite in the Hongan Group of the Dabie Mountains. The deposit length ranges from 80 m to 115 m, the deposit width ranges from 60 m to 120 m, and the deposit thickness ranges from 5 m to 18 m. The ore mainly contains quartz, and the size of the quartz grain is about 0.8 mm, while a part of the quartz grain is larger than 1 mm.

2.2. Mineralogical Analysis

Examples of the typical bulk hydrothermal quartz were sliced up and polished into sections for microscopy observation. Mineral phases and their distribution in quartz ore were determined by polarizing microscope (DLMP, Leica Microsystems, Wetzlar, Germany). Element components of gangue minerals were analyzed by electron probe microanalysis (EPMA, JXA-8230, JEOL Ltd., Tokyo, Japan) and X-ray energy spectrometer (EDS).

The electron optical system of EPMA consists of a LaB6 electron gun with a centered cartridge filament; acceleration voltage: 0~30 kV; beam ranging: 10^{-5}~10^{-12} A; and image magnification: ×40~×300,000. Analysis precision of EDS: analysis element: 5B-92U, 2%~3% for common element, detection limit: 0.1 wt %~0.5 wt %. Electron images were obtained at an acceleration voltage of 20.0 kV and a working distance of 11.2 mm. The GBT15617-2002 standard is used for EPMA [25].

2.3. Chemical Analysis

The chemical composition of the hydrothermal quartz was analyzed by inductively coupled plasma (ICP) techniques according to national standards SJ/T 11554-2015 [26] and GB/T 32650-2016 [27] in China. The analytical method and procedure are as follows.

2.0000 g per sample were dissolved by 12 cm^3 hydrofluoric acid (40 wt %, guarantee reagent) in a pressure-tight reaction kettle with a lining of polytetrafluoretyhylene (PTFE, 50 cm^3), while digestion temperature and time were 200 °C and 3 h, respectively. The produced fluid was heated and evaporated in a PTFE beaker at 150 °C. When the fluid shrank into droplets, 3 cm^3 ultrapure water was added for further evaporation, while the operation was repeated three times, and the remained fluid was commonly diluted to 20 cm^3, 500 cm^3, or 1000 cm^3, according to the different contents of impurity elements.

The element contents of diluted fluids were analyzed by inductively coupled plasma-optical emission spectroscopy (ICP-OES, Prodigy 7, LEEMAN LABS INC., Hudson, NH, USA) and inductively coupled plasma mass spectrometry (ICP-MS, iCAP-Qc, THERMO FISHER SCIENTIFIC, Waltham, MA, USA).

The analytical conditions include:

Wavelength coverage of ICP-OES: 165~1100 nm, optical resolution: 0.007 nm at 200 nm, relative standard deviation: <2%.

Mass spectrum of ICP-MS: 4~290 amu, sensitivity: >150 Mcps·ppm^{-1} for common elements, detection limit: <0.1 ppt for common element, relative standard deviation: <2% in short time, <3% in long time.

The computational formula to obtain the element contents is as follows:

$$\varphi = \frac{V(\rho - \rho_0)}{m} \tag{1}$$

where φ is the mass of element content in 1 g of quartz sand, µg·g^{-1}; V is the volume of the diluted fluid, cm^3; ρ is the mass concentration of the ananlyzed element by ICP, ρ_0 is the mass concentration of the blank sample, mg·dm^{-3} or µg·dm^{-3}; m is the mass of the analyzed sample, 2.0000 g.

2.4. Processing and Characterization of Quartz Sand

Mining, washing, crushing, and grinding of quartz ore were accomplished by Kaidi Building Materials Co., Ltd. (Huangshi, China). The mined quartz ore was washed and air-dried. Then, the bulk quartz was crushed by a jaw crusher (PE-Φ100 × 125), and ground by a Raymond mill (3R2115). Produced quartz sand ranging from 106 µm (140 mesh) to 212 µm (70 mesh) was selected by standard sieve, and used for experiments. The quartz sand was calcinated at 900 °C for 5 h in a muffle furnace (KSY-12-16A), and then cooled by ultrapure water (18.2 MΩ·cm). The calcinated quartz sand was leached by mixed leaching agents in a pressure-tight reaction kettle. Leached quartz sand was washed by ultrapure water, and the leach liquor and washing liquor were mixed to wash the calcinated quartz sand in turn. Quartz and gangue minerals during processing were observed by scanning electron microscope (SEM, JSM-IT300) and EPMA after carbon coating.

Secondary electron images were obtained at 20.0 kV of acceleration voltage, 16.2~16.6 mm of working distance, and 30 nm of spot size.

3. Results and Discussion

3.1. Mineralogy of Hydrothermal Quartz

Bulk quartz is milky-white as a whole, and its size ranges from 20 cm to 40 cm. Scattered red materials are distributed in quartz fractures.

3.1.1. Impurity Elements in Quartz Ore

Elemental concentrations of the quartz sand are shown in Table 1. The main impurity elements determined by ICP are Al (353 $\mu g \cdot g^{-1}$), K (118 $\mu g \cdot g^{-1}$), Fe (61.2 $\mu g \cdot g^{-1}$), P (15.5 $\mu g \cdot g^{-1}$), Na (13.4 $\mu g \cdot g^{-1}$), Mg (11.8 $\mu g \cdot g^{-1}$), B (10.8 $\mu g \cdot g^{-1}$), Ti (8.31 $\mu g \cdot g^{-1}$) and Ca (8.04 $\mu g \cdot g^{-1}$) by ICP analysis, and proportions of these elements in total impurities are 57.0 wt %, 19.1 wt %, 9.89 wt %, 2.50 wt %, 2.16 wt %, 1.91 wt %, 1.74 wt %, 1.34 wt % and 1.30 wt %, respectively.

Table 1. Contents of main impurity elements within the hydrothermal quartz analyzed by inductively coupled plasma (ICP) analysis.

Element	Al	Fe	Na	S	P	Li	K	Ca	Ti
Content [1] ($\mu g \cdot g^{-1}$)	353	61.2	13.4	5.64	15.5	2.20	118	8.04	8.31
Element	Mg	Ni	Zr	Zn	As	B	In Total		
Content ($\mu g \cdot g^{-1}$)	11.8	1.01	6.46	0.567	3.16	10.8	619		

[1] Relative standard deviations (RSD) are less than 4%.

3.1.2. Optical Microscope Analysis

The hydrothermal quartz mainly contains solid inclusions of muscovite and hematite, as well as fluid inclusions (Figure 2). The muscovite platelets are widely distributed along the quartz grain boundaries (Figure 2a), but a small quantity of muscovite is included in the quartz grains (Figure 2b). Muscovite is commonly thought to be an important source of impurity elements, including Al, K, Mg, Fe and Ti [28].

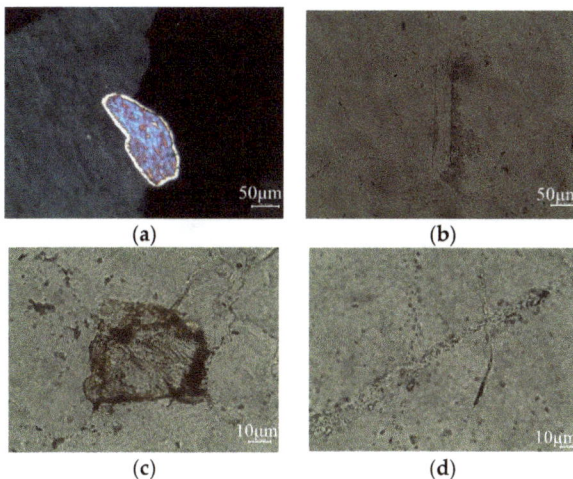

(a) (b)

(c) (d)

Figure 2. Transmitted-light microscopy images of muscovite, hematite, and fluid inclusions within quartz thin sections: (**a**) muscovite along a grain boundary (XPL); (**b**) muscovite included in quartz crystal (PPL); (**c**) hematite in compression fractures of quartz (PPL); (**d**) two generations of fluid inclusions (PPL).

Hematite is the main iron carrier in these veins, and mainly occurs filling thin cracks (Figure 2c). In hydrothermal quartz, most of the fluid inclusion generations (Figure 2d) traverse the whole quartz grains, so these secondary fluid inclusions commonly containing gas and liquid phases are a possible source of fluid impurities [29].

3.1.3. Electron Probe Microanalysis

Figure 3 shows secondary electron images of muscovite, hematite, apatite, and quartz sections. Figure 3a is a micrograph of muscovite inclusions within quartz. The analysis (Table 2) shows that Al, K, Mg, and Fe are the major elements obtained in the muscovite. These impurities occurring in aluminum silicate minerals could reduce the quality of quartz production [30]. Figure 3b is an image of hematite (the mineral phase was determined by polarizing microscope within an electron probe microanalyzer) in quartz, and its composition (Table 2) shows that the secondary mineral contains elevated Fe (the Cr is from polishing powder). Generally, the iron in high-grade quartz is typically less than 1 ppm [31]. It is difficult to remove some filmy iron on the surface of quartz particles by conventional physical methods [32]. Figure 3c is an image of apatite, which is associated with hematite. Its chemical analysis (Table 2) shows that P, Ca, Fe, Mg, and Na occur in the mixed minerals. The apatite is included in quartz, and it is a diagnostic mineral of hydrothermal quartz [33].

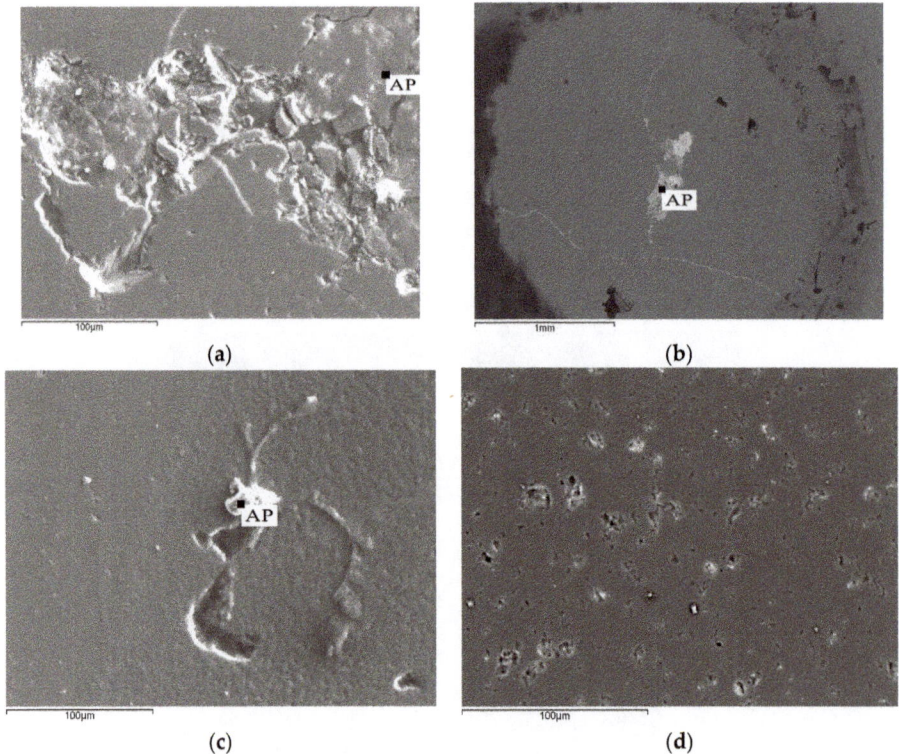

Figure 3. Details of solid inclusions in quartz veins using EMP (electron microprobe) images in SE (secondary electron) mode: (**a**) muscovite inclusions in quartz; (**b**) hematite in quartz fracture; (**c**) apatite and hematite included in quartz; (**d**) a typical microarea used for area-scan analysis of the electron probe; AP—analyzed point by X-ray energy spectrometer (EDS).

Table 2. Elemental components of analyzed points in Figure 3a–c by EDS analysis of electron probe.

Element	Na	Mg	Al	Si	P	S	K	Ca	Mn	Cr	Fe	O
Figure 3a (wt %)		0.31	19.16	24.69			8.40				0.28	47.17
RSD [1] (%)		2.43	1.10	0.96			1.22				1.96	
Figure 3b (wt %)			2.24	2.48	0.63					2.89	65.91	25.85
RSD (%)			2.23	1.32	2.64					1.08	0.68	
Figure 3c (wt %)	3.31	5.06	1.57	6.63	10.38	2.01	0.87	14.50	0.91		14.52	40.24
RSD (%)	2.88	2.09	2.11	1.16	1.67	1.34	2.89	1.44	1.57		0.83	

[1] Relative standard deviations (RSD) are less than 3%.

Besides impurities in these inclusions of the quartz, some lattice elements in the hydrothermal quartz cannot be neglected. As shown in Figure 4, Götze (2009) [34] reported four substitutions between Si and impurity elements in a quartz lattice. In addition, substitution with interstitial charge compensator is suitable to be used for explaining occurrences of impurity elements in the hydrothermal quartz, because P mainly occurs in the apatite, and Ge is not detected by ICP analysis.

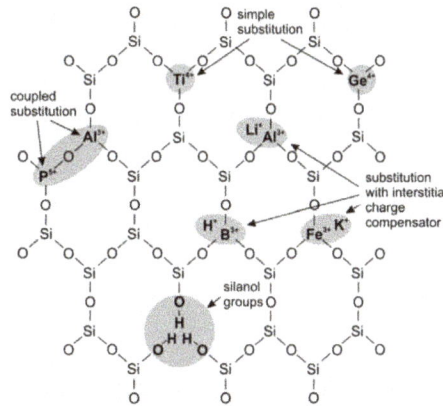

Figure 4. Schematic quartz structure showing the configuration of trace elements in the quartz lattice (modified from Götze 2009) [34]. Mclaren et al. (1983) proposed that the substitution of Si^{4+} by four H^+ is also possible (silanol groups) [35]. Since the illustration is two-dimensional, the fourth H^+ is not shown on the figure.

Figure 5 shows the surface distributions of Si, O, Al, and Na in the microarea of Figure 3d. Al and Na are uniformly distributed across the microarea. This indicates that the two elements replace Si in a quartz lattice in substitution with the interstitial charge compensator, although Li and B cannot be effectively detected by electron probe.

Based on a combination of optical microscope analysis and electronic microscope analysis, the main mineral inclusions in the quartz are muscovite, hematite and apatite. The muscovite contains Al, K, Mg, and Fe. The hematite contains Fe, while some hematite is associated with apatite. The apatite contains P and Ca. The associated Na, Mg, and Al might be from the wall rock of biotite plagioclase gneiss. For the main impurities Al, K, and Fe, the microscopic examination and component analysis indicate that Al and K mainly occur in muscovite, and Fe mainly exists in hematite. In addition, some Al and Na possibly occur in the quartz lattice. In addition, fluid inclusions can be possible sources for H, Na, and othe alkali and alkali Earth elements (and probably anions such as Cl). Most of the fluid inclusions contain gas and liquid phases, and are secondary. The quartz ore mainly contains quartz, muscovite, hematite, apatite, and secondary fluid inclusions, which are common features of hydrothermal deposits [36].

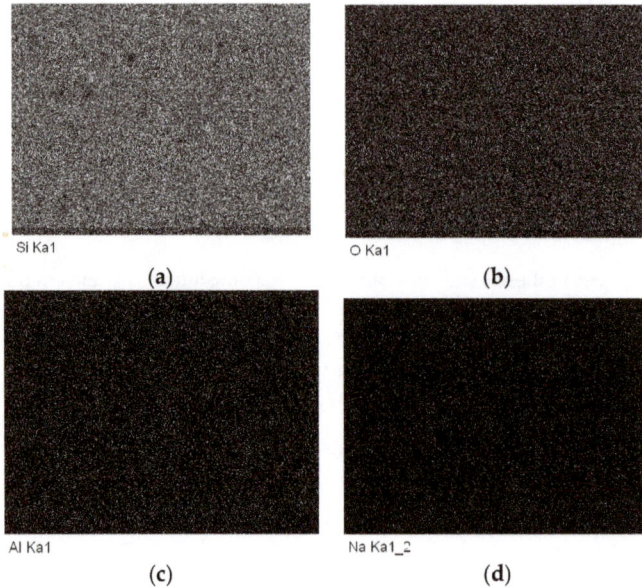

Figure 5. Surface distributions of Si, O, Al, and Na in the microarea of Figure 3d (analyzing for 12 min): luminance was enhanced for showing clearly: (**a**) the surface distribution of Si; (**b**) the surface distributions of O; (**c**) the surface distribution of Al; and (**d**) the surface distributions of Na. White spots represent elements at the corresponding positions of Figure 3d, and the scale bars are the same as Figure 3d.

3.2. Quartz Processing

Xiong et al. used pressure acid leaching with hydrofluoric acid as an efficient purification process for purifying the vein quartz [10]. However, in the conventional acid leaching process of quartz ore, quartz purification depends on an acid attack to gangue minerals, which results in high acid consumption and a long leaching time [10,11,22]. The two problems significantly reduce equipment life, increase production costs, and decrease production efficiency. Due to the pollution of the fluoric leaching agent [16,17], it is therefore necessary to develop an efficient fluorine-free leaching system to purify vein quartz. The objective of the recommended process is to realize a fluoride-free technology of pressure acid leaching at the same time as ascertaining quartz quality.

Calcination at 900 °C for 5 h is deemed to be effective to remove organic impurities and H_2O within fluid inclusions, and increase the exposure probabilities of mineral inclusions, including muscovite, hematite, and apatite, according to different thermal expansibilities [10,11,37]. In addition, calcination could destroy the crystal structures of gangue minerals, especially muscovite [37]. When combined with the calcination process, pressure acid leaching has the power to effectively purify quartz sand without any fluorides.

3.2.1. Recommended Process

Figure 6 shows two recommended processes for purifying the hydrothermal quartz. Quartz sand was calcinated at 900 °C for 5 h, and the calcinated quartz sand was washed by leaching and washing liquors. In a pressure-tight reaction kettle [10], the washed quartz sand was either leached by acid solutions containing H_2SO_4 and NH_4Cl, or HCl and NH_4Cl. Leached quartz sand was again washed by ultrapure water 5–10 times, and then dried at 105 °C for 3 h.

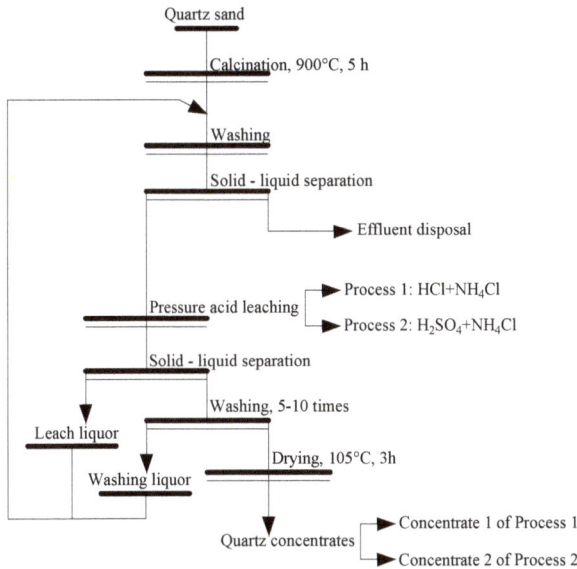

Figure 6. Engineering flow sheet of the recommended processes for purifying the hydrothermal quartz.

Following calcination, the first washing process is used to make full use of acids remained in leaching and washing liquors, and cut the cost effluent disposal. Using H_2SO_4 or HCl, calcinated gangue minerals can be efficiently dissolved and removed from quartz at high pressure and temperature. The NH_4Cl is used to provide hydrogen ions with a more stable and higher concentration during the hydrolysis of NH_4^+.

Based on a combination of calcination and pressure acid leaching, the quartz sand was efficiently purified. The optimal process conditions and contents of impurity elements in quartz concentrates for two leaching processes are shown in Tables 3 and 4, respectively. The effective removal of Al, K, and Fe indicates that the main gangue minerals, including muscovite and hematite, are efficiently removed by the two processes. Parts of P and Ca were removed, but the removal rates were not so high. The possible reason is that some primary apatite is included in the quartz particle, and some P occurs in the quartz lattice. S, Ni, Zn, and As are not detected in quartz concentrates, which indicates that these impurity elements are liberated from quartz particles after ore grinding. Zr occurs in hydrothermal zircon due to the ultra-high pressure action in the Dabie Mountains [38], but it is difficult to be dissolved by acid solution [39], so its removal rates are only 0.155 wt %. The removal rates of some elements, including Li, B, Na, P, and Ti, are not so high because parts of them occur in the quartz lattice. As a result, the fluoride-free leaching processes have little influence on removing them.

Table 3. Optimal conditions of leaching processes 1 and 2.

Leaching Process	Temperature (°C)	Acid Concentration (HCl or H_2SO_4) (mol·dm^{-3})	NH_4Cl Concentration (mol·dm^{-3})	Liquid/Solid Ratio (cm^3·g^{-1})	Leaching Time (h)
HCl + NH_4Cl	280	0.8	0.8	10	6
H_2SO_4 + NH_4Cl	250	0.3	0.45	5	7

Table 4. Contents and removal rates of major elements at optimal conditions in Table 3.

Element	Ore ($\mu g \cdot g^{-1}$)	Concentrate 1 [1] ($\mu g \cdot g^{-1}$)	Removal Rate 1 (wt %)	Concentrate 2 [1] ($\mu g \cdot g^{-1}$)	Removal Rate 2 (wt %)
Al	353	41.5	88.2	44.1	87.5
Fe	61.2	1.14	98.1	1.12	98.2
Na	13.4	11.3	15.7	12.2	8.96
S	5.64	-	-	-	-
P	15.5	5.00	67.7	5.39	65.2
Li	2.20	2.19	0.455	2.04	7.27
K	118	1.21	99.0	2.24	98.1
Ca	8.04	4.54	43.5	4.40	45.3
Ti	8.31	4.89	41.2	5.38	35.3
Mg	11.8	4.88	58.6	7.15	39.4
Ni	1.01	-	-	-	-
Zr	6.46	6.45	0.155	6.45	0.155
Zn	0.567	-	-	-	-
As	3.16	-	-	-	-
B	10.8	8.77	18.8	8.77	18. 8
In total	619	91.9	85.2	99.2	84.0

[1] Relative standard deviations (RSD) of element contents in quartz concentrates are less than 4%, and not shown in the Table; (-) = below the limit of detection.

3.2.2. Effects of the Calcination Process

Muscovite, as a main gangue mineral in the hydrothermal quartz, can be transformed into amorphous phases during the calcination process [37]. Figure 7 shows the surface topographies of muscovite in calcinated quartz sand. The planar water and constitution water were removed from the muscovite during calcination (Formula (2)), and the water evaporation increased the interlayer spacing of the muscovite [37], and further promoted a seperation of different muscovite sheets along the cleavage plane (Figure 7a). Muscovite flakes close to its surface were separated from the host, and some of them were damaged into fragments with micron sizes (Figure 7b). The fragments consisting of active K_2O, Al_2O_3, and SiO_2 (Formula (3)) could be easily dissolved by acid solutions. This demonstrates that calcination is a necessary procedure for removing muscovite by a fluoride-free acid leaching process.

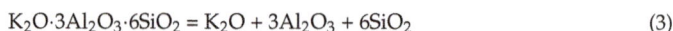

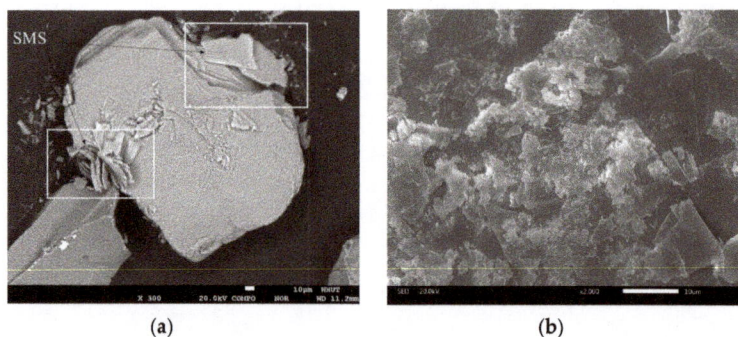

$$K_2O \cdot 3Al_2O_3 \cdot 6SiO_2 \cdot 2H_2O = K_2O \cdot 3Al_2O_3 \cdot 6SiO_2 + 2H_2O\uparrow \tag{2}$$

$$K_2O \cdot 3Al_2O_3 \cdot 6SiO_2 = K_2O + 3Al_2O_3 + 6SiO_2 \tag{3}$$

(a)   (b)

Figure 7. Surface topographies of muscovite in calcinated quartz sand: calcination temperature 900 °C, calcination time 5 h; (a) backscattered electron image obtained by electron probe microanalysis (EPMA); (b) secondary electron image obtained by scanning electron microscope (SEM); SMS—seperated muscovite sheets along cleavage plane.

The calcination process is also effective for removing other mineral inclusions, such as hematite and apatite. The exposure probabilities of gangue minerals included in quartz grains have great influences on impurities removal. The impurities included in quartz particles are hardly removed unless a calcination process is used, especially in case of a fluoride-free purification process of quartz sand. The different expansion coefficients between quartz and hematite during calcination promoted the exposure of hematite. The calcination is also effective for apatite when it is associated with hematite.

Calcination is also an effective method for removing fluid impurities within quartz fluid inclusions. Schmidt–Mumm [40] thought that the decrepitation of the fluid inclusions could be classified as several processes: microcracks opening and propagating, the rupture of grain boundaries, transgranular fracturing, intragranular fracturing, and the decrepitation of large/small fluid inclusions. Figure 8 shows a decrepitation pit on the surface of a calcinated quartz particle. The small pit is about 7 μm long and 1μm wide. It indicates that calcination at 900 °C for 5 h can even remove micron-size fluid inclusions. Volatile materials occurring in the fluid inclusions were removed during the calcination process, and inorganic salts dissolved in the fluids could be leached in the following acid leaching process.

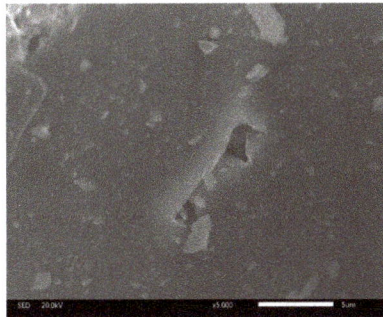

Figure 8. Decrepitation pit on the surface of a calcinated quartz particle: calcination temperature 900 °C, calcination time 5 h.

At the optimal leaching conditions of processes 1 and 2 in Table 3, the element contents in quartz sand only after pressure acid leaching (rather than using a calcination process) were compared with those of quartz concentrates so as to describe the effect of the calcination process on the removal of impurity elements. As shown in Table 5, the calcination process can effectively promote the leaching of Al, Fe, S, P, K, and As. The contents of Al and K were respectively reduced from 167 and 188 $\mu g \cdot g^{-1}$ to 41.5 and 44.1 $\mu g \cdot g^{-1}$ by an additional calcination process. The differences indicate that the calcination process activates some gangue minerals and reduces the chemical reaction resistance of the leaching process by destroying their crystal structures. In addition, the decrease of the Fe and P contents may be caused by exposure of the hematite and apatite during calcination. S and As could easily be removed during calcination due to their volatility. Moreover, the decrease of Ca and Na contents might be caused by the decrepitation of fluid inclusions (compare to Figure 8).

Table 5. Contents of major elements in quartz sand with or without a calcination process.

Element	Ore ($\mu g \cdot g^{-1}$)	C and PHAL ($\mu g \cdot g^{-1}$)	PHAL [1] ($\mu g \cdot g^{-1}$)	C and PSAL ($\mu g \cdot g^{-1}$)	PSAL [1] ($\mu g \cdot g^{-1}$)
Al	353	41.5	167	44.1	188
Fe	61.2	1.14	7.42	1.12	13.4
Na	13.4	11.3	13.5	12.2	12.9
S	5.64	-	4.00	-	4.52
P	15.5	5.00	13.6	5.39	10.4
Li	2.20	2.19	2.20	2.04	2.19

Table 5. *Cont.*

Element	Ore (μg·g^{-1})	C and PHAL (μg·g^{-1})	PHAL [1] (μg·g^{-1})	C and PSAL (μg·g^{-1})	PSAL [1] (μg·g^{-1})
K	118	1.21	42.6	2.24	45.9
Ca	8.04	4.54	7.50	4.40	8.01
Ti	8.31	4.89	6.07	5.38	5.91
Mg	11.8	4.88	8.60	7.15	8.12
Ni	1.01	-	-	-	-
Zr	6.46	6.45	6.46	6.45	6.46
Zn	0.567	-	-	-	-
As	3.16	-	1.47	-	2.15
B	10.8	8.77	9.80	8.77	8.47
In total	619	91.9	290	99.2	316

[1] Relative standard deviations (RSD) of the element contents in quartz concentrates are less than 4%, and not shown in the Table; C—calcination process (900 °C for 5 h), PHAL—pressure acid leaching (HCl + NH$_4$Cl), PSAL—pressure acid leaching (H$_2$SO$_4$ + NH$_4$Cl); (-) = below the limit of detection.

3.2.3. Effects of Fluoride-Free Pressure Acid Leaching

Figure 9 shows the surface topographies of muscovite in leached quartz sand. Compared with the surface topographies of muscovite (Figure 7) in calcinated quartz sand, the fragments with micron sizes were dissolved and removed from the muscovite surface (Figure 9a). This process can be described by the dissolution of the active fragments containing K$_2$O and Al$_2$O$_3$ (Formulas (4) and (5)). In addition, the muscovite in calcinated quartz sand was further dissolved along its edge. More importantly, some fractures were developed in the vertical direction of the long side of muscovite when concentrations of leaching agents were doubled (Figure 9b). Moreover, expanded interlayer spacing provided favorable conditions for leaching reactions, and promoted a faster process of muscovite dissolution. The several effects of acid leaching led to the structure collapsing and the dissolution of muscovite.

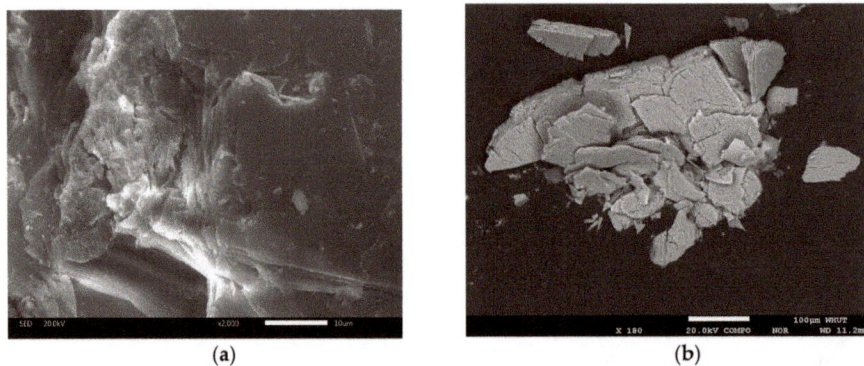

$$K_2O + 2H^+ = 2K^+ + H_2O \tag{4}$$

$$Al_2O_3 + 6H^+ = 2Al^{3+} + 3H_2O \tag{5}$$

(a)   (b)

Figure 9. Surface topographies of muscovite in leached quartz sand: (**a**) secondary electron image obtained by SEM, leaching conditions: 0.05 mol·dm^{-3} H$_2$SO$_4$, 0.10 mol·dm^{-3} NH$_4$Cl, liquid/solid ratio 5 cm^3·g^{-1}, leaching time 7 h, and leaching temperature 200 °C; (**b**) backscattered electron image obtained by EPMA, leaching conditions: 0.10 mol·dm^{-3} H$_2$SO$_4$, 0.20 mol·dm^{-3} NH$_4$Cl, liquid/solid ratio 5 cm^3·g^{-1}, leaching time 7 h, and leaching temperature 200 °C.

During calcination, most of the mineral inclusions in quartz sand were transformed into mineral intergrowths of quartz and gangue minerals due to heat stress. Then they were exposed in the leaching solution. With further calcination, the mineral intergrowth was promoted into separated mineral monomers. Figure 10 shows the surface topographies of calcinated and leached quartz sands. Although associated minerals were liberated from quartz particles in Figure 10a, some impurities unavoidably remained in the pores of the calcinated quartz surface. The impurities that remained on the surface of calcinated quartz were leached by mixed leaching agents. As a result, the surface of the leached quartz sand in Figure 10b became smoother than that of the calcinated quartz sand in Figure 10a. Common intergrowths of quartz–hematite and quartz–hematite–apatite were leached by mixed leaching agents, and the leaching processes are as follows:

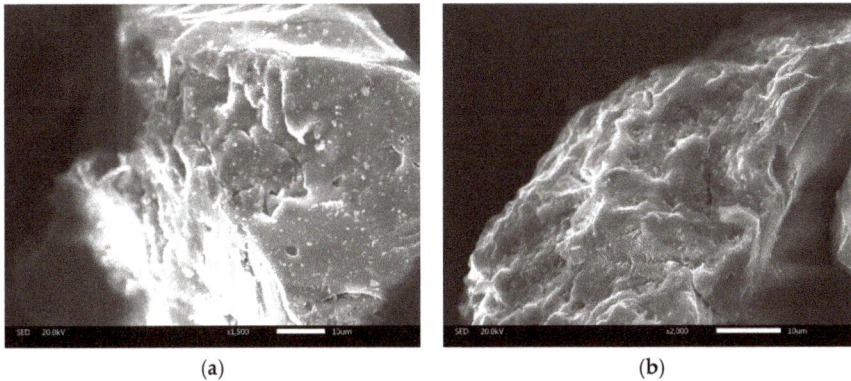

$$Fe_2O_3 + 6H^+ = 2Fe^{3+} + 3H_2O \tag{6}$$

$$Ca_5[PO_4]_3(OH) + H^+ \rightarrow Ca^{2+} + H_3PO_4 + H_2PO_4{}^- + HPO_4{}^{2-} + PO_4{}^{3-} + H_2O \tag{7}$$

(a)   (b)

Figure 10. Surface topographies of quartz sand: (**a**) calcinated quartz sand (900 °C for 5 h); (**b**) quartz concentrate of process 2.

By a combined process of calcination and fluoride-free acid leaching, gangue minerals including muscovite, hematite, and apatite were effectively separated from the hydrothermal quartz. The calcination process not only destroyed the crystal structures of the gangue minerals; it also removed volatile compounds. Meanwhile, the exposure probabilities of the impurities were enhanced due to the calcination process. In general, the calcination process had a positive influence on the removal of impurity elements with the fluoride-free pressure acid leaching process.

4. Conclusions

The present study analyzed the geological formation and mineralogy of the hydrothermal quartz deposit from Hengche in Qichun County, of the Hubei Province in China, A combined purification process consisting of calcination and fluoride-free pressure acid leaching was developed based on two fundamental investigations. Moreover, the effects of this processing procedure on the raw quartz material were discussed, and the following conclusions obtained:

(1) Impurity elements in the hydrothermal quartz are related to fluid inclusions and associated gangue minerals such as muscovite, hematite, and apatite. The muscovite not only occurs along the grain boundaries of quartz particles; it is also included in quartz particles. The hematite is mainly of secondary origin, and occurs along crystal fractures in quartz. However, some of the hematite is also associated with apatite.

(2) The gangue minerals such as muscovite, hematite, and apatite were effectively removed by a combined process of calcination and fluoride-free pressure acid leaching, while the removal rates

of the most important impurity elements Al, K, and Fe were 88.2 wt %, 99.0 wt % and 98.1 wt %, respectively, for the HCl-NH$_4$Cl system, and 87.5 wt %, 98.1 wt % and 98.2 wt % for the H$_2$SO$_4$-NH$_4$Cl system. The removal rates of some elements, including Li, B, Na, P and Ti were not so high, because they might be structurally incorporated in the quartz lattice.

(3) The calcination process not only destroyed the crystal structure of the gangue minerals, it also removed volatile compounds. The liberation probabilities of impurities were enhanced due to the calcination process. In conclusion, the calcination process had a positive influence on the removal of impurity elements during the fluoride-free pressure acid leaching process.

Acknowledgments: This research was supported by the Open Foundation of Engineering Center of Avionics Electrical and Information Network of Guizhou Province Colleges and Universities of Anshun University (HKDZ201404).

Author Contributions: Min Lin and Zhneyu Pei conceived and designed the experiments; Min Lin performed the experiments; Min Lin, Zhneyu Pei and Shaomin Lei analyzed the data; Shaomin Lei contributed reagents/materials/analysis tools; Min Lin wrote the paper.

Conflicts of Interest: The authors declare no conflict of interest. The founding sponsors had no role in the design of the study; in the collection, analyses, or interpretation of data; in the writing of the manuscript, and in the decision to publish the results.

References

1. Götze, J. Classification, mineralogy and industrial potential of SiO$_2$, minerals and rocks. In *Quartz: Deposits, Mineralogy and Analytics*, 1st ed.; Götze, J., Möckel, R., Eds.; Springer: Berlin/Heidelberg, Germany, 2012; pp. 1–27.

2. Howm, R.A. Silica: Physical behavior, geochemistry and materials applications. *Mineral. Mag.* **1996**, *60*, 390–391. [CrossRef]

3. Armington, A.F.; Larkin, J.J. Purification and analysis of α-quartz. *J. Cryst. Growth* **1986**, *75*, 122–125. [CrossRef]

4. Li, J.S.; Li, X.X.; Shen, Q.; Zhang, Z.Z.; Du, F.H. Further purification of industrial quartz by much milder conditions and a harmless method. *Environ. Sci. Technol.* **2010**, *44*, 7673. [CrossRef] [PubMed]

5. Bayaraa, B.; Greg, B.; Noriyoshi, T. Hydrothermal quartz vein formation, revealed by coupled SEM-CL imaging and fluid inclusion microthermometry: Shuteen Complex, South Gobi, Mongolia. *Resour. Geol.* **2010**, *55*, 1–8. [CrossRef]

6. Johnson, G.R. History of the industrial production and technical development of single crystal cultured quartz. In Proceedings of the IEEE International Frequency Control Symposium & Exposition, Montreal, QC, Canada, 23–27 August 2004.

7. Mavrogenes, J.A.; Bodnar, R.J. Hydrogen movement into and out of fluid inclusions in quartz: Experimental evidence and geologic implications. *Geochim. Cosmochim. Acta* **1994**, *58*, 141–148. [CrossRef]

8. Tomlinson, E.L.; Mcmillan, P.F.; Zhang, M.; Jones, A.P.; Redfern, S.A.T. Quartz-bearing C–O–H fluid inclusions diamond: Retracing the pressure–temperature path in the mantle using calibrated high temperature IR spectroscopy. *Geochim. Cosmochim. Acta* **2007**, *71*, 6030–6039. [CrossRef]

9. Sayilgan, A.; Arol, A.I. Effect of carbonate alkalinity on flotation behavior of quartz. *Int. J. Miner. Process.* **2004**, *74*, 233–238. [CrossRef]

10. Xiong, K.; Lei, S.M.; Zhong, L.L.; Pei, Z.Y.; Yang, Y.Y.; Zang, F.F. Thermodynamic mechanismand purification of hot press leaching with vein quartz. *China Min. Mag.* **2016**, *25*, 129–132. [CrossRef]

11. Lei, S.M.; Pei, Z.Y.; Zhong, L.L.; Ma, Q.L.; Huang, D.D.; Yang, Y.Y. Study on the technology and mechanism of reverse flotation and hot pressing leaching with vein quartz. *Nonmet. Mines* **2014**, *37*, 40–43. [CrossRef]

12. Wang, L.; Li, C.X.; Wang, Y.; Yin, D.Q. China technologies present of high-purity quartz processing and the development propositions. *J. Mineral. Petrol.* **2011**, *31*, 110–114. [CrossRef]

13. Haßler, S.; Kempe, U.; Monecke, T.; Götze, J. Trace Element Content of Quartz from the Ehrenfriedersdorf Sn-w Deposit, Germany: Results of an Acid-Wash Procedure. In *Mineral Deposit Research: Meeting the Global Challenge, Proceedings of the Eighth Biennial SGA Meeting, Beijing, China, 18–21 August 2005*; Mao, J.W., Bierlein, F.P., Eds.; Society of Economic Geologists, Inc.: Littleton, CO, USA, 2006.

14. Haus, R.; Prinz, S.; Priess, C. Assessment of high purity quartz resources. In *Quartz: Deposits, Mineralogy and Analytics*, 1st ed.; Götze, J., Möckel, R., Eds.; Springer: Berlin/Heidelberg, Germany, 2012; pp. 45–49.

15. Vidyadhar, A.; Hanumantha, R.K. Adsorption mechanism of mixed cationic/anionic collectors in feldspar-quartz flotation system. *J. Colloid Interface Sci.* **2007**, *306*, 195–204. [CrossRef] [PubMed]

16. An, J.; Lee, H.A.; Lee, J.; Yoon, H.O. Fluorine distribution in soil in the vicinity of an accidental spillage of hydrofluoric acid in Korea. *Chemosphere* **2015**, *119*, 577–582. [CrossRef] [PubMed]

17. Dasgupta, P.K. Comment on "hydrofluoric acid in the Southern California atmosphere". *Environ. Sci. Technol.* **1998**, *31*, 427. [CrossRef]

18. Zhou, Y.H. Study on refining quartz powder by leaching in HF acid solution. *J. Mineral. Petrol.* **2005**, *25*, 23–26. [CrossRef]

19. Scott, H.S. The decrepitation method applied to minerals with fluid inclusions. *Econ. Geol.* **1948**, *43*, 637–654. [CrossRef]

20. Knotter, D.M. Etching mechanism of vitreous silicon dioxide in HF-Based solutions. *J. Am. Chem. Soc.* **2000**, *122*, 4345–4351. [CrossRef]

21. Su, Y.; Zhou, Y.H.; Huang, W.; Gu, Z.A. Study on reaction kinetics between silica glasses and hydrofluoric acid. *J. Chin. Ceram. Soc.* **2004**, *32*, 287–293. [CrossRef]

22. Xue, N.N.; Zhang, Y.M.; Liu, T.; Huang, J.; Zheng, Q.S. Effects of hydration and hardening of calcium sulfate on muscovite dissolution during pressure acid leaching of black shale. *J. Clean Prod.* **2017**, *149*, 989–998. [CrossRef]

23. Mao, X.W.; Ye, Q.; Liao, M.F.; Yang, J.X.; Zhang, H.J.; Wang, Z.Y. Division and discussion of geotectonic units in Hubei Province. *Resour. Environ. Eng.* **2014**, *1*, 6–15. [CrossRef]

24. Zhang, P.C.; Liu, Y.F.; Li, J.F.; Deng, M.; Liu, S.T. Study on high-purity quartz mineral resource engineering. *J. Mineral. Petrol.* **2012**, *32*, 38–44. [CrossRef]

25. China Technical Committee for Standardization of Microbeam Analysis. *GB/T 15617-2002 Methods for Quantitative Analysis of Silicate Minerals by Electron Probe*; Standards Press of China: Beijing, China, 2002.

26. China Technical Committee for Standardization of Semiconductor Equipment and Materials. *SJ/T 11554-2015 Determination of the Metals' Concentration of Hydrofluoric Acid by ICP-OES*; Standards Press of China: Beijing, China, 2015.

27. China Technical Committee for Standardization of Semiconductor Equipment and Materials. *GB/T 32650-2016 Determining the Content of Trace Elements in Arenaceous Quartz by Inductively Coupled Plasma Mass Spectrometry (ICP-MS)*; Standards Press of China: Beijing, China, 2016.

28. Lei, S.M.; Lin, M.; Pei, Z.Y.; Wang, E.W.; Zang, F.F.; Xiong, K. Occurrence and removal of mineral impurities in quartz. *China Min. Mag.* **2016**, *25*, 79–83. [CrossRef]

29. Shmulovich, K. An experimental study of phase equilibria in the systems H_2O-CO_2-$CaCl_2$ and H_2O-CO_2-NaCl at high pressures and temperatures (500–800 °C, 0.5–0.9 GPa): Geological and geophysical applications. *Contrib. Mineral. Petrol.* **2004**, *146*, 450–462. [CrossRef]

30. Yan, F.L. Distribution properties and hosting conditions and purification methods of baneful impurity elements in quartz. *J. Geol.* **2009**, *33*, 277–279. [CrossRef]

31. Zhang, Z.Z.; Li, J.S.; Li, X.X.; Huang, H.Q.; Zhou, L.F.; Xiong, T.T. High efficiency iron removal from quartz sand using phosphoric acid. *Int. J. Miner. Process.* **2012**, *114–117*, 30–34. [CrossRef]

32. Bai, J.X.; Li, S.Q.; Yang, C.Q.; Kong, J.W. Study on the influence of ultrasound on iron removal by acid leaching for quartz sand. *Nonmet. Mines* **2016**, *39*, 69–71. [CrossRef]

33. Tang, Q.; Sun, X.M.; Xu, L.; Zhai, W.; Liang, J.L.; Liang, Y.H.; Shen, K. U-Th-Pb Chemical dating of monazite exsolutions in apatite aggregates in quartz veins of UHP rocks from the Chinese Continental Scientific Drilling (CCSD) Project. *Acta Petrol. Sin.* **2006**, *22*, 1927–1932.

34. Götze, J. Chemistry, textures and physical properties of quartz-geological interpretation and technical application. *Mineral. Mag.* **2009**, *73*, 645–671. [CrossRef]

35. Mclaren, A.C.; Cook, R.F.; Hyde, S.T.; Tobin, R.C. The mechanisms of the formation and growth of water bubbles and associated dislocation loops in synthetic quartz. *Phys. Chem. Miner.* **1983**, *9*, 79–94. [CrossRef]

36. Zhou, L.G. *The Basic of Ore Petrology*, 3rd ed.; Metallurgical Industry Press: Beijing, China, 2007; pp. 206–309.

37. Liu, C.; Lin, J. Influence of calcination temperature on dielectric constant and structure of the micro-crystalline muscovite. *China Nonmet. Min. Ind. Her.* **2008**, *5*, 38–46. [CrossRef]

38. Liu, X.C.; Wu, Y.B.; Gong, H.J.; Yang, S.H.; Wang, J.; Peng, M.; Jiao, W.F. Zircon age and Hf isotopic composition of quartz veins in UHP eclogites from western Dabie Mountains. *Chin. Sci. Bull.* **2009**, *54*, 1449–1454. [CrossRef]

39. Lu, H.P.; Wang, R.C.; Lu, X.X.; Xu, S.J.; Chen, J.; Gao, J.F. Study on dissolution behavior of zircon in hydrothermal solution of 180 °C. *Prog. Nat. Sci.* **2003**, *13*, 1042–1047. [CrossRef]

40. Schmidt-Mumm, A. Low frequency acoustic emission from quartz upon heating from 90 to 610 °C. *Phys. Chem. Miner.* **1991**, *17*, 545–553. [CrossRef]

minerals

MDPI

Article

Dissolution Behaviors of Trace Muscovite during Pressure Leaching of Hydrothermal Vein Quartz Using H_2SO_4 and NH_4Cl as Leaching Agents

Zhenyu Pei, Min Lin *, Yuanyuan Liu and Shaomin Lei

School of Resources and Environmental Engineering, Wuhan University of Technology, Wuhan 430070, China; 108344@whut.edu.cn (Z.P.); liuyuanyuan@whut.edu.cn (Y.L.); shmlei@163.com (S.L.)
* Correspondence: 208726@whut.edu.cn

Received: 26 December 2017; Accepted: 7 February 2018; Published: 11 February 2018

Abstract: Dissolution behaviors of trace muscovite during pressure leaching of hydrothermal vein quartz using H_2SO_4 and NH_4Cl as leaching agents have been studied by means of optical and electronic microscopes. Phase transformations of pure muscovite during calcination and the pressure leaching were analyzed by powder X-ray diffraction (XRD) and thermal analysis (TG-DSC), which are used for indirectly discussing dissolution mechanisms of the trace muscovite. Structure damages of trace muscovite are caused by calcination, and further developed during pressure leaching of the quartz sand using H_2SO_4 and NH_4Cl as leaching agents. The trace muscovite is dissolved, and then efficiently separated from quartz sand by coupling effects of calcination and fluorine-free pressure leaching.

Keywords: muscovite; dissolution behaviors; vein quartz; pressure leaching

---

1. Introduction

Vein quartz, as an industrial substitute of crystal quartz, usually contains some mica minerals, such as muscovite and biotite [1]. These mica minerals usually contain impurity elements of Al, K, Fe, and Ti, etc., which could evidently reduce quality of quartz products [2,3]. High-efficiency separation of the muscovite from quartz ore has been focused on by researchers for a long time [4,5]. In recent years, fluorine-free flotation has been developed to separate muscovite with quartz [6,7], but the flotation technique is only suitable for separating liberated ores, although separation efficiency of muscovite and quartz is not high enough [8,9]. Moreover, fluoric acid leaching shows great effects on removing muscovite (less than 0.2 wt %) within quartz [10,11]. However, fluorides used in the leaching process commonly lead to severe environmental pollution [12].

In the conventional process of oxygen pressure acid leaching of quartz ore, purification of quartz sand depends on acid attack to mica minerals, resulting in high acid consumption and a long leaching time [13]. The two problems lead to significant decreases in equipment life and production efficiency, and an obvious increase in production costs. In addition, pressure acid leaching with mixed agents consisting of acids and inorganic salts is deemed to be an effective method to dissolve and remove muscovite without using any fluorides [13,14]. Metallic ions in the inorganic salts are unacceptable in quartz purification because they would unavoidably drag in metallic impurities during processing, but the use of NH_4Cl avoids the problem. Mixed leaching agents consisting of H_2SO_4 and NH_4Cl show great effects on removing trace muscovite from hydrothermal vein quartz with high leaching pressure and extremely low acid consumptions [15].

Table 1 presents the impurity contents of ore and concentrate, and detailed process conditions [15]. Separation efficiency of the trace muscovite and quartz can reach 98.10% when total removal rate of impurity elements is about 84.0%.

Table 1. Contents of main impurity elements in quartz sand (μg/g) [15].

Element	Fe	Li	Mg	Ni	Ti	Ca	K	Na	Al	Zr	Others	In Total
Ore	61.2	2.20	11.8	1.01	8.34	8.05	118	13.5	353	6.46	35.1	619
Concentrate [1]	1.12	2.04	7.15	-	5.38	4.40	2.24	12.2	44.1	6.45	14.16	99.2

[1] Leaching conditions: 0.30 mol/L H_2SO_4, 0.45 mol/L NH_4Cl, 5 mL/g of L/S ratio, leaching temperature of 250 °C, leaching time of 6 h.

Although the trace muscovite has been efficiently removed by the fluorine-free pressure leaching process, dissolution behaviors of trace muscovite during the pressure leaching of hydrothermal vein quartz have not been studied in detail. Meanwhile, the removal rate of Al in high-grade quartz is not suitable to be used for the calculation of the separation efficiency of trace muscovite and quartz because substitution of Al and Si is widely found in hydrothermal quartz. Hence, a quantitative calculation method is necessary. When the $c(NH_4Cl)/c(H_2SO_4)$ ratio was 2/1, Al content obviously decreased with H_2SO_4 concentration (0.025 mol/L to 0.300 mol/L), and remained approximately constant from 0.300 mol/L to 0.500 mol/L [16]. Hence, the quartz samples leached by the NH_4Cl-H_2SO_4 solution with low concentrations can be used for microscopic analysis so as to investigate dissolution behaviors of trace muscovite.

Based on the previous research [15,16], this study is to elaborate the dissolution behaviors of trace muscovite during pressure leaching of hydrothermal vein quartz using H_2SO_4 and NH_4Cl as leaching agents, and investigating occurrences of main lattice elements in leached quartz so as to provide evidence for calculating the separation efficiency of trace muscovite and quartz using the removal rate of trace K. The research is focused on analyzing the removal mechanism of trace muscovite from hydrothermal vein quartz by characterizing the phase transformations and structure modifications of the muscovite.

2. Materials and Methods

2.1. Materials

Hydrothermal vein quartz used in this study is from Hengche Town in Qichun County, Hubei Province, China. The sampled quartz ore was washed, dried, crushed by a Raymond mill (3R2115), and grain-size separated with standard sieves. Separated samples ranging in size from 106 μm (140 mesh) to 212 μm (70 mesh) was used for experiments. Muscovite is the main Al-K host mineral in the studied quartz ore. The main element impurities in the powdered quartz obtained by [15] are presented in Table 1.

2.2. Methods

Sieved samples were calcinated (at 900 °C for 5 h) and leached (5 g/sample) by H_2SO_4-NH_4Cl solutions (50 cm^3) at different agent concentrations and temperatures in an airtight reaction kettle for 6 h.

Microstructures of trace muscovite occurred in quartz sand at different leaching conditions were analyzed by biological microscope (ALPHAPHOT-2 YS-2, Nikon, Tokyo, Japan). Muscovite was then subjected to electron microprobe (JXA-8230/INCAX-ACT, JEOL Ltd., Tokyo, Japan) for morphology observation of the BSE model (backscattered electron image) after selection from the calcined and leached quartz sands by a biological microscope due to the distinctive colors and light transmittances of muscovites. The morphology and structure of the selected muscovites were analyzed by a polarizing microscope (DLMP, Leica Microsystems, Wetzlar, Germany) and an electron microprobe (JXA-8230/INCAX-ACT). Electron images of electron probe microanalysis were obtained at an acceleration voltage of 20.0 kV and a working distance of 11.2 mm based on the GBT15617-2002 standard [17]. Leaching conditions and numbers of micro-images are shown in Table 2.

Table 2. Design of the microscopic analysis.

Leaching Condition [1]	0.025 mol/L H$_2$SO$_4$ 0.050 mol/L NH$_4$Cl	0.050 mol/L H$_2$SO$_4$ 0.100 mol/L NH$_4$Cl	0.100 mol/L H$_2$SO$_4$ 0.200 mol/L NH$_4$Cl
150 °C	Figures 2a and 3a	Figures 2d and 3d	
200 °C	Figures 2b and 3b	Figures 2e and 3e	Figure 4b
250 °C	Figures 2c, 3c and 6	Figures 2f and 3f	

[1] Liquid/solid ratio—5 mL/g, Leaching time—6 h.

Occurrences of the main metallic elements in leached quartz sand were analyzed by X-ray photoelectron spectroscopy (XPS, ESCALAB 250Xi, THERMO FISHER SCIENTIFIC, Waltham, MA, USA). Narrow spectrum analyses (XPS) of Al and Na were respectively measured 11 times. The binding energy scale was corrected based on a C1s peak from contaminations (around 284.79 eV) as the internal binding energy standard [18].

Pure natural muscovite was calcinated (at 900 °C for 5 h) and leached (2 g/sample) by H$_2$SO$_4$-NH$_4$Cl solutions (0.3 mol/L H$_2$SO$_4$, 0.6 mol/L NH$_4$Cl, 10 mL/g of L/S ratio) at 250 °C for 4 h. Mineral phases of pure natural muscovite, calcinated muscovite, and leached muscovite were analyzed by powder X-ray diffraction (RU-200B/D/MAX-RB). The powder X-ray diffraction (XRD) used a rotation anode high-power X-ray diffractometer (RU-200B/D/MAX-RB, Rigaku Corporation, JPN, Karlsruhe, Germany) employing CuKα radiation (λ = 0.154 nm, 40 kV, 50 mA) over scanning range 2θ = 5°–70° with step width 2°·min^{-1}.

The results of thermal analyses (TG-DSC) of pure natural muscovite were obtained by simultaneous thermal analysis (STA449F3) (NETZSCH, Selb, Germany). The experiment of thermal analyses of pure muscovite was carried out under air atmosphere from room temperature to 1000 °C at a heating rate of 6 °C/min.

3. Results and Discussion

3.1. Effect of Calcination Process on Muscovite Structure

The calcination process, as a pretreatment technique, is commonly used to destroy crystal structure of muscovite so as to provide more active sites during later pressure leaching [16]. Structural damages around surface, edge, interior, and cleavage plane of muscovite in quartz sand are caused by high-temperature calcination (Figure 1). Surface oxidation, volatilization of interlayer water, and thermal dilation of muscovite during the calcination process led to the structural damage. In leaching processing, leaching agents dissolve the muscovite along its surface, but after calcination, the leaching sites of the muscovite are increased, and could be dissolved along fracture and cleavage planes, additionally.

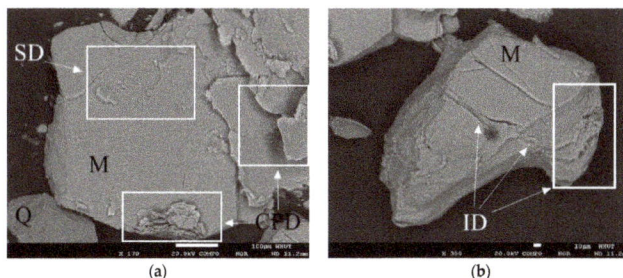

(a)   (b)

Figure 1. Surface topographies of muscovite in calcinated quartz sand: surface damage (SD), cleavage plane damage (CPD), interior damage (ID), quartz (Q), and muscovite (M): (**a**) flaky muscovite, (**b**) platelike muscovite.

3.2. Effect of Pressure Leaching on Muscovite

Calcinated quartz sand was leached by mixed agents consisting of H_2SO_4 and NH_4Cl. During the leaching process, some factors, including the leaching temperature and agent concentration, have great influences on the dissolution and separation of trace muscovite from quartz. The dissolution behaviors of the trace muscovite in the H_2SO_4-NH_4Cl leaching system were also investigated.

3.2.1. Effects of Leaching Conditions on Muscovite

The optical morphology of muscovite in calcinated quartz sand (Figure 2) shows its dissolution along the surface and fracture planes. Light transmittances of the muscovite increase with the leaching temperature. This indicates that leaching agents can directly dissolve the surface {001} of the muscovite. Observably different from Figure 2a–c, the muscovites in Figure 2d–f are efficiently dissolved from its edges. This shows that the pressure leaching process tends to dissolve the weakened of calcinated muscovite. The "gulf" in leached muscovite is caused by the dissolution tendency because the "gulf" area shows higher chemical activity than the others. Directional fractures are developed and expended to interlaced fracture when the leaching temperature was raised from 150 °C to 250 °C. The depth and width of the interlaced fractures are commonly deeper than the directional fractures, so the leaching process tends to further destroy the structure of the muscovite along the fractures caused by calcination.

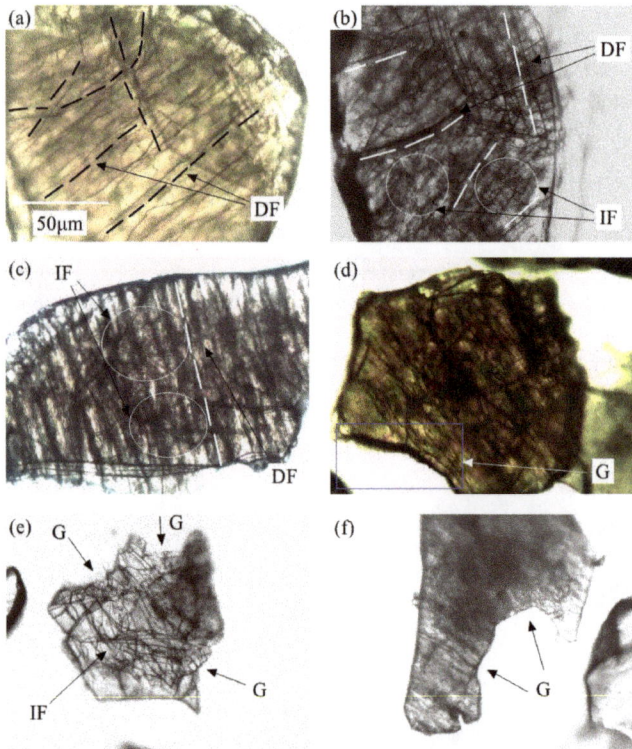

Figure 2. Optical morphology of muscovite within leached quartz sand: (1) 0.025 mol/L H_2SO_4, 0.050 mol/L NH_4Cl (a) −150 °C; (b) −200 °C; (c) −250 °C; (2) 0.050 mol/L H_2SO_4, 0.100 mol/L NH_4Cl (d) −150 °C; (e) −200 °C; and (f) −250 °C; DF—directional fracture, IF—interlaced fracture. G—"gulf"; the scale bars are the same.

Similar results are obtained by electron probe micro-analysis. As shown in Figure 3, surface etching, edge damage, directional and interlaced fractures are the main results of the pressure leaching process. Structural damage is widely developed around muscovite surfaces (Figure 3a). With the rise of the leaching temperature, the structural damage is developed from the surface into the interior of the muscovite. Some directional fractures are produced in the internal layer of the muscovite when the leaching temperature reaches 250 °C (Figure 3c). With higher agent concentrations (Figure 3d–f), directional and interlaced fractures are quickly developed around the surface and the interior of the muscovite by the pressure leaching process. Flaky muscovite is transformed into active fragments (Figure 3d–f) with its structural damage. This indicates that the directional and interlaced fractures not only occur on the surface, but are also extended into the interior of the muscovite. The directional and interlaced fractures further lead to a whole disintegration of muscovite with their further extending. As shown in Figure 3c,f, the muscovite is destroyed into several parts along these fractures. Leaching agents can, therefore, enter into the interior of the muscovite along the fracture. The active fragments have a large specific area so as to promote leaching reactions. This indicates that the H_2SO_4-NH_4Cl leaching system can accelerate the formation of the directional and interlaced fractures, and ultimately lead to the whole disintegration of the muscovite. Chemical dissolution of the edge of the muscovite further leads to a comminuted disintegration (Figure 3f). Leaching agents, therefore, enter into the cleavage planes of the muscovite, and further increases the cleavage plane spacing. The effects of calcination and pressure leaching on the fractures and cleavage planes lead to structural damage of the trace muscovite associated in quartz sand.

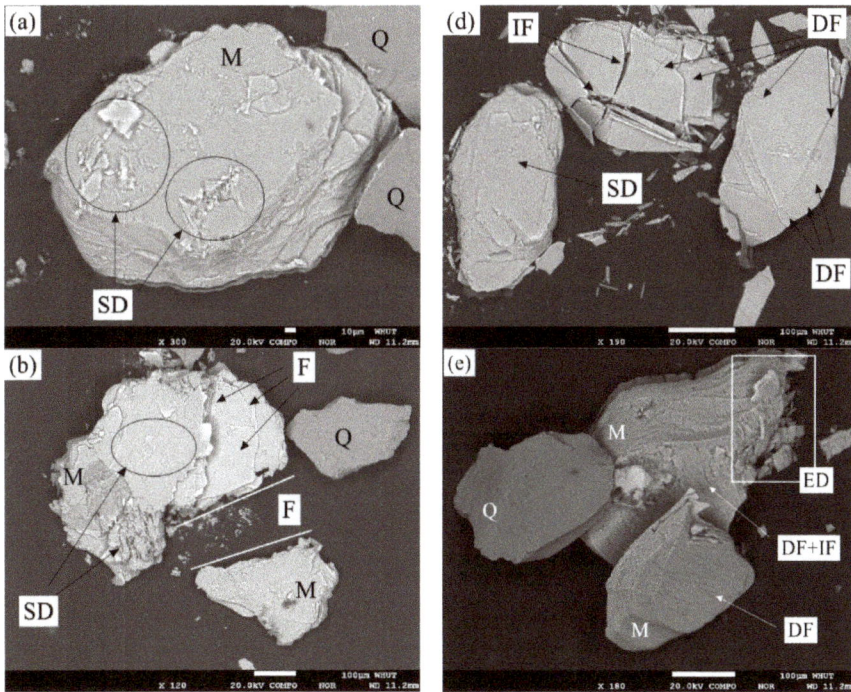

Figure 3. *Cont.*

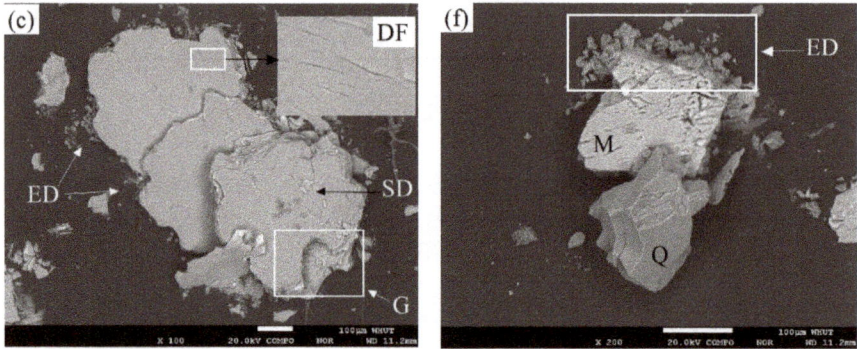

Figure 3. Surface topography of muscovite in leached quartz sand: (1) 0.025 mol/L H_2SO_4, 0.050 mol/L NH_4Cl (**a**) −150 °C; (**b**) −200 °C; (**c**) −250 °C; (2) 0.050 mol/L H_2SO_4, 0.100 mol/L NH_4Cl (**d**) −150 °C; (**e**) −200 °C; and (**f**) −250 °C; SD—surface damage, F—fracture, DF—directional fracture, IF—interlaced fracture, ED—edge damage, M—muscovite, and Q—quartz.

Integrated effects of the process on optical and surface morphologies are shown in Figure 4. Fractures not only exist on the surface of the muscovite, but also inside, because transmitted, crossed and polarized light can synthetically reflect the influences of this process on muscovite structures. The interference color (sky blue, an inexistent color in natural muscovite) of muscovite is caused during calcination. Polarized microscopic analysis shows that directional and interlaced fractures in muscovite are caused by the calcination process, and further developed during the pressure leaching process. The two processes, including calcination and pressure leaching, lead to significant changes in space structures of muscovite because the refractivity of natural muscovite is absolutely changed. These directional fractures, perpendicular to the longer side of the muscovite, are supposedly caused by heat stress, and transformed to the interlaced form by chemical dissolution during pressure leaching. Created fractures during the calcination process, and developed during the leaching process, provide new channels for internal diffusion of leaching agents and more active sites for dissolving muscovite.

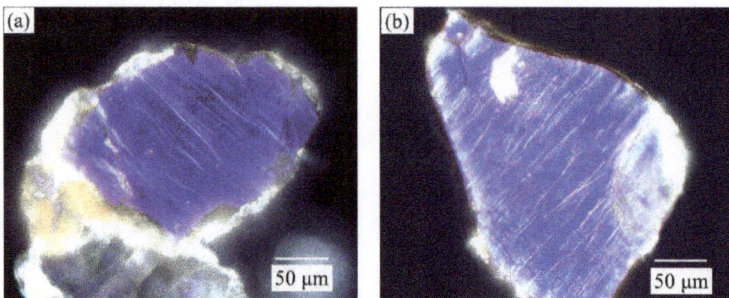

Figure 4. Polarizing luminescence of muscovite at transmitted, crossed, and polarized light: (a) muscovite within calcinated quartz sand (900 °C for 5 h); and (b) muscovite within leached quartz sand (0.100 mol/L H_2SO_4, 0.200 mol/L NH_4Cl, 5 mL/g of L/S ratio, 200 °C).

Structural damage of muscovite is mainly distributed around the surface, edge, interior, and cleavage planes. The structural damage caused by the calcination process could reduce chemical reaction resistances, and provide more chemically active sites. During the pressure leaching process of calcinated quartz sand, the muscovite is further destroyed into several fragments with the formation of micro-fractures around the surface, edge, interior, and cleavage planes. The structural damage not only provides more chemically active sites to reduce the chemical reaction resistances, but also increases specific surface areas so as to reduce internal diffusion resistances of leaching agents. In general, the muscovite is dissolved and separated from quartz sand by coupling the effects of calcination and fluorine-free acid leaching.

3.2.2. Mechanism Analysis of Muscovite Dissolution

Natural pure muscovite (99.9 wt %) is used for analyses of TG-DSC and XRD. The DSC curve (Figure 5) shows a wide exothermic peak around 600–800 °C when the TG curve shows obvious mass loss due to dehydroxylation above 700 °C [19]. Results of thermal analyses of pure muscovite (TG-DSC) show that muscovite structure is destroyed by calcination above 600 °C. The DSC curve also shows two obvious exothermic peaks at 895.9 °C and 957.6 °C. The two exothermic peaks are due to recrystallizations of spinel and γ-Al$_2$O$_3$ [19,20]. Below 800 °C, distributions of most atomics, especially for Al and Si, are not changed [21]. Around 800–900 °C, Al$_{VI}$–O octahedron close to the interlayer is destroyed with the break of Al–O bonds when the Al$_{IV}$–O tetrahedron in the muscovite layer remains unchanged [20,21]. Thus, calcination at 900 °C triggers a dehydroxylation of muscovite, and further promotes the structural damage of Al$_{VI}$–O octahedrons close to the interlayer. Since the trace muscovites are distributed uniformly in the quartz sand, the minor phases are difficult to be produced during calcination of actual minerals [15,21,22].

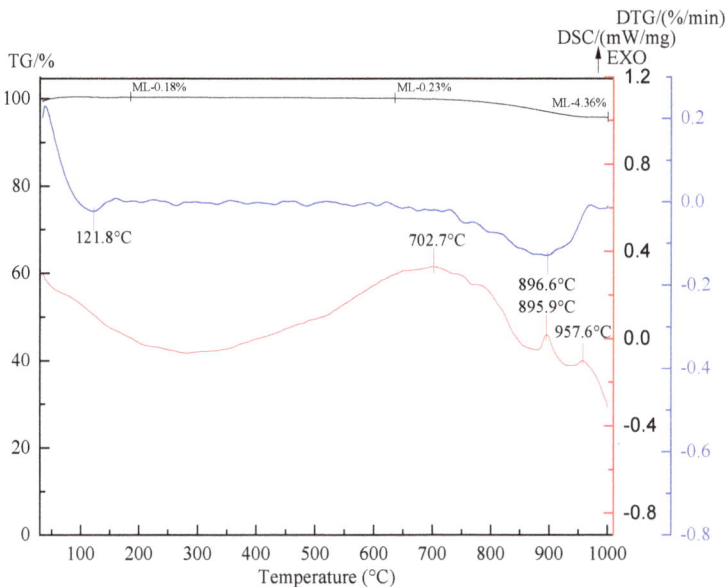

Figure 5. Thermal analysis curves of pure muscovite (99.9 wt %): mass loss (ML), exothermic (EXO).

XRD patterns for natural, calcinated (at 900 °C), and leached muscovite (Figure 6) show crystal structural destruction during calcination. Diffraction peaks of crystal faces, including {001} and {003}, are seriously impaired by calcination at 900 °C, and further weakened after fluorine-free pressure leaching. Furthermore, diffraction peaks of crystal faces including {002}, {131}, {005}, and {151} almost disappear after calcination. This shows that the calcination process provides favorable conditions for pressure leaching of trace muscovite by transforming crystal muscovite into active structures. The active structures mainly include Si–O–K, Al–O–K, and Si–O–Al (Equation (1)) [23]. The active structures are disordered in atomic arrangement, but hold a certain shape (Figure 1) based on XRD analysis. Although effects of the pressure leaching process on the structure of calcinated muscovite are not so obvious in Figure 6, small differences in diffraction peaks of {001} and {003} may indicate great effects on dissolution of trace muscovite in hydrothermal vein quartz.

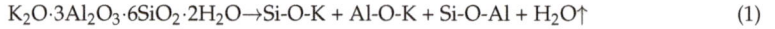

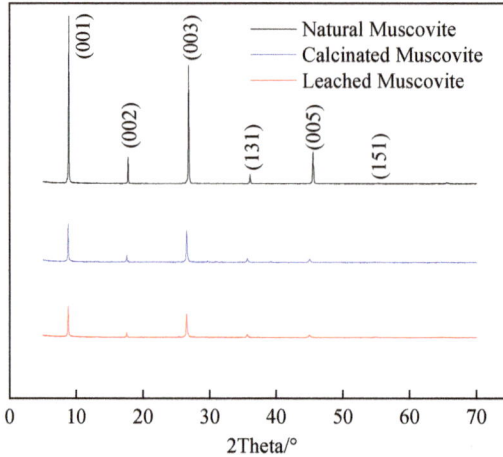

$$K_2O \cdot 3Al_2O_3 \cdot 6SiO_2 \cdot 2H_2O \rightarrow Si\text{-}O\text{-}K + Al\text{-}O\text{-}K + Si\text{-}O\text{-}Al + H_2O\uparrow \tag{1}$$

Figure 6. XRD patterns of natural, calcinated and leached muscovites: a pure muscovite (>99.9 wt%) is marked as "natural muscovite", the muscovite calcinated at 900 °C for 5 h is marked as "calcinated muscovite", and the calcinated muscovite (2 g) is leached by H_2SO_4-NH_4Cl solutions (0.3 mol/L H_2SO_4, 0.6 mol/L NH_4Cl, 10 mL/g of L/S ratio) at 250 °C for 4 h. The leaching residue is marked as "leached muscovite".

Analyses of XRD and TG-DSC shows that a small amount of calcinated muscovite only loses planar water so as to hold a metastable state (Equation (2)). The space between muscovite cleavage planes increases due to the evaporation of planar water. The expended cleavage planes provide an important channel for leaching of interlayer cations, especially for K^+ (Equation (3)):

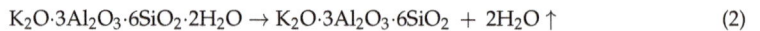

$$K_2O \cdot 3Al_2O_3 \cdot 6SiO_2 \cdot 2H_2O \rightarrow K_2O \cdot 3Al_2O_3 \cdot 6SiO_2 + 2H_2O \uparrow \tag{2}$$

$$K_2O \cdot 3Al_2O_3 \cdot 6SiO_2 + 2H^+ \rightarrow H_2O \cdot 3Al_2O_3 \cdot 6SiO_2 + 2K^+ \tag{3}$$

Furthermore, active Al–O–Si skeleton is also damaged during calcination. The compounds can easily be dissolved by H_2SO_4 without HF (Equation (4)). Meanwhile, the active structures are preferentially dissolved (Equation (5)):

$$H_2O \cdot 3Al_2O_3 \cdot 6SiO_2 + 18H^+ \rightarrow 6SiO_2 + 6Al^{3+} + 10H_2O \tag{4}$$

$$Si\text{-}O\text{-}K + Al\text{-}O\text{-}K + Si\text{-}O\text{-}Al + H^+ \rightarrow K^+ + Al^{3+} + Si-OH + H_2O \tag{5}$$

Since H_2SO_4 is completely ionized in dilute leaching solution (Equation (6)), the concentration of H^+ decreases with leaching time in leaching system of H_2SO_4 solution. However, NH_4^+ could provide a more stable leaching environment by its hydrolysis [24,25]. With the consumption of H^+, the hydrolysis balance of NH^{4+} moved to the right for maintaining the concentration of H^+ so as to reduce the chemical reaction resistances caused by the decreasing concentration of leaching agents (Equation (7)):

$$H_2SO_4 \rightarrow HSO_4^- + H^+ \rightarrow SO_4^{2-} + 2H^+ \tag{6}$$

$$NH^{4+} + H_2O \rightleftharpoons NH_3 \cdot H_2O + H^+ \tag{7}$$

In addition, the NH_4Cl is seen as an inhibitor and catalyzer by inhibiting a hydrolysis of Al^{3+}, which is caused by the triggering of elevated temperatures and the Si–O$^-$ structure, and further promotes the dissolution of Al by introducing more H^+ [16].

Surface morphologies of different layers of the muscovite in leached quartz sand are shown in Figure 7. The directional fracture not only exists in the muscovite surface, but also occurs in the internal layers. Some elements, including Si and K, in certain areas of the muscovite are preferentially leached, but leaching of Al is hysteretic. This indicates that chemical dissolution of muscovite edges are achieved by damaging Si–O–Al bonds. The possible reaction equation is shown in below:

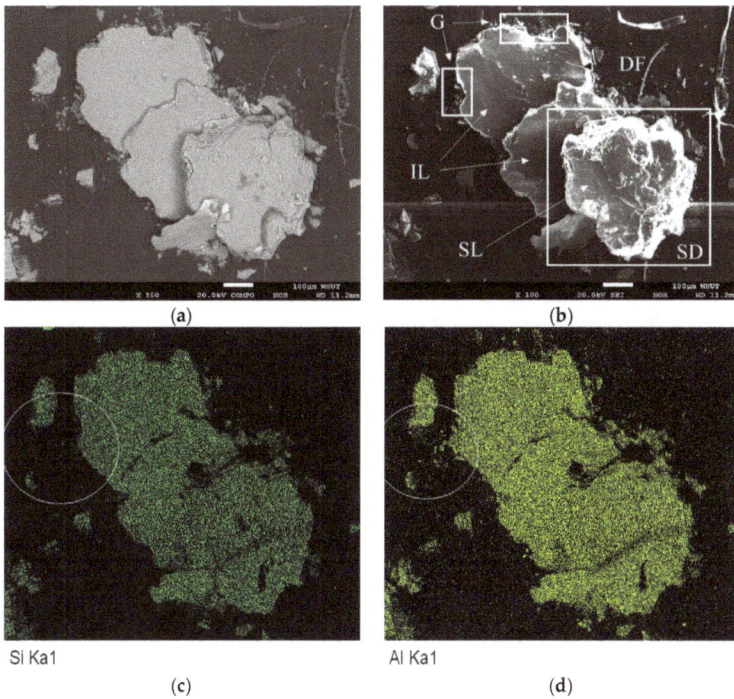

$$K_2O \cdot 3Al_2O_3 \cdot 6SiO_2 + 5H_2O + 2H^+ \rightarrow 3Al_2O_3 + 6H_2SiO_3 + 2K^+ \tag{8}$$

(a)

(b)

Si Ka1

(c)

Al Ka1

(d)

Figure 7. *Cont.*

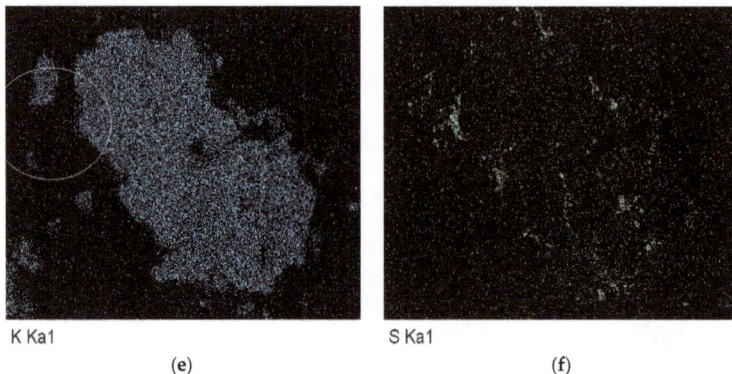

Figure 7. Morphology and element distribution of muscovite in leached quartz sand: leaching conditions (0.025 mol/L H_2SO_4, 0.050 mol/L NH_4Cl, 5 mL/g of L/S ratio, 250 °C); G—"gulf", DF—directional fracture, IL—internal layer, SL—surface layer, and SD—surface damage: (**a**) component image, (**b**) morphology image, (**c**) distribution of Si, (**d**) distribution of Al, (**e**) distribution of K, (**f**) distribution of S.

Sulfur, a representational element of the main leaching agent (H_2SO_4), was just distributed around the edge of muscovite. This indicates that muscovite dissolution is a chemical process achieved by destroying the Si–O–Al bonds along the edge of muscovite. Active Al_2O_3 could be produced during the hydrolysis of leached Al^{3+}, which is caused by the triggering of elevated temperatures and the Si–O$^-$ structure:

$$Al^{3+} + Si-O^- \rightarrow Al(OH)_3 + Si-OH \rightarrow Al_2O_3 + H_2O + Si-OH \tag{9}$$

Dissolving Al^{3+} in strong acid is most likely to be hydrated in the muscovite hydrated layer. The shearing of S shows that S–OH in H_2SO_4 could react with muscovite Si–O–Al so as to realize H^+ and Al^{3+} cation exchange. The Al would be dissolved again once the S–O–Al ($Al_2(SO_4)_3$) enters into diffusion layer:

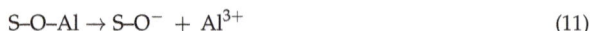

$$S-OH(S-O^-) + Si-O-Al \rightarrow Si-OH(Si-O^-) + S-O-Al \tag{10}$$

$$S-O-Al \rightarrow S-O^- + Al^{3+} \tag{11}$$

3.3. Removal Efficiency of Trace Muscovite from Vein Quartz

In conventional processing of low-grade silicate minerals, Al_2O_3 content is used for ascertaining the separation efficiency of aluminosilicate minerals and quartz [26]. However, the method error cannot be accepted in processing of high-grade quartz as Al is a major trace element in the quartz lattice [27–29]. The Al and K contents are reduced from 353 μg/g and 118 μg/g to 44.1 μg/g and 2.24 μg/g, respectively, when the calcinated quartz sand was leached by H_2SO_4-NH_4Cl solution (0.30 mol/L H_2SO_4 and 0.45 mol/L NH_4Cl) at 250 °C for 6 h (liquid/solid ratio = 5 mL/g) [15]. As shown in Figure 8, contents of Al and K cannot be further reduced even using excess leaching agents containing HF. This indicates that muscovite has been efficiently dissolved and separated. Moreover, Al (about 44 μg/g) remained in quartz concentrate could not exist in gangue mineral. As shown in Figure 9, quartz concentrate shows some Al and Na by XPS analysis. The K removal rate is 98.1% when that of Al and other elements are 87.5% and 84.0%, respectively. Obvious differences of K and Al removal rates shows that it is inadvisable to only use Al_2O_3 content to ascertain the separation efficiency of trace muscovite (Al–K host mineral) and quartz.

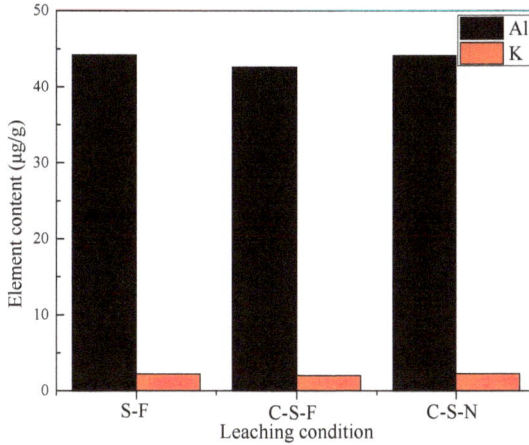

Figure 8. Contents of Al and K at different conditions: S-F (pressure leaching: 2.00 mol/L H_2SO_4, 1.00 mol/L HF, 5 mL/g of L/S ratio, leaching temperature of 250 °C, leaching time of 6 h), C-S-F (calcination: 900 °C for 5 h; pressure leaching: 2.00 mol/L H_2SO_4, 1.00 mol/L HF, 5 mL/g of L/S ratio, leaching temperature of 250 °C, leaching time of 6 h), C-S-N (calcination: 900 °C for 5 h; pressure leaching: 0.30 mol/L H_2SO_4, 0.45 mol/L NH_4Cl, 5 mL/g of L/S ratio, leaching temperature of 250 °C, leaching time of 6 h).

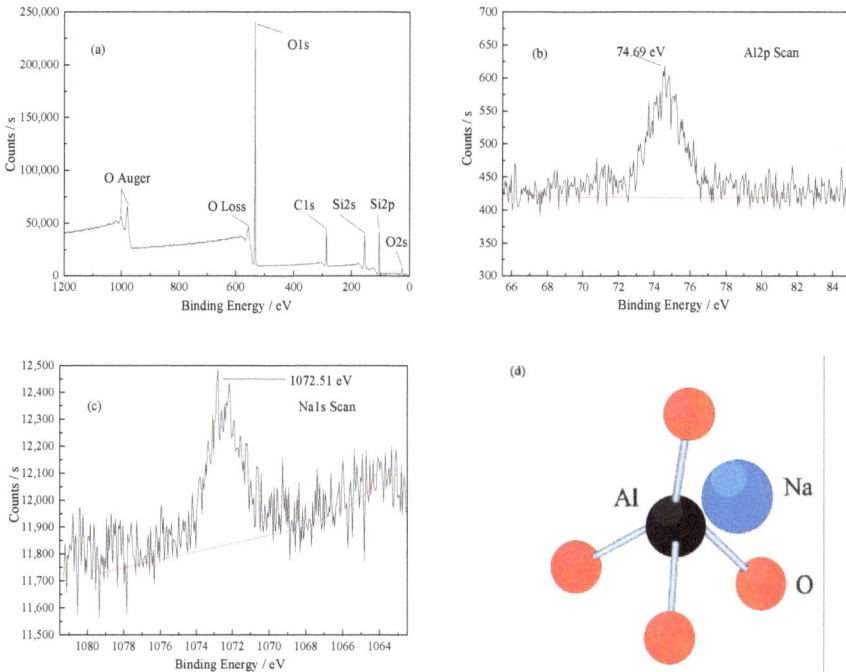

Figure 9. X-ray photoelectric spectroscopy and structure simulation of quartz concentrate: (**a**) broad spectrum of quartz concentrate; (**b**) narrow spectrum of Al; (**c**) narrow spectrum of Na; and (**d**) structural simulation of lattice substitution.

Narrow Al and Na spectra are shown in Figure 9b,c, and compared with those in the XPS handbook and native oxides, the two peaks are those of Al_2O_3 (74.7 eV) and Na (OX-1072.5 eV) [18,30]. Combined with replacement theory of Al and Si, the peak (74.69 eV) binding energy is the characteristic peak of the Al–O bond. As shown in Figure 9d, Al and Na replace Si in the tetrahedron [30–35]. Narrow patterns of possible lattice elements, including Fe, Li, Mg, Ti, and Ca, are not obtained. The phenomenon might be due to asymmetry distribution and low concentration of lattice impurities. XPS data show that Al is the main lattice trace element substitute in quartz and K in muscovite. Thus, the removal rate of K is more suitable for calculating the separation efficiency of trace muscovite and quartz:

$$\text{Separation efficiency (muscovite)} \geq \text{Removal rate (K)} = (118 - 2.24)/118 = 98.1\%$$

where 118 μg/g is the content of element K in the calcinated quartz, 2.24 μg/g is the content of element K in the leached quartz sand.

4. Conclusions

(1) Calcination leads to structural damage of the trace muscovite around surface, edge, interior, and cleavage planes. Destroyed sites provide larger specific area and higher chemical activity so as to reduce internal diffusion resistances of leaching agents and chemical reaction resistances. Structural damage of trace muscovite are caused by high-temperature calcination, and further developed during pressure leaching of the quartz sand using H_2SO_4 and NH_4Cl as leaching agents. The trace muscovite is dissolved and separated from quartz sand by coupling effects of calcination and fluorine-free pressure leaching.

(2) Si and K within muscovite are preferentially leached before Al during fluorine-free pressure leaching of the hydrothermal vein quartz. S–OH in H_2SO_4 react with the Si–O–Al structure of calcinated muscovite so as to realize a cation exchange of H^+ and Al^{3+}. The remaining active Al_2O_3 is finally dissolved when $Al_2(SO_4)_3$ enters into diffusion layer. The reason why the removal rate of Al is limited as 87.5% is that the remaining trace elements Al and Na replace Si in the quartz lattice.

Acknowledgments: This research was supported by the Open Foundation of Engineering Center of Avionics Electrical and Information Network of Guizhou Province Colleges and Universities (HKDZ201404). Special thanks to the Materials Research and Testing Center of Wuhan University of Technology.

Author Contributions: Zhenyu Pei and Min Lin conceived and designed the experiments; Zhenyu performed the experiments; Zhenyu Pei, Min Lin, and Shaomin Lei analyzed the data; Shaomin Lei contributed reagents/materials/analysis tools; and Zhenyu Pei wrote the paper.

References

1. Wang, L.; Sun, W.; Hu, Y.H.; Xu, L.H. Adsorption mechanism of mixed anionic/cationic collectors in muscovite-quartz flotation system. *Miner. Eng.* **2014**, *64*, 44–50. [CrossRef]
2. Marques, F.O.; Burlini, L.; Burg, J.P. Microstructure and mechanical properties of halite/coarse muscovite synthetic aggregates deformed in torsion. *J. Struct. Geol.* **2011**, *33*, 624–632. [CrossRef]
3. Wang, L.; Liu, R.; Hu, Y.; Liu, J.; Sun, W. Adsorption behavior of mixed cationic/anionic surfactants and their depression mechanism on the flotation of quartz. *Powder Technol.* **2016**, *302*, 15–20. [CrossRef]
4. Vegló, F.; Passariello, B.; Barbaro, M.; Plescia, P.; Marabini, A.M. Drum leaching tests in iron removal from quartz using oxalic and sulphuric acids. *Int. J. Miner. Process.* **1998**, *54*, 183–200. [CrossRef]
5. Iuga, A.; Cuglesan, I.; Samuila, A.; Blajan, M. Electrostatic separation of muscovite mica from feldspathic pegmatites. *IEEE Trans. Ind. Appl.* **2004**, *40*, 422–429. [CrossRef]

6. Marion, C.; Jordens, A.; Mccarthy, S.; Grammatikopoulos, T.; Waters, K.E. An investigation into the flotation of muscovite with an amine collector and calcium lignin sulfonate depressant. *Sep. Purif. Technol.* **2015**, *149*, 216–227. [CrossRef]

7. Jordens, A.; Marion, C.; Grammatikopoulos, T.; Waters, K.E. Understanding the effect of mineralogy on muscovite flotation using QEMSCAN. *Int. J. Miner. Process.* **2016**, *155*, 6–12. [CrossRef]

8. Vieira, A.M.; Peres, A.E.C. The effect of amine type, pH, and size range in the flotation of quartz. *Miner. Eng.* **2007**, *20*, 1008–1013. [CrossRef]

9. Xu, L.; Wu, H.; Dong, F.; Wang, L.; Wang, Z.; Xiao, J. Flotation and adsorption of mixed cationic/anionic collectors on muscovite mica. *Miner. Eng.* **2013**, *41*, 41–45. [CrossRef]

10. Monk, D.J.; Soane, D.S.; Howe, R.T. A review of the chemical reaction mechanism and kinetics for hydrofluoric acid etching of silicon dioxide for surface micromachining applications. *Thin Solid Films* **1993**, *232*, 1–12. [CrossRef]

11. Knotter, D.M. Etching mechanism of vitreous silicon dioxide in HF-based solutions. *J. Am. Chem. Soc.* **2000**, *122*, 4345–4351. [CrossRef]

12. Lei, S.M.; Guo, Z.H. Hazards of fluoride pollution and technical research progress of treating fluoride-containing waste water. *Metal Mine* **2012**, *41*, 152–155.

13. Xue, N.N.; Zhang, Y.M.; Liu, T.; Huang, J.; Zheng, Q.S. Effects of hydration and hardening of calcium sulfate on muscovite dissolution during pressure acid leaching of black shale. *J. Clean. Prod.* **2017**, *149*, 989–998. [CrossRef]

14. Xue, N.N.; Zhang, Y.M.; Liu, T.; Huang, J. Study of the dissolution behavior of muscovite in stone coal by oxygen pressure acid leaching. *Metall. Mater. Trans. B Proc. Metall. Mater. Proc. Sci.* **2016**, *47*, 694–701. [CrossRef]

15. Lin, M.; Pei, Z.Y.; Lei, S.M. Mineralogy and Processing of Hydrothermal Vein Quartz from Hengche, Hubei Province (China). *Minerals* **2017**, *7*, 161. [CrossRef]

16. Lin, M.; Pei, Z.Y.; Lei, S.M.; Liu, Y.Y.; Xia, Z.J.; Xie, F.X. Trace muscovite dissolution separation from vein quartz by elevated temperature and pressure acid leaching using sulphuric acid and ammonia chloride solutions. *Physicochem. Probl. Mineral Process.* **2017**. [CrossRef]

17. China Technical Committee for Standardization of Microbeam Analysis. *GB/T 15617-2002 Methods for Quantitative Analysis of Silicate Minerals by Electron Probe*; Standards Press of China: Beijing, China, 2002.

18. Crist, B.V. *Handbook of Monochromatic XPS Spectra, the Elements and Native Oxides*; John Wiley & Sons: New York, NY, USA, 2000; pp. 20–106.

19. Lin, L.; An, L.Y.; Liu, S.R. Study on extracting potassium from sericite by roasting. *Ind. Miner. Process.* **2012**, *41*, 12–15.

20. He, D.S.; Li, Q.S.; Qang, X.C.; Liu, X.; Qin, F. Dissolution behavior of muscovite in mixed solution of sulfuric acid and hydrofluoric acid. *Nonferrous Met. (Extr. Metall.)* **2016**, *12*, 37–39.

21. Liu, C.; Lin, J. Influence of calcination temperature on dielectric constant and structure of the micro-crystalline muscovite. *China Non-Met. Min. Ind. Her.* **2008**, *5*, 38–46.

22. Lin, M.; Pei, Z.Y.; Liu, Y.Y.; Xia, Z.J.; Xiong, K.; Lei, S.M.; Wang, E.W. High-efficiency trace Na extraction from crystal quartz ore used for fused silica—A pretreatment technology. *Int. J. Min. Met. Mater.* **2017**, *24*, 1075–1086. [CrossRef]

23. Lei, S.M.; Pei, Z.Y.; Zhong, L.L.; Ma, Q.L.; Huang, D.D.; Yang, Y.Y. Study on the technology and mechanism of reverse flotation and hot pressing leaching with vein quartz. *Non-Met. Mines* **2014**, *2*, 40–43.

24. Ranjitham, A.M.; Khangaonkar, P.R. Leaching behaviour of calcinated magnesite with ammonium chloride solutions. *Hydrometallurgy* **1990**, *23*, 177–189. [CrossRef]

25. Xu, L.J.; Qu, G.; Zhou, Z.G. A process for recovery leaching of strontium from strontium waste residues using ammonia chloride. *J. Chongqin Univ. (Nat. Sci. Ed.)* **2008**, *31*, 1174–1177.

26. El-Salmawy, M.S.; Nakahiro, Y.; Wakamatsu, T. The role of alkaline earth cations in flotation separation of quartz from feldspar. *Min. Eng.* **1993**, *6*, 1231–1243. [CrossRef]

27. Gotze, J. Classification, mineralogy and industrial potential of SiO$_2$, minerals and rocks. In *Quartz: Deposits, Mineralogy and Analytics*, 1st ed.; Möckel, R., Ed.; Springer-Verlag: Berlin/Heidelberg, Germany, 2012; pp. 1–27.

28. Beall, G.H. Industrial applications of silica. *Silica Phys. Behav. Geochem. Mater. Appl.* **1994**, *29*, 469–505.

29. Gotze, J. Chemistry, textures and physical properties of quartz—Geological interpretation and technical application. *Mineral. Mag.* **2009**, *73*, 645–671. [CrossRef]
30. Shea, J.J. Handbook of monochromatic XPS spectra—The elements and native oxides [book review]. *IEEE Electr. Insul. Mag.* **2003**, *19*, 73. [CrossRef]
31. Zang, F.F.; Lei, S.M.; Zhong, L.L.; Pei, Z.Y.; Yang, Y.Y.; Xiong, K. Purification of vein quartz by mixed acid thermal pressure leaching and it's mechanism. *China Min. Mag.* **2016**, *25*, 106–110.
32. Liu, Y.F.; Huang, Z.L.; Yang, N.; Sun, J.J. The research of preparation of ultra-pure quartz sand. *Non-Met. Mines* **2016**, *39*, 84–86.
33. Lei, S.M.; Lin, M.; Pei, Z.Y.; Wang, E.W.; Zang, F.F.; Xiong, K. Occurrence and removal of mineral impurities in quartz. *China Min. Mag.* **2016**, *25*, 79–83.
34. Pankrath, R.; Florke, O.W. Kinetics of Al-Si exchange in low and high quartz: Calculation of al diffusion coefficients. *Eur. J. Mineral.* **1994**, *6*, 435–457. [CrossRef]
35. Botis, S.M.; Pan, Y. Theoretical calculations of $[AlO_4/M^+]^0$ defects in quartz and crystal-chemical controls on the uptake of Al. *Mineral. Mag.* **2009**, *73*, 537–550. [CrossRef]

minerals

MDPI

Article

The Use of Infrared Spectroscopy to Determine the Quality of Carbonate-Rich Diatomite Ores

Adriana Guatame-Garcia * and Mike Buxton

Resource Engineering Section, Department of Geoscience and Engineering, Delft University of Technology, Stevinweg 1, 2628 CN Delft, The Netherlands; m.w.n.buxton@tudelft.nl
* Correspondence: l.a.guatame-garcia@tudelft.nl; Tel.: +31-15-278-3425

Received: 28 February 2018; Accepted: 17 March 2018; Published: 20 March 2018

Abstract: Diatomite, a rock formed by the accumulation of opaline diatom frustules, is a preferred raw material for the manufacturing of filters. Its uniqueness relies on the high porosity and inertness of the frustules. The presence of carbonates in some diatomite ores hinders these properties. The purpose of this study was to identify the type of carbonates and their association with the ore in a diatomite deposit, and to assess the suitability of determining the quality of the ore using techniques with potential for in-pit implementation. For this, run-of-mine samples were analysed using environmental scanning electron microscopy (ESEM) and infrared spectroscopy. The ESEM images showed that carbonate is present as cement and laminae. The infrared data revealed that the carbonate minerals correspond to aragonite and calcite, and that their occurrence is linked to the total amount of carbonate in the sample. By using a portable spectral instrument that uses diffuse reflectance, it was possible to classify the spectra of the ore samples based on the carbonate content. These results indicate that infrared technology could be used on-site for determining the quality of the ore, thus providing relevant information to assist the optimisation of mining and beneficiation activities.

Keywords: diatomite ore; opal; carbonate; environmental scanning electron microscopy (ESEM); infrared spectroscopy

1. Introduction

Diatomite is a sedimentary rock formed mainly by the accumulation of the skeletal remains of diatoms. Diatoms are a kind of unicellular algae that have exoskeletons (frustules) made of amorphous opaline silica with an intricate porous structure [1,2]. These characteristics make of diatomite a unique material with high porosity, low density and low chemical reactivity. Due to these properties, diatomite has great importance as an industrial raw material and is widely used as filtration media, adsorbent, filler and functional mineral additive. From these applications, filtration represents the majority of the consumption of the diatomite production [3].

The purity of diatomite is especially important to ensure the adequate performance of filter grade products. Industry standards, such the Food Chemical Codex (FCC), determine that the amount of non-siliceous substances in diatomite filters should not be more than 25 wt % [4]. Filters used for beverages pay particular attention to impurities that can be soluble in the filtered media [5]. However, during the formation of diatomite deposits, other minerals deposit along with the diatom frustules, among which carbonate minerals are common [6]. The high reactivity of CO_3 in these minerals affects the chemical purity of the diatomite ore. Furthermore, the intimate association of the frustules and carbonates can affect the porosity of diatomite. Conventionally, to generate filter-grade diatomite products, diatomite deposits are mined through open-pit methods, the extracted ore is then treated by disagglomeration, drying, degritting and sizing. Special care is taken through all these processes to preserve the delicate diatom structure [2]. Finally, to adjust the pore structure and particle size required for filtration, the beneficiated ore is calcined in a rotary kiln [7,8]. At this stage, the presence of CO_3 in the ore is also detrimental to the process, since it increases the energy consumption for the kiln and increases the CO_2 emissions [9].

In the processing of filter-grade diatomite products, the objective is to generate a grade or blend with a controlled amount of impurities. Such product must provide an acceptable flow-rate and clarity by the industrial standards and customers' requirements, at the time that maintains the essential pore structure of the diatom frustules. In this context, a considerable number of beneficiation techniques have been developed to remove the carbonate minerals and improve the purity of the diatomite finished products. Some methods use physical separation of ore and impurities such as sequential classification [10], electric field-based separation [11,12] and centrifugal separation [13]. These are sometimes complemented with enrichment methods such as flotation [14] and acid leaching [15,16]. Even though these techniques have been proven successful in increasing the purity of diatomite products, they are not ideal for the beneficiation process. The physical techniques can destroy the diatom structure [15,16], reducing the porosity of the diatomite, whereas acid leaching and flotation generate large amounts of acid wastewater, increasing the environmental impact of the process. As a consequence, it is relevant to reduce the effect of the beneficiation techniques in the quality of the diatomite products and the environment. This could be achieved by feeding the processing with a pre-upgraded ore, allowing the optimisation of the beneficiation techniques.

The author's recent research [17] suggested that the in-pit classification of the diatomite ore based on the carbonate content could support a controlled feed to the processing plant. Following this approach, prior knowledge about the characteristics of the diatom frustules and their association with the type and amount of carbonate minerals would also influence the decision-making for the beneficiation process. For this information to be useful for operational feedback, it is necessary to be able to acquire the data of the ore synchronously with the mining and processing activities, that is to say, on-time and in (near) real-time. In this context, infrared (IR) spectroscopy can detect the minerals present in the diatomite ore (opal and carbonates) and has the technology maturity to be used as an on-site technique.

In this work, a diatomite deposit in Spain was chosen to perform a detailed characterisation of a carbonate-rich diatomite ore. This study analyses how the characteristics of the ore are related to the carbonate content and how this information can be used for optimising mining and processing. For this, microscopic techniques were used to characterise the morphology and preservation of the diatom frustules and their association with the carbonate minerals and to analyse their possible influence in the porosity of the ore. Additionally, laboratory and portable infrared spectroscopy techniques were used to determine the type and amount of carbonate minerals in the ore. The observations focused on the capability of the portable techniques to be used as an on-site sensor for the characterisation of the diatomite ore. The results are further discussed in the context of offering new insights for the optimised beneficiation of diatomite ores.

2. Materials and Methods

2.1. Geological Background and Samples

The samples for this study were obtained from a diatomite deposit in the Elche de la Sierra basin (SE Spain). The deposit forms part of a system of continental Neogene basins located on the external side of the Betic Ranges Figure 1 [6]. During the Upper Miocene, the basins were filled mainly with lacustrine sediments. Towards the end of the evolution of the basins, active volcanism generated a surplus of silica in the water, favouring the thriving of diatoms. The deposition of diatomites occurred together with marls and carbonates [18], generating diatomite deposits with a variable amount of carbonates and some terrigenous intercalations through the stratigraphic sequence [19]. The formation of the carbonate minerals and the type and abundance of diatoms was affected by seasonal variations of composition and volume of the water in the lakes [6]. The diatom flora indicates a depositional environment in a shallow lacustrine basin with sporadic marine inputs [20].

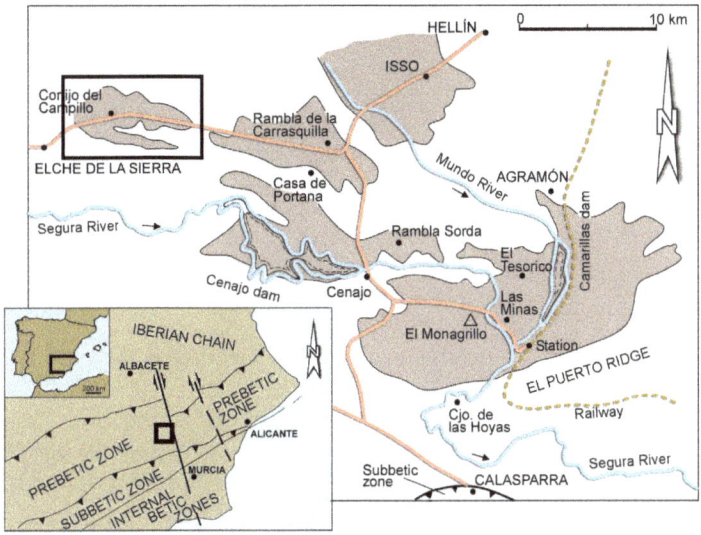

Figure 1. Neogene basins on the external side of the Betic Ranges, and location of the Elche de la Sierra basin (black rectangle in the upper-left corner) (modified from [21]).

Locally, Jurassic and Tertiary limestones and faulted Quaternary alluvium enclose the Elche de la Sierra diatomite deposit. Northwards, diatomite transforms progressively to clays, marls and conglomerate. Southwards, the edge of the body thins and transforms into opal-CT (cristobalite/tridymite). In the central part of the basin, the diatomite deposit has a maximum thickness of 80 m. The stratigraphic sequence consists of a basal conglomerate, overlain by the diatomaceous ore body, lacustrine limestones, and a Quaternary polymictic conglomerate in the top. The diatomaceous ore body consists of 8 to 11 tabular diatomite layers, interbedded with limestones, marls, sand and clay horizons.

The material for this study was composed of run-of-mine (ROM) samples collected from a mine face in the deposit at different stratigraphic positions. Figure 2 shows a section of the deposit where diatomite layers of 1 m to 3 m thickness are recognisable by their white colour, whereas a light grey colour identifies the limestone layers. Intercalation of thin layers of diatomite, limestones, marls and clay horizons are also present in this part of the deposit. Only diatomite layers thicker than 1 m are mineable. From these layers, 13 samples were taken at different stratigraphic positions and different points in the pit upon accessibility, taking special care that all the diatomite layers were represented in the sample set. The samples were named consecutively D1 to D13, although the numbering does not necessarily relate to the stratigraphic position.

Based on the threshold established by the FCC of maximum 25 wt % of non-siliceous substances in diatomite filters, for this research, a cut-off value of 25 wt % $CaCO_3$ was established as the maximum threshold for a sample to be regarded as diatomite ore. For an analogue with the actual processing of diatomite ore, the samples were classified into different quality grade (QG) levels based on the $CaCO_3$ content. The cut-off values for every QG assigned in this study do not exactly correspond to those used in the industry (disclosure not possible due to commercial sensitivity), but yet they attempt to resemble the classification used by some mining companies. The cut-off values used in this study are as follows: QG1 < 10 wt %, QG2 = 10–16 wt %, QG3 = 16–22 wt % and QG4 = 22–25 wt %. To ensure that all the QG levels were represented in the sample set, four samples with known CaCO$_3$ values were added, namely samples D14 to D17.

Figure 2. Active mine face section of the diatomite deposit where some of the ore samples were obtained: (**a**) limestone layer with small intercalations of diatomite; (**b**) diatomite layer with small intercalations of limestones; (**c**) limestone layer; (**d**) intercalation of diatomite, limestones, marls and clay horizons; (**e**) diatomite layer with small intercalations of limestones.

2.2. Analytical Methods

The association between carbonates and diatom frustules as well as the variations in the diatom morphology and preservation were investigated using environmental scanning electron microscopy (ESEM). The observations were conducted on rock chips to avoid any alteration of the diatom frustules induced by sample preparation. The instrument used was a Philips XL30 ESEM (Amsterdam, The Netherlands) and supported with element analysis. The amount of CaCO$_3$ and the concentration of major elements in the samples to support the mineralogical analyses were determined using X-ray fluorescence (XRF) using a PANalytical Magix Pro (Almelo, The Netherlands). The mineralogical content of the samples was identified by using infrared (IR) spectroscopy. This technique was used firstly due to its ability to detect amorphous phases in minerals, which is adequate for the detection of opaline silica [22], and secondly due to its demonstrated capabilities for operating in mining and industrial environments. The samples used for all the spectral analyses were gently powdered using pestle and mortar, trying to preserve the delicate diatom structure, and later homogenised. Laboratory data were collected using attenuated total reflection (ATR), with an ATR-Diamond accessory in a Perkin Elmer Spectrum 100 FTIR spectrometer (Shelton, CT, USA). The measuring conditions included compression of 1.5 kbar, spectral range 7.5 to 15 µm (1333 to 667 cm^{-1}), resolution of 2 cm^{-1} and 16 scans per second.

Spectra were also collected using portable spectrometers to assess the optimum measuring and analysing conditions required for an actual on-site application. Bidirectional reflectance spectra were collected using a PANalytical ASD FieldSpec spectrometer covering the 0.3 to 2.5 µm (33,333 to 4000 cm^{-1}) range, integrated with a contact probe and an internal light source; the spectral resolution was 0.01 µm, using 50 scans to generate one spectrum per measurement. The measuring time per spectrum (50 scans) was on average 60 s. Diffuse reflectance spectra were measured with an Agilent 4300 hand-held FTIR spectrometer (Edinburgh, UK) using coarse silver calibration, ranging from 1.9 to 15 µm (5263 to 666 cm^{-1}), spectral resolution of 4 cm^{-1}, using 128 scans to generate one spectrum per measurement. The measuring time per spectrum (128 scans) was in average 90 s.

The bidirectional and diffuse spectral measurements were carried out using Petri dishes as sample containers. The surface was first compacted and flattened with a spatula to minimise void spaces and then slightly roughened with the edge of the spatula to maximise the direction of the reflections. It is important to note that, with this method, it is not possible to achieve the same roughness in all samples, causing variations in the overall reflectance from sample to sample. Five spectral measurements per sample were recorded. The spectral files were processed in the R environment. First, noise reduction was conducted by averaging the spectra recorded for every sample. The bidirectional reflectance spectra were processed using continuum removal and derivatives. The diffuse reflectance spectra were smoothed using the Savitzky–Golay filter with polynomial order of 3 and window size of 55 data points. Further principal component analysis (PCA) was also performed in the R environment using the Chemometrics With R package [23].

3. Results

3.1. Ore ESEM Microscopy

Environmental Scanning Electron Microscope images, displayed in Figure 3, were used to characterise the texture of the diatomite ore. These images allowed the identification of the morphological characteristics of the frustules, as well as their relationship with the carbonate minerals. Most of the observed frustules are centric with radial symmetry and a variable size between 10 to 40 μm. The majority of the diatoms correspond to the genus *Cyclotella*, as Focault et al. [19] and Servant-Vildary et al. [20] reported previously, although diatoms of the genus *Tertiarius* might be present as well. Sporadic occurrences of diatoms with bilateral symmetry were also observed, and they correspond to the genera *Navicula* (Figure 3b), *Cocconeis* and *Cymatopleura*. The microscopic images showed that most of the valves are separated but complete, and occur together with fragments of the loose girdle bands (Figure 3b,d,f).

The carbonate was determined to be cement, coating the diatom frustules and in some cases filling the pores (Figure 3d), but also occurs as carbonate laminae in sharp contact with diatom laminae (Figure 3e). Biogenic aragonite with rod-like structure [24] was also present in some samples mixed with the diatoms frustules (Figure 3g–i) or forming the carbonate laminae (Figure 3j). In addition, the diatom frustules and the carbonate minerals, sponge spicules (Figure 3a,c) and detrital fragments were found in a very small proportion.

In order to identify any relationship between the previous observations and the QG levels, the samples from the same quality grade were compared. As Figure 3 shows, the morphological characteristics of the diatoms do not show any particular pattern regarding the QG level. *Cyclotella*, the most common of the diatoms, appears in all the QG levels. The other types of diatoms seem to occur randomly regardless of the type or amount of impurities in the samples. Concerning the amount and type of carbonate, it was found that, in samples from QG1 and QG2, the carbonate is most commonly found as cement, whereas laminae were almost exclusive from QG3 and QG4. Moreover, the occurrence of rod-like aragonite was also restricted to the levels with high amount of carbonate.

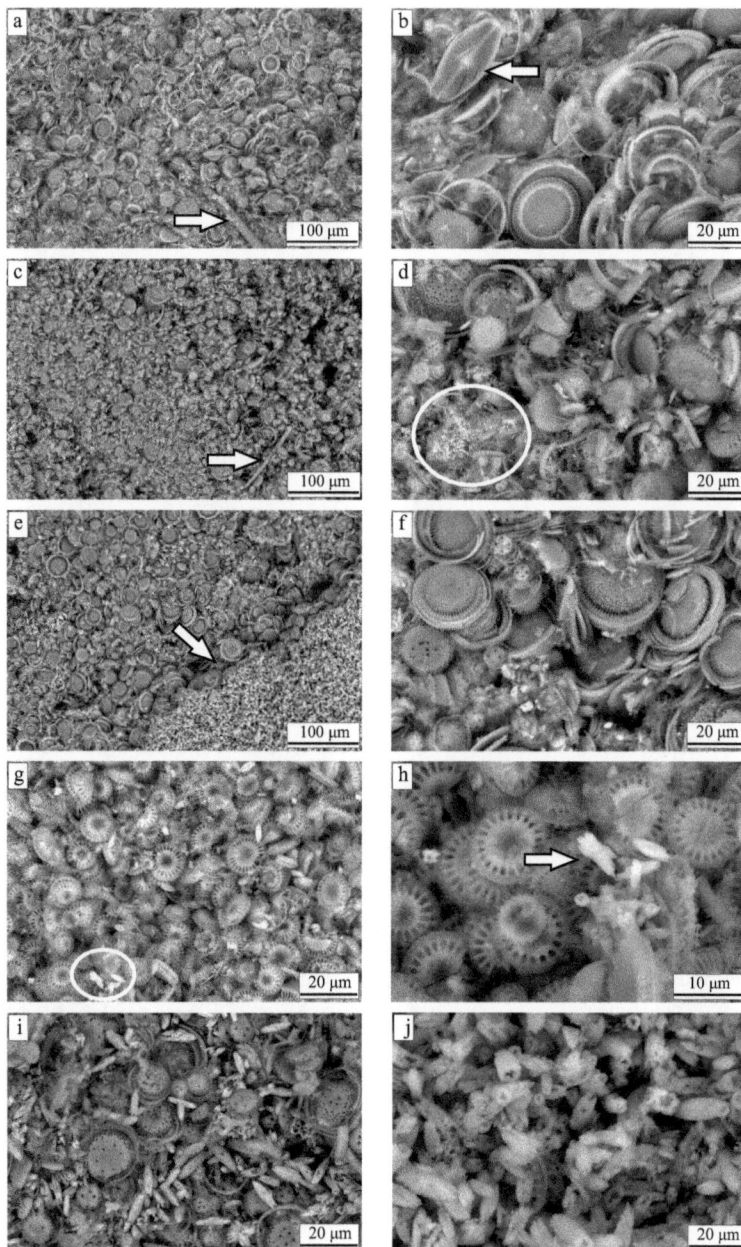

Figure 3. Environmental scanning electron microscopy (ESEM) microphotographs of the diatomite ore: (**a**) QG1 with fragments of sponge spicules; (**b**) QG1 with the occurrence of the genus *Navicula*; (**c**) QG2 with fragments of sponge spicule; (**d**) QG2 with carbonate cement; (**e**) QG3 with sharp contact between silica and carbonate laminae; (**f**) QG3; (**g–i**) QG4 with rod-like aragonite crystals; (**j**) carbonate lamina composed mainly of rod-like aragonite crystals (zoom in from microphotograph (**e**)).

3.2. XRF Characterisation

The chemical composition of the samples, summarised in Table 1, reveals that the most abundant component in the samples is SiO_2, comprising an average of 80% in the samples. Based on the ESEM microphotographs, most of the SiO_2 can be attributed to the diatom frustules, which are made of opaline silica ($SiO_2 \cdot H_2O$). The second component in abundance, and therefore the principal impurity in the ore, is $CaCO_3$, which constitutes on average 18% of the composition of the samples. It can be assigned to calcite, aragonite or vaterite, which are calcium carbonate polymorphs. The remaining 2% consist of Al_2O_3, Cl, Na_2O, SrO and SO_3, indicating that the other mineral impurities correspond to clay minerals and evaporites, such as halite, celestine and gypsum, expected from evaporitic shallow lacustrine environments, as is the case of the Elche de la Sierra basin [21]. The Mg content in the samples could be either associated with clay minerals, such as montmorillonite, or with carbonates, in the form of dolomite. However, the low concentration of Mg in the samples makes, for this study, the likely presence of dolomite negligible.

Table 1. X-ray fluorescence (XRF) analysis of the diatomite ore samples (concentrations expressed in wt %).

Sample ID	Al_2O_3	$CaCO_3$	Cl	Fe_2O_3	K_2O	MgO	Na_2O	P_2O_5	SiO_2	SO_3	SrO	TiO_2
D1	0.56	19.9	0.06	0.43	0.10	0.19	0.12	0.07	68.4	0.04	0.17	0.03
D2	0.42	18.6	0.07	0.14	0.06	0.16	0.08	0.07	70.9	0.03	0.17	0.04
D3	0.46	10.9	0.09	0.33	0.08	0.14	0.12	0.04	82.2	0.08	0.08	0.05
D4	0.64	16.4	0.07	0.34	0.11	0.19	0.09	0.06	73.7	0.05	0.10	0.03
D5	0.45	22.1	0.07	0.14	0.07	0.16	0.09	0.06	65.5	0.04	0.17	0.06
D6	0.54	22.1	0.04	0.22	0.09	0.16	0.08	0.06	65.3	0.05	0.22	0.06
D7	0.77	22.2	0.03	0.34	0.13	0.28	0.06	0.08	64.7	0.04	0.14	0.07
D8	0.67	20.8	0.04	0.69	0.14	0.21	0.08	0.08	66.7	0.03	0.13	0.05
D9	0.67	24.3	0.06	0.25	0.12	0.20	0.14	0.04	61.8	0.04	0.19	0.08
D10	0.96	23.1	0.12	0.40	0.18	0.45	0.13	0.07	62.7	0.05	0.21	0.04
D11	0.38	9.61	0.03	0.47	0.04	0.16	0.04	0.04	84.3	0.02	0.07	0.03
D12	0.68	10.4	0.02	0.32	0.09	0.25	0.06	0.05	82.8	0.02	0.02	0.05
D13	0.84	21.5	0.03	0.50	0.15	0.32	0.05	0.08	65.5	0.03	0.13	0.06
D14	0.51	6.29	0.03	0.28	0.07	0.17	0.05	0.04	89.3	0.02	0.02	0.04
D15	0.73	10.9	0.02	0.45	0.11	0.25	0.06	0.07	81.8	0.02	0.05	0.06
D16	0.74	16.4	0.05	0.50	0.11	0.29	0.11	0.07	73.3	0.03	0.12	0.07
D17	0.72	24.4	0.03	0.42	0.10	0.28	0.05	0.07	61.4	0.05	0.17	0.01
Mean	0.63	17.6	0.05	0.36	0.10	0.23	0.08	0.06	80.6	0.04	0.13	0.05
SD	0.16	5.9	0.03	0.14	0.03	0.08	0.03	0.01	6.08	0.02	0.06	0.02
Min	0.38	6.3	0.02	0.14	0.04	0.14	0.04	0.04	73.6	0.02	0.02	0.01
Max	0.96	24.2	0.12	0.69	0.18	0.45	0.14	0.08	92.4	0.08	0.22	0.08
Error	0.02	0.08	0.006	0.02	0.009	0.01	0.008	0.007	0.10	0.006	0.009	0.007

3.3. Mineralogical Characterisation Using Laboratory Infrared Spectroscopy

The infrared spectra provide a more precise characterisation of the mineralogical content in the diatomite ore. Figure 4 presents the spectra of the ore reference samples with different amounts of $CaCO_3$. The most prominent spectral features are those that are characteristic of opaline silica. The main feature, centred at 9.4 μm, is caused by the asymmetric stretching of the $[SiO_4]$ tetrahedron; the feature around 12.6 μm is associated with the Si–O–Si bending vibration mode; and the shoulder at 10.2 μm is due to Si–OH molecular vibrations [25]; this feature is in particular distinctive for diatoms [22,26,27]. Because the silica polymorphs have similar spectra in the analysed range, it was not possible to identify any other silica minerals besides opal. A particular feature of quartz at 14.4 μm [26] was not detected in any of the samples, suggesting that quartz is either absent or in a very low concentration.

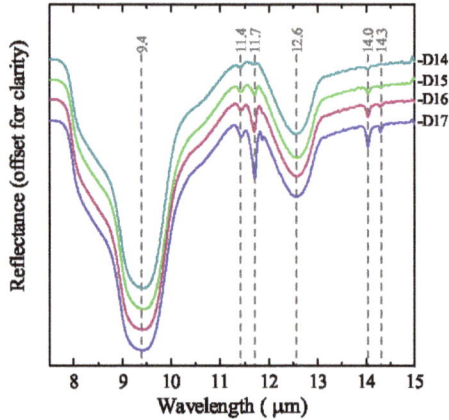

Figure 4. Attenuated total reflection (ATR) infrared (IR) spectra of diatomite ore samples with variable content of $CaCO_3$: D14 = 6.29 wt %, D15 = 10.9 wt %, D16 = 16.4 wt %, D17 = 24.4 wt %.

The identification of carbonate minerals was based first on the spectral feature at 14.0 μm, caused by the OCO bending (in-plane deformation) mode. Calcite and aragonite are differentiated by the CO_3 out-of-plane deformation mode feature located at 11.4 μm for calcite and 11.7 μm for aragonite; in addition, aragonite present as well a second feature associated with the OCO bending mode at 14.3 μm [28,29]. The characteristic feature of vaterite at 13.4 μm [24] was not detected in any of the samples. It is expected that the intensity of the spectral features of the carbonate minerals decreases with lower $CaCO_3$ content. However, it is noticeable that the calcite features, even if weak, are present in the spectra of all the samples, whereas the aragonite features are only present at relatively high $CaCO_3$ content (above ~11 wt %). The IR spectra did not detect any of other minerals likely to be present in the samples, namely clays and evaporites, since their concentration in the samples is too low.

3.4. Mineral Identification Using Portable Infrared Devices

Two portable spectral techniques and ranges were tested to assess the suitability of detecting on-site the purity of the diatomite ore in relation to the carbonate content. Each technique makes use of a different type of infrared reflection and a different spectral range. The potential of these techniques for an on-site implementation was assessed based on their ability to differentiate the carbonate minerals from the opaline silica and to make a qualitative assessment of the amount of impurities in the ore.

The spectra recorded using bidirectional reflectance were constrained to the 1.9 to 2.5 μm wavelength range, where the absorption features of opaline silica and carbonates occur. In this range, opaline silica presents features at 2.21 μm and 2.26 μm caused by the stretching mode of isolated Si–OH and bending mode of the H-bound silanol [27]. Aragonite and calcite present strong absorption features at 2.33 μm and 2.26 μm due to the combination tones of the carbonate vibrations; for these two minerals, the features occur at the same or slightly shorter wavelengths [30]. Other carbonate features are present between 1.7 to 2.1 μm, but they are usually not present in mixed spectra. In Figure 5 (top), the spectra of the diatomite ore were processed using the Continuum Removal (CR) technique to enhance the differences between spectral features. However, only the features of opaline silica are visible. The overlap of opal and carbonate features at 2.26 μm hinders the use of this wavelength as a diagnostic mark, and the 2.33 μm feature is masked. To amplify the possible differences in the spectra, the first derivative was calculated. Figure 5 (bottom) reveals the 2.33 μm carbonate feature in the D16 and D17 samples, indicating that, despite the overwhelming spectra of opaline silica, carbonate minerals can be detected as long as the amount of $CaCO_3$ is no less than 11 wt %.

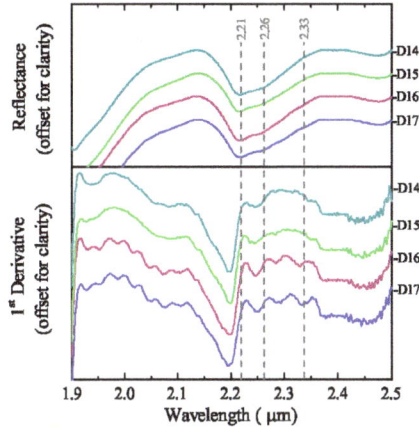

Figure 5. Bidirectional reflectance spectra of diatomite ore samples with variable content of CaCO₃ after continuum removal (**top**) and 1st derivative (**bottom**): D14 = 6.29 wt %, D15 = 10.9 wt %, D16 = 16.4 wt %, D17 = 24.4 wt %.

In the diffuse reflectance, the spectra were noisier than using bidirectional reflectance, and long wavelengths were influenced by the volume scattering of the diatomite powders [31], generating low-quality spectra. For these data, the most useful spectral region was the one that covers the opal and carbonate between 6.0 to 8.0 μm. In this range, the features of opaline silica are detected at 6.1 μm, caused by the H–O–H bend in isolated molecular water, and at 7.4 μm, corresponding to the Christiansen Feature (CF) for silicates [27]. The features for carbonate minerals in this region are due to the asymmetric C–O stretching; absorptions at 6.3 μm and 6.6 μm are characteristic of aragonite, whereas those at 6.9 μm and 7.2 μm correspond to calcite [28,29]. In the spectra of the diatomite ore, shown in Figure 6, the opal Christiansen Feature in the D14 sample is broadened to shorter wavelengths due to the 7.2 μm calcite feature. With the increasing CaCO₃ content in the other samples, the calcite absorptions of calcite become better defined along with the aragonite ones. In the particular case of aragonite, the broad bands in Figure 6 are an indication of a distorted amorphous lattice [28].

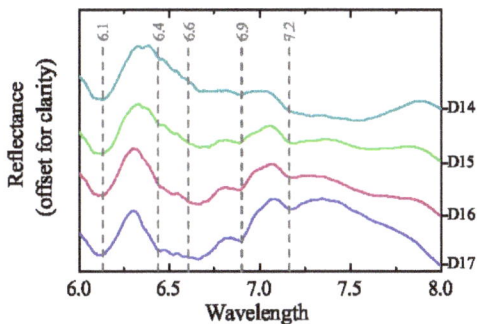

Figure 6. Diffuse reflectance spectra of diatomite ore samples with variable content of CaCO₃: D14 = 6.29 wt %, D15 = 10.9 wt %, D16 = 16.4 wt %, D17 = 24.4 wt %.

The results of the characterisation of the diatomite ore using diffuse reflectance spectra are in general in agreement with those achieved by using the laboratory ATR data. Furthermore, they provide more complete information than that acquired by using the bidirectional reflectance set-up. Based on this outcome, further analysis was carried out for the diffuse reflectance data. Principal component analysis (PCA) was used to assess whether it is possible to classify the ore according to the QG levels using the infrared spectra. PCA analysis presents the advantage of identifying the largest possible variance in the sample set, giving the most meaningful information in the first components. Other sources of variability related to noise, for example, are relegated to other components. In the case of the spectra used in this study, the PCA analysis is convenient since it identifies not only the spectral variability related to the mineral composition, but also because it minimises the influence of spectral variations in the overall reflectance due to the sample preparation.

Figure 7 shows the scores and loadings plots of the first principal component (PC1). The scores plot is coloured based on the quality grade levels. Even though the size of the sample set is rather small, it is possible to identify trends in the PCA scores. Positive scores are associated with QG1 and QG2, whereas negative scores are related to QG3 and QG4. Moreover, the dimension of the positive scores separates QG1 from QG2. In contrast, in the negative scores, there is not a clear separation between QG3 and QG4. The variability in the spectra described by PC1 is related to the silica and the carbonate features, as the loadings plot shows. Positive loadings belong to the region between 6.3 to 6.8 µm, where the aragonite features occur. In contrast, negative loadings correspond to the regions that host the opal features, between 6.0 to 6.3 µm and 7.0 to 7.7 µm. It is not surprising that the PCA does not differentiate sharply between quality grade levels since, on the one hand, the cut-off values for the quality grades are arbitrary; on the other hand, the composition and consequently the spectra of the samples vary continuously along quality grades.

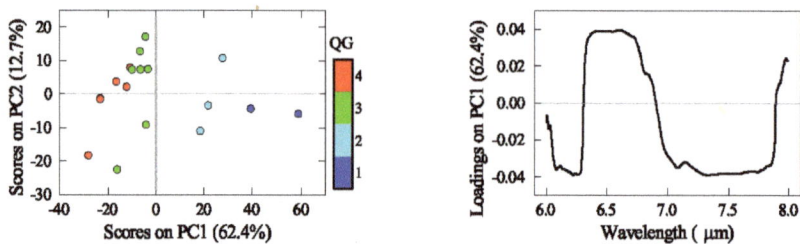

Figure 7. Principal component analysis (PC1 vs. PC2) of diatomite ores (**left**) and corresponding loadings plot (**right**).

4. Discussion

4.1. Influence of the Ore Characteristics in Mineral Processing

The observations about the texture of the diatomite ore revealed that, even though the girdles are commonly detached from the diatom frustule, complete valves and complete shells make up the majority of the diatom remains. This implies that the natural porosity of the diatomite ore is not affected by the disaggregation of the frustules. However, it was also observed that the carbonate present as cement fills the pores of some of the frustules, decreasing the natural porosity of the ore, as shown in Figure 3d. In contrast, biogenic aragonite crystals do not affect the porosity, due to their particle size (10 to 20 µm), and carbonate laminae have a reduced surface of contact with the frustules.

The association between the carbonate and the frustules, together with the type of carbonate present in the ore, might influence the performance of certain beneficiation techniques [12]. Removal of carbonate present as cement would be more effective by using wet beneficiation techniques that dissolve the $CaCO_3$, whereas large aragonite crystals and carbonate laminae, in general, could be removed by using physical beneficiation techniques. Since the presence of carbonate as a cement is usually found in ores of higher quality (QG1 and QG2), and laminae and aragonite crystals are more common in high carbonate ores (QG3 and QG4), high and low quality ores could be beneficiated separately using adequate methods for the type of mineral association.

4.2. Potential for the Use of Infrared Sensors as a Tool for Optimising Mining and Mineral Processing

The characterisation of the diatomite using infrared spectroscopy enable the identification of calcite and aragonite as the main ore impurities. Moreover, it established a correspondence between the type of carbonate mineral and the quality of the diatomite, where aragonite is restricted to high carbonate ores. The identification of the type of carbonate was possible by using not only a well-established laboratory set-up, but also by using a portable instrument that uses diffuse reflectance spectroscopy. Even though the quality of the data collected with the device equipped with diffuse reflectance is lower than the laboratory and bidirectional reflectance ones, the acquired spectra were shown to be adequate to perform a qualitative analysis of the composition of the diatomite ore.

The main advantage of the diffuse over the bidirectional reflectance is based on the possibility of using a spectral range where the features of carbonates and opal do not overlap, enabling a straightforward identification of the two mineral phases present in the ore. Moreover, in the spectral range used, it was also possible to differentiate the features of calcite from those of aragonite. Given these factors, it was feasible to perform a qualitative analysis using principal component analysis (PCA). The PCA results demonstrated that, by using the spectra above, it is possible to classify the diatomite ore based on the carbonate content. These results also open up the opportunity for conducting more precise quantitative analysis with the aid of chemometric tools, such as partial least squares regression (PLS), as suggested by Meyer et al. [32], or relational models based on the inherent parameters of the carbonate absorption features (e.g., wavelength position, depth) and the actual $CaCO_3$ content following the approach used by Zaini et al. [33].

Besides the analytic advantages provided by the use of diffuse reflectance spectroscopy, this technique also represents a suitable sensor for the in-pit characterisation of the diatomite ore. The spectral range used is not affected by particle size, and, therefore, can be used either for rock or powder samples. The availability of portable instruments that work in the mentioned spectral range enables the collection of data on-site, with little or no sample preparation. The spectral acquisition takes a few minutes per sample only, and the data does not require extensive processing or high computing capabilities, making it possible to generate information in (near) real-time.

Despite these advantages, to the knowledge of the authors, the technological advances to date in the 6.0 to 8.0 μm range allow collecting point data only, leaving applications such as sample and face imaging out of reach. Hyperspectral imagers are available only in the Short Wavelength IR (SWIR) (1.0 μm to 2.5 μm) and in the thermal IR (TIR) (8.0 μm to 12.0 μm). The results of this study discouraged the use of the SWIR range due to the overwhelming spectra of opal (Figure 5) and the use of the TIR range due to issues related to particle size. Nevertheless, for the applications that work for whole rock rather than for powders, this constraint is not valid anymore. In this case, the features that were studied in this work using laboratory spectra (Figure 4) could be possibly used to establish the relationship between the spectra and the $CaCO_3$ content. However, unlike the diffuse reflectance data, the collection and processing of spectral images, either as mine face or drill core scanners, are time-consuming, and therefore, they could not be considered for processes that require real-time data.

The use of diffuse reflectance spectroscopy, using a portable device ranging around 6.0 to 8.0 μm, would enable the on-site determination of the quality grade of the diatomite ore. Even though it works as a point measurement technique, results can be obtained on-site and (near) real-time offering the possibility of collecting several data points that characterise either a section of the pit or high volumes of stocked piled ore. The results of the spectral analysis could be used, for example, to aid selective mining, by discriminating sections of the pit with a given quality grade. The spectral classification of the ore could also assist with the upgrading of the ore before mineral processing, aiding in the selection of convenient and optimised beneficiation techniques.

5. Conclusions

This study presented the main characteristics of the diatomite ore that are relevant for generating filter-grade finished products. In addition, it assessed the potential of using infrared spectroscopy as a technique for the on-site characterisation of the diatomite ore with a view to the optimisation of mining and beneficiation.

The textural observations of the ore revealed that the diatom frustules are excellently preserved and the disaggregation state does not diminish the porosity of the diatomite. The presence of carbonate minerals was identified as the only factor that affects the quality of the ore. For this case study, carbonate was found as calcite cement in ores with relatively low carbonate content, and as cement, loose aragonite crystals and laminae in ores with relatively high carbonate content. These minerals were identified by using laboratory and portable infrared spectroscopy. From the portable techniques tested, diffuse reflectance using the 6.0 to 8.0 µm spectral range provided a better mineral identification that could be used for classifying the diatomite ore according to the pre-determined quality grades.

These results show that the quality of the diatomite ore can be determined by using infrared spectroscopy. This information integrated with the acquired knowledge about the type of associations between the carbonate and opaline silica phases can be used for optimising the beneficiation process. Consequently, this study offers insights on the advantages and the potential of using infrared spectroscopy as an on-site tool for the characterisation of the diatomite ore. An infrared-based determination of the quality of the ore would provide relevant and timely information at mining and pre-beneficiation stages that can support proactively the decision-making process. Ultimately, the implementation of infrared ore characterisation would be reflected in the more efficient use of the diatomite resources, optimised mineral processing and generation of high-quality diatomite products.

Acknowledgments: The authors wish to thank Imerys Ltd., Cornwall, UK and Alicante, Spain, for access to the samples, and to Emiel Klifman for the ESEM analyses. This work was financially supported by the European FP7 project "Sustainable Technologies for Calcined Industrial Minerals in Europe" (STOICISM), grant NMP2-LA-2012-310645.

Author Contributions: Adriana Guatame-Garcia and Mike Buxton conceived and designed the experiments; Adriana Guatame-Garcia performed the experiments analysed the data; Adriana Guatame-Garcia wrote the paper and Mike Buxton reviewed it before submission.

Conflicts of Interest: The authors declare no conflict of interest. The founding sponsors had no role in the design of the study; in the collection, analyses, or interpretation of data; in the writing of the manuscript, and in the decision to publish the results.

References

1. Korunic, Z. Review—Diatomaceous earths, a group of natural insecticides. *J. Stored Prod. Res.* **1998**, *34*, 87–97.
2. Breese, R.O.Y.; Bodycomb, F.M. Diatomite. In *Industrial Minerals and Rocks*, 7th ed.; Kogel, J.E., Trivedi, N.C., Barker, J.M., Krukowski, S.T., Eds.; Society for Mining, Metallurgy, and Exploration: Englewood, CO, USA, 2006; pp. 433–450.
3. Crangle, R.D. 2015 Minerals Year Book–Diatomite [Advance Release]. 2016. Available online: https://minerals.usgs.gov/minerals/pubs/commodity/diatomite/myb1-2015-diato.pdf (accessed on 30 January 2018).
4. USPC. Food Chemicals Codex. 2016. Available online: https://app.knovel.com/hotlink/toc/id:kpFCCE0031/food-chemicals-codex/food-chemicals-codex (accessed on 14 March 2018).
5. Braun, F.; Hildebrand, N.; Wilkinson, S.; Back, W.; Krottenthaler, M.; Becker, T. Large-Scale Study on Beer Filtration with Combined Filter Aid Additions to Cellulose Fibres. *J. Inst. Brew.* **2011**, *117*, 314–328.
6. Bellanca, A.; Calvo, J.P.; Censi, P.; Elizaga, E.; Neri, R. Evolution of Lacustrine Diatomite Carbonate Cycles of Miocene Age, Southeastern Spain: Petrology and Isotope Geochemistry. *J. Sediment. Petrol.* **1989**, *59*, 45–52.
7. Martinovic, S.; Vlahovic, M.; Boljanac, T.; Pavlovic, L. Preparation of filter aids based on diatomites. *Int. J. Miner. Process.* **2006**, *80*, 255–260.
8. Ediz, N.; Bentli, İ.; Tatar, İ. Improvement in filtration characteristics of diatomite by calcination. *Int. J. Miner. Process.* **2010**, *94*, 129–134.
9. Moffat, W.; Walmsley, M.R.W. Understanding lime calcination kinetics for energy cost reduction. In Proceedings of the 59th Appita Conference, Auckland, New Zealand, 16–19 May 2005.

10. Al-Wakeel, M.I. Characterization and process development of the Nile diatomaceous sediment. *Int. J. Miner. Process.* **2009**, *92*, 128–136.

11. Jung, K.W.; Jang, D.; Ahn, K.H. A novel approach for improvement of purity and porosity in diatomite (diatomaceous earth) by applying an electric field. *Int. J. Miner. Process.* **2014**, *131*, 7–11.

12. Moradi, S.; Moseley, D.; Hrach, F.; Gupta, A. Electrostatic beneficiation of diatomaceous earth. *Int. J. Miner. Process.* **2017**, *169*, 142–161.

13. Sun, Z.; Mao, J.; Hu, Z.; Zheng, S. Study on pilot-scale centrifugal separator for low-grade diatomite purification using response surface methodology. *Part. Sci. Technol.* **2017**, *35*, 119–126.

14. Rezai, B. The beneficiation studies of diatomite by flotation and hydrocyclone. In *Mineral Processing Technology Mpt-2005*; Venugopal, R., Sharma, T., Saxena, V.K., Mandre, N.R., Eds.; McGraw Hill: New York, NY, USA, 2005; pp. 214–227.

15. Şan, O.; Gören, R.; Özgür, C. Purification of diatomite powder by acid leaching for use in fabrication of porous ceramics. *Int. J. Miner. Process.* **2009**, *93*, 6–10.

16. Zhang, G.; Cai, D.; Wang, M.; Zhang, C.; Zhang, J.; Wu, Z. Microstructural modification of diatomite by acid treatment, high-speed shear, and ultrasound. *Microporous Mesoporous Mater.* **2013**, *165*, 106–112.

17. Guatame-García, A.; Buxton, M. Detection of mineral impurities in diatomite ores. In Proceedings of the 2nd International Conference on Applied Mineralogy & Advanced Materials AMAM-ICAM 2017, Taranto, Italy, 5–9 June 2017; Fiore, S., Ed.; Digilabs: Bari, Italy, 2017; Volume 4, pp. 19–24.

18. Elizaga, E.; Calvo, J.P. Evolución sedimentaria de las cuencas lacustres neógenas de la zona prebética (Albacete, España). Relación, posicióny efectos del vulcanismo durante la evolución. Interés minero. *Boletín Geológicoy Minero* **1988**, *99*, 837–846.

19. Foucault, A.; Calvo, J.; Elizaga, E.; Rouchy, J.; Servant-Vildary, S. Situation of the late Miocene lacustrine deposits from Hellin—Province of Albacete, Spain—in the geodynamic evolution of the Betic Cordilleras. [Place des depots lacustres d'age miocene superieur de la region de Hellin (Province de Albacete, Espagne) dans l'evolution geodynamique des Cordilleres betiques.]. *Comptes Rendus Acad. Sci. Ser. II* **1987**, *305*, 1163–1166.

20. Servant-Vildary, S.; Rouchy, J.; Pierre, C.; Foucault, A. Marine and continental water contributions to a hypersaline basin using diatom ecology, sedimentology and stable isotopes: An example in the Late Miocene of the Mediterranean (Hellin Basin, southern Spain). *Palaeogeogr. Palaeoclimatol. Palaeoecol.* **1990**, *79*, 189–204.

21. Ortí, F.; Rosell, L.; Gibert, L.; Moragas, M.; Playà, E.; Inglès, M.; Rouchy, J.M.; Calvo, J.P.; Gimeno, D. Evaporite sedimentation in a tectonically active basin: The lacustrine Las Minas Gypsum unit (Late Tortonian, SE Spain). *Sediment. Geol.* **2014**, *311*, 17–42.

22. Chester, R.; Elderfield, H. The Infrared Determination of Opal in Siliceous Deep-sea Sediments. *Geochim. Cosmochim. Acta* **1968**, *32*, 1128–1140.

23. Wehrens, R. *Chemometrics with R*; Springer: Berlin/Heidelberg, Germany, 2011.

24. Gopi, S.P.; Subramanian, V. Polymorphism in $CaCO_3$—Effect of temperature under the influence of EDTA (di sodium salt). *Desalination* **2012**, *297*, 38–47.

25. Gendron-Badou, A.; Coradin, T.; Maquet, J.; Fröhlich, F.; Livage, J. Spectroscopic characterization of biogenic silica. *J. Non-Cryst. Solids* **2003**, *316*, 331–337.

26. Moenke, H.H.W. Silica, the Three-dimensional Silicates, Borosilicates and Beryllium Silicates. In *The Infrared Spectra of Minerals*; Monograph 4, Book Section 16; Farmer, V.C., Ed.; Mineralogical Society: Chantilly, VA, USA, 1974; pp. 365–382.

27. Goryniuk, M.C. The reflectance spectra of opal-A (0.5–25 microns) from the Taupo Volcanic Zone: Spectra that may identify hydrothermal systems on planetary surfaces. *Geophys. Res. Lett.* **2004**, *31*, doi:10.1029/2004GL021481.

28. Andersen, F.A.; Brečević, L. Infrared Spectra of Amorphous and Crystalline Calcium Carbonate. *Acta Chem. Scand.* **1991**, *45*, 1018–1024.

29. Gunasekaran, S.; Anbalagan, G.; Pandi, S. Raman and infrared spectra of carbonates of calcite structure. *J. Raman Spectrosc.* **2006**, *37*, 892–899.

30. Gaffey, S. Spectral reflectance of carbonate minerals in the visible and near infrared (0.35–2.55 microns): Calcite, aragonite, and dolomite. *Am. Mineral.* **1986**, *71*, 151–162.

31. Cooper, B.L.; Salisbury, J.W.; Killen, R.M.; Potter, A.E. Mid-infrared spectral features of rocks and their powders. *J. Geophys. Res. E Planets* **2002**, *107*, doi:10.1029/2000JE001462.
32. Meyer-Jacob, C.; Vogel, H.; Boxberg, F.; Rosén, P.; Weber, M.E.; Bindler, R. Independent measurement of biogenic silica in sediments by FTIR spectroscopy and PLS regression. *J. Paleolimnol.* **2014**, *52*, 245–255.
33. Zaini, N.; van der Meer, F.; van Ruitenbeek, F.; de Smeth, B.; Amri, F.; Lievens, C. An Alternative Quality Control Technique for Mineral Chemistry Analysis of Portland Cement-Grade Limestone Using Shortwave Infrared Spectroscopy. *Remote Sens.* **2016**, *8*, 950, doi:10.3390/rs8110950.

minerals

MDPI

Article

Silica Colloid Ordering in a Dynamic Sedimentary Environment

Moritz Liesegang * and Ralf Milke

Institut für Geologische Wissenschaften, Freie Universität Berlin, Malteserstrasse 74-100, 12249 Berlin, Germany; milke@zedat.fu-berlin.de
* Correspondence: limo@zedat.fu-berlin.de; Tel.: +49-(0)30-838-70323

Received: 1 December 2017; Accepted: 4 January 2018; Published: 7 January 2018

Abstract: The formation of ordered particle arrays plays an essential role in nanotechnology, biological systems, and inorganic photonic structures in the geosphere. Here, we show how ordered arrays of amorphous silica spheres form in deeply weathered lithologies of the Great Artesian Basin (central Australia). Our multi-method approach, using optical and scanning electron microscopy, X-ray microdiffraction, Raman spectroscopy, and electron probe microanalysis, reveals that particle morphologies trace the flow of opal-forming colloidal suspensions and document syn- and post-depositional deformation. The micromorphology of amorphous silica pseudomorphs suggests that the volume-preserving replacement of non-silicate minerals proceeds via an interface-coupled dissolution precipitation process. We conclude that colloid flow and post-depositional shearing create but also destroy natural photonic crystals. Contrary to previous studies, our results indicate that purely gravitational settling/ordering is the exception rather than the rule during the formation of three-dimensional periodic sphere arrays in the highly dynamic colloidal suspensions of chemically weathered clastic sediments.

Keywords: silica; opal-A; common opal; precious opal; silica colloid; photonic crystal; particle deformation; SEM; interface-coupled dissolution-precipitation; Australia

1. Introduction

The hydrated mineralogical assemblages of the Great Artesian Basin (central Australia) record the acidic oxidative weathering of volcaniclastic sediments on a multimillion-year time scale [1]. Chemical weathering of the silicate rocks induces the in situ formation of amorphous silica at silicate mineral surfaces [2] and the release of silica into solution, followed by amorphous nanoparticle precipitation through inorganic processes [3]. Scanning electron microscopy shows that nanosphere-based amorphous silica (opal-A) in central Australia consists of subparticles tens of nanometers in size, indicating aggregative particle growth [4–7]. Ordering of the final silica spheres leads to the formation of a natural photonic crystal that modulates visible light due to Bragg diffraction. In precious opal-A, uniform spheres form a regular three-dimensional array that diffracts visible light, giving the characteristic play-of-color [8,9], which is absent in common opal-A. So far, the formation of ordered sphere lattices in fractures and pores of the host rocks and during mineral replacement has been attributed to gravitational sphere settling [1,4,5,10,11]. All of these models agree that gravity and short-range particle interaction potentials create structural order when uniform spheres form in a gel and sediment from or through it at quiescent conditions in sealed environments. This classical theory contrasts considerably with the abundant presence of opal-bearing hydraulic fractures, viscous colloid flow textures, opal reactivation structures, and nanoscale replacement processes in the subsurface precious opal deposits [1,6,7,11]. These characteristics demonstrate the highly dynamic environment that produces ordered arrays of uniform, X-ray amorphous silica spheres. In this study, we use a multi-method analytical approach to identify the mineralogical and micromorphological characteristics of these ordered arrays in two different contexts: fracture infilling and replacement. These new data allow us to provide a coherent interpretation of particle ordering processes that are consistent with the dynamic environment. Our results contradict the classical theory of gravitational sphere settling at quiescent conditions in sealed environments.

2. Materials and Methods

2.1. Sample Material

Australian opals and their host rocks were collected from the precious opal fields (Figure 1) at Andamooka and Mintabie (South Australia) and analyzed using petrographic microscopy, X-ray microdiffraction, Raman spectroscopy, scanning electron microscopy, and electron probe microanalysis. From a total of 82 samples from Andamooka and 13 samples from Mintabie, we selected 22 and 6 samples, respectively, based on textural and mineralogical characteristics, for in-depth analyses. Andamooka opal samples are associated with the deeply weathered early Cretaceous sediments of the Marree Subgroup [1,12]. Samples were extracted from the bleached Early Cretaceous Bulldog Shale (~20 m beneath the surface) in the Teatree Flat field, which is located about 15 km northwest of the Andamooka Township.

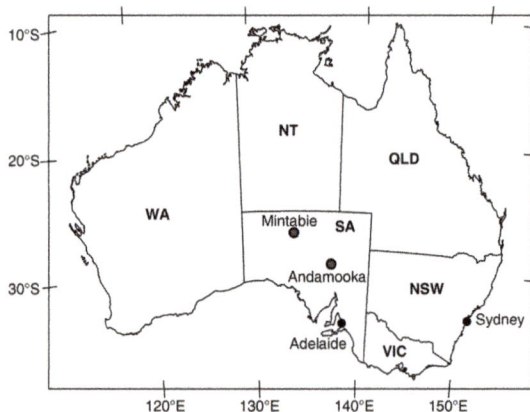

Figure 1. Location map of the sample sites in Andamooka and Mintabie (South Australia, Australia).

Silicified sample material from Andamooka comprises silt- and sandstones, conglomerates, and cherts with oolitic structures (Figure 2a). Opaline material fills extensional fractures and pseudomorphically replaces rhombohedral and twinned crystals. The fracture-filling opals from Mintabie are exceptional because they are hosted by tightly cemented, microcline-rich Ordovician sandstone of the Mintabie beds (Figure 2b; [2]). These samples are typical for the deepest parts of the opal-bearing profiles in the Mintabie field [2]. Opals analyzed in this study include gray, milky, white, brown, and transparent samples. Precious opals are transparent with a play-of-color covering the visible spectrum.

2.2. Analytical Methods

Diamond-polished thin sections (30 μm thick) and thick sections were prepared using standard procedures. For polarized light microscopy, a Zeiss Axio Lab.A1 petrographic microscope (Carl Zeiss, Jena, Germany) was used. At least 40 intersection angles between thin straight lines in single photonic crystals were measured digitally and averaged.

Figure 2. Photographs of polished thick sections of samples from Andamooka (**a**) and raw, unprepared sample material from Mintabie (**b**). (**a**) Andamooka samples, from left to right: opal-filled extensional fractures crosscut the bedding in a bleached sandstone; precious opal in an extensional fracture in sandstone with clay clasts; conglomerate with abundant opal cement and large clasts containing pseudomorphs after carbonates; laminated chert sample with oolitic structures and opal pseudomorphs after rhombohedral crystals. (**b**) Mintabie samples, from left to right: bleached sandstone with precious and common (white) vein opal; bleached sandstone with predominantly common (white and translucent) vein opal; sandstone fragments in a sample composed of varicolored common opal with minor precious opal domains.

The micromorphology of opals and their host rocks was investigated on the surfaces of thin and thick sections and freshly fractured material by SEM. Specimens were etched in 10 vol % hydrofluoric acid (HF) solution for 15 s, dried, and sputter-coated with ~15 nm W. Secondary electron (SE) images were obtained in a Zeiss Supra 40 VP Ultra SEM instrument (Carl Zeiss, Jena, Germany), at an acceleration voltage of 5 kV and a beam current of 10 nA. Sphere diameters were determined by sizing over 1000 particles from secondary electron images. The sphere diameter dispersity has been calculated as relative standard deviation.

Quantitative element concentrations were determined on carbon-coated, polished thin sections using a JEOL JXA 8200 Superprobe operated at 15 kV accelerating voltage, 20 nA beam current, and a beam diameter of 10 μm. For each opal specimen, 30 point analyses were measured. The acquisition time for Na analysis was 5 s on peak and 5 s on background. The peak and background of other elements were measured for 10 s each. The instrument was internally calibrated using natural silicate, oxide, and basalt (VG-2) and rhyolite (VG-568) glass. Elemental maps were acquired using the wavelength dispersive spectrometer (WDS) detectors. The operating conditions were a 15 kV accelerating voltage and a 20 nA beam current (on Faraday cup), with a beam diameter of 1 μm and a 60 ms counting time per 0.5–1 μm pixel size.

Non-destructive X-ray microdiffraction was used on polished thick sections, at the Eberhard Karls Universität Tübingen (Tübingen, Germany), with a Bruker AXS micro-X-ray diffractometer D8 Discover (Bruker AXS GmbH, Karlsruhe, Germany) with focusing X-ray optics (IfG Berlin, Berlin, Germany; incidence angle of 10°), a HOPG-monochromator, and a large VÅNTEC-500 2D-detector (μ-XRD^2). Diffractograms were recorded for 300 s at a beam diameter of 50 μm in the 2θ range of 7–67°, using CoKα radiation ($\lambda\alpha_1$ = 1.78897 Å) at 30 kV and a tube current of 30 mA. The step size of the diffractogram was 0.05° 2θ. We analyzed X-ray diffraction patterns with PeakFit 4 (Systat Software, San Jose, CA, USA). After a manual baseline subtraction, we used a five-point Savitzky–Golay moving filter [13] to smooth the diffraction pattern and minimize human bias in peak maxima determination. Diffractograms, peak positions, and the full width at half the maximum intensity (FWHM) are expressed as d-spacing (in Å) calculated from the diffraction angles (° 2θ) to facilitate the comparison of diffractograms recorded with different anode material.

Raman spectroscopy analyses were conducted on a Horiba Jobin Yvon LabRAM HR 800 instrument (Horiba Jobin Yvon, Bensheim, Germany) coupled to an Olympus BX41 microscope (Olympus, Hamburg, Germany) at the Museum für Naturkunde (Berlin, Germany). A 785 nm air-cooled diode laser was used to excite the sample with a 100× objective, a spectral integration time of 60 s, and three accumulations. With the Peltier-cooled charge-coupled device (CCD) detector (1024 × 256 pixels), a spectral resolution of ~0.2 cm^{-1}/pixel is achieved. Scattered Raman light was collected in backscattering geometry and dispersed by a grating of 600 grooves/mm after passing through a 100 µm entrance slit. The confocal hole size was set to 1000 µm. Unpolarized spectra were collected with the Labspec 6 software over a range from 100 to 1200 cm^{-1}. An internal intensity correction (ICS, Horiba) was used to correct detector intensities. The instrument was calibrated using the Raman band of silica at 520.7 cm^{-1}.

3. Results

3.1. Opal Mineralogy

We used X-ray microdiffraction and Raman spectroscopy to identify the mineralogy of fracture-filling and replacive opals. The typical X-ray microdiffraction pattern of the studied opals is shown in Figure 3a. The diffractograms, expressed as a function of d-spacing, generally show a broad asymmetric peak with a maximum ranging between 3.97 and 4.06 Å and a high d-spacing side shoulder. A secondary broad reflection of low intensity occurs at ~2 Å. The full width at half-maximum intensity (FWHM) of all samples scatters unsystematically from 1.07 to 1.19 Å and is independent of the main peak position. The maximum intensity of the main peak varies unsystematically by less than 5% between samples. According to the Jones and Segnit [14] opal classification scheme, the diffractograms of all samples are consistent with those of opal-A. The peak shapes and positions are uniform within individual samples, including those composed of intermingled composites of precious and common opal. Overall, we found no co-variation between peak shapes/positions and the visual appearance, micromorphology, or chemical aspects of the studied material.

Figure 3b shows a typical Raman spectrum of the studied opals. The spectra and band maximum positions of opals at all sites are similar and display a prominent band at 425 ± 3 cm^{-1} and bands of lower intensity at 791 ± 1, 961 ± 2, and 1067 ± 3 cm^{-1}. The bands at 425 and 961 cm^{-1} show a positive skewness, while the band at 791 cm^{-1} shows a negative skewness. These asymmetries are likely due to the superposition of separate bands with different shapes and positions. Previous studies assigned the Raman bands at ~425, 791, and 1067 cm^{-1} to fundamental vibrations of the SiO amorphous silica framework [15]. The band at ~961 cm^{-1} indicates a Si-OH stretching mode due to silanol groups [16]. The Raman spectra of all samples are consistent with those of opal-A [17] and lack a correlation between chemical composition, micromorphological features, and peak shapes and positions obtained from the X-ray diffraction analyses.

3.2. Colloid Flow Structures in Fractures

SEM images of vein-filling opal-A show that macroscopic opal characteristics directly link to micromorphological features (Figure 4). Hydraulic extensional fractures contain alternating zones of precious and common opal with variable color and parabolic interfaces pointing toward fracture tips. In these geometries, transparent opals with a play-of-color consist of regularly arranged, uniform spheres ranging in size from 160 to 440 nm. Irregularly arranged spheres with a size of 100–300 nm and minor nonspherical particles form the adjacent translucent common opals. The sphere diameters in translucent common and transparent precious opal are similar (±10 nm) within each sample. The white and milky common opal consists of non-uniform, round to ellipsoidal particles up to 1 µm in length, with a maximum aspect ratio of 0.25. These particles are elongated parallel to the fracture surfaces and parabolic interfaces between adjacent opals. Areas with accumulated, oriented ellipsoidal particles frequently display a moderate first-order grey birefringence between crossed polarizers (Figure 4d). The sequences of vein-filling opal with parabolic interfaces lack chemical or structural (determined by X-ray microdiffraction and Raman spectroscopy) gradients.

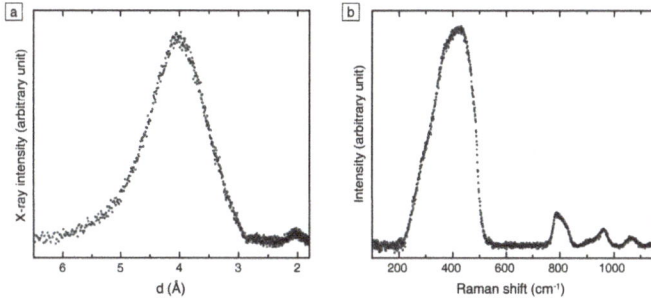

Figure 3. A typical X-ray diffractogram (**a**) and Raman spectrum (**b**) of the studied samples. (**a**) X-ray diffraction patterns (plotted as a function of d-spacing calculated from diffraction angles) generally show a broad asymmetric peak with a maximum ranging between 3.97 and 4.06 Å. A smaller, secondary reflection occurs at ~2 Å. (**b**) The Raman spectra of opals studied here show broad asymmetric bands with a maximum at 425 ± 3, 791 ± 1, and 961 ± 2 cm^{-1}, respectively. A more symmetrical band exists at 1067 ± 3 cm^{-1}. All spectra are consistent with opal-A [17].

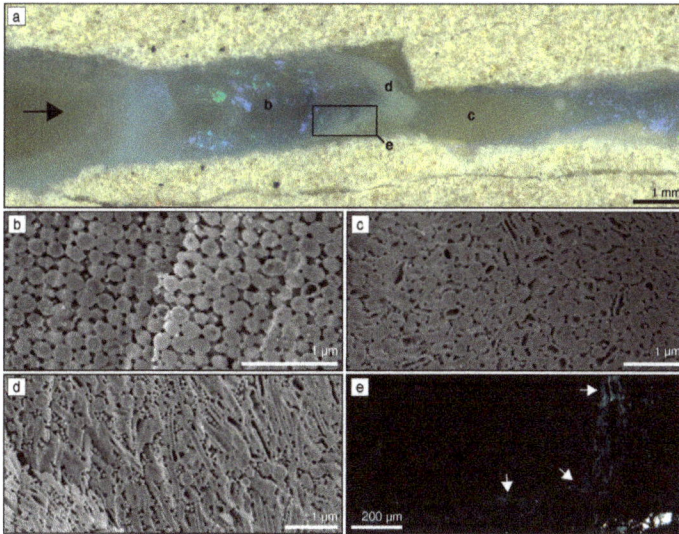

Figure 4. Textures and microstructural features of Andamooka vein opal. The SE images (**b,d**) were obtained after hydrofluoric acid (HF) etching. (**a**) Photograph of alternating zones of transparent precious and translucent to milky common opal. The parabolic interfaces point toward the fracture tip (arrow). (**b**) SE image of precious opal composed of ordered uniform spheres. (**c**) SE image of irregularly arranged spheres and minor nonspherical particles in translucent common opal. (**d**) White to milky common opal composed of ellipsoidal particles with a preferred orientation ~N–S and a maximum aspect ratio of 0.25. (**e**) Microphotograph (crossed polarizers) of the area highlighted in (**a**) depicting the faint grey birefringence of ellipsoidal, ordered particles in white to milky common opal-A.

The fractures in opal-indurated sandstones of the Mintabie field commonly contain veins of transparent opal intermingled with wave-shaped, layered white opal (Figure 5). These wave features are concentric and resemble unflattened roll-up structures [18]. Minute-sized (<1 μm) kaolinite flakes are embedded in the material and remnants of the host rock are located on its surface. The white opal consists of irregularly arranged, non-uniform spheres <300 nm in diameter and contains randomly scattered, sphere-lined pores, ~0.4 to 3.5 μm in diameter (Figure 5b). These pore architectures are also widespread in the white opal from Andamooka. Wedge-shaped photonic crystals with thin parallel striations and spindle-shaped extensional fractures are located between roll-up structures (Figure 5c). A sharp interface separates the regularly ordered regions from transparent opal composed of dominantly uniform but irregularly arranged spheres (Figure 5d). The chemical composition of the white and transparent opal and the host rock cement varies insignificantly (Table 1).

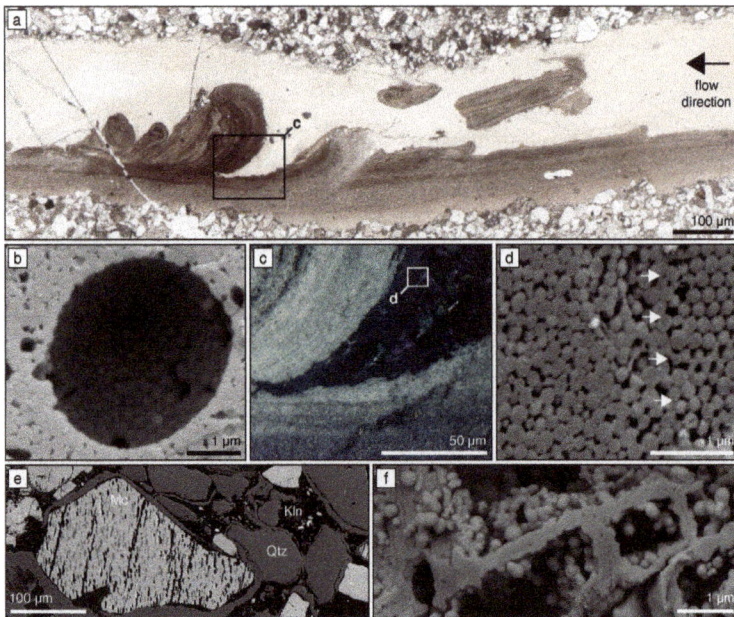

Figure 5. Textures and microstructural features of Mintabie vein opal (**a–d**) and its bleached sandstone host rock (**e–f**). (**a**) Microphotograph of a partially fragmented, brown opal (white in hand specimen) with roll-up structures overlain by transparent common and precious opal. The arrow indicates the flow direction of the concentrated colloid solution that deformed the brown opal. (**b**) SE image of sphere-lined pore space in the brown opal, typical of white opal at Mintabie and Andamooka. The sample was HF-etched prior to imaging. (**c**) Microphotograph of the area highlighted in (**a**), showing a wedge-shaped photonic crystal embedded in transparent common opal, between crossed polarizers. (**d**) SE image on HF-etched material of the area highlighted in **c**, showing the contact between transparent common and precious opal. Arrows emphasize the interface between disordered (left) and ordered (right) regions of similar-sized spheres. (**e**) Backscattered electron image of bleached Mintabie sandstone with microcline (Mc), quartz (Qtz), and kaolinite (Kln) in their typical spatial context. Microcline and quartz grains are surrounded by an opal-A cement. Na-rich regions (dark) in microcline dissolve preferentially during alteration. (**f**) SE image of an opal-A pseudomorph after microcline with dissolution cavities containing non-uniform spheres <300 nm in diameter. The opal-replaced cavity walls trace microcline cleavage planes.

Table 1. Chemical composition of opal-A samples shown in Figures 4–6 analyzed by electron probe microanalysis (EPMA) (in wt %). The data are averaged from 30 point analyses per sample. LLD denotes lower limit of detection. Vein opal from Andamooka refers to Figure 4. Mintabie opals are shown in Figure 5. Replacement opal-A from Andamooka comprises twinned and untwinned crystals shown in Figure 6.

Oxide	Andamooka					Mintabie	
	Vein Opal	Twinned Crystal [1]	Rhombs [1]	Inverse Opal	Cement	Transparent	White Roll-Ups
SiO_2	90.60	92.63	87.90	84.45	90.25	90.72	89.98
TiO_2	0.05	0.06	<LLD	<LLD	0.01	0.03	0.03
Al_2O_3	0.93	1.23	1.40	1.42	1.17	1.13	1.18
Fe_2O_3-total	0.09	0.29	0.10	0.09	0.04	0.07	0.08
MgO	0.04	0.08	0.04	0.04	0.01	0.04	0.04
CaO	0.06	0.12	0.09	0.08	0.14	0.12	0.13
SrO	0.03	0.03	0.02	0.02	0.04	0.03	0.03
BaO	0.09	0.10	0.08	0.06	0.10	0.08	0.09
Na_2O	0.10	0.13	0.29	0.30	0.18	0.17	0.18
K_2O	0.09	0.12	0.09	0.05	0.27	0.26	0.32
SUM Total	92.08	94.79	90.01	86.50	92.21	92.65	92.07
SUM impurities	2.19	2.16	2.11	2.06	1.96	1.93	2.08
$(H_2O + OH)$ [2]	10.73	5.21	9.99	13.50	7.79	7.35	7.93

[1] Precious opal, [2] Calculated by balance of SUM total with 100 wt %.

3.3. Carbonate Replacement Structures

Transmitted light microscopy reveals that fractures in silicified siltstones contain an- to subhedral, randomly oriented, optical photonic crystals up to 1.5 mm in size (Figure 6a). Observation between crossed polarizers reveals parallel striations (3 to 180 μm wide) that mimic polysynthetic twin lamellae on trigonal {018} indicative of recrystallized calcite. The twinned condition of the lamellae is visible as a periodic color variation that changes both color and intensity upon rotation on the microscope stage. The color variations arise from Bragg diffraction effects from close-packed, uniform silica spheres and air-filled pores with rhythmically changing orientation (Figure 6b). The average sphere diameter varies between samples from 200 to 320 nm (size dispersion <4%). X-ray microdiffraction, Raman spectroscopy, and elemental mapping confirm the amorphous nature of the opaline material and indicate the absence of crystalline precursor remnants.

Randomly oriented, opal-filled rhombohedra (<120 μm) are abundant in quartz- and opal-cemented cherts containing round to ellipsoidal oolitic structures (Figure 6c–e). These samples are likely reworked fragments of Cambrian age Andamooka Limestone [19]. The ooids range in size between 200 and 600 μm and are composed of a uniform mosaic of up to 20 μm-sized, interlocked quartz crystals. Small (<2 μm), anhedral dolomite crystals ($Ca_{1.1}Mg_{0.9}(CO_3)_2$ on average) randomly scatter through the quartz mosaic. In transmitted light, opal rhombohedra resembling dolomite rhombs consist of inclusion-free transparent to brown opal-A with variable micromorphology. A homogeneous amorphous silica veneer (~60 nm) covers the inner surface of the rhombs (Figure 6d). Transparent common opal inside the rhombohedra consists of 200–300 nm large, uniform spheres in a predominantly irregular arrangement. Precious opal is composed of similar-sized, but close-packed spheres. Dark brown rhombohedra consist of inverse opal with a regular arrangement of spherical void spaces surrounded by solid walls of opal-A (Figure 6e).

The sandstones of the Mintabie field contain abundant microperthitic microcline partially or entirely altered to translucent opal and/or kaolinite flakes (Figure 5e,f). Aggregates of submicrometer-sized kaolinite platelets and non-uniform silica spheres 120 to 270 nm in diameter fill the dissolution cavities in relictic feldspar grains. The opal-replaced cavity walls resemble fused spheres and occasionally contain subparticles ~50 nm in diameter.

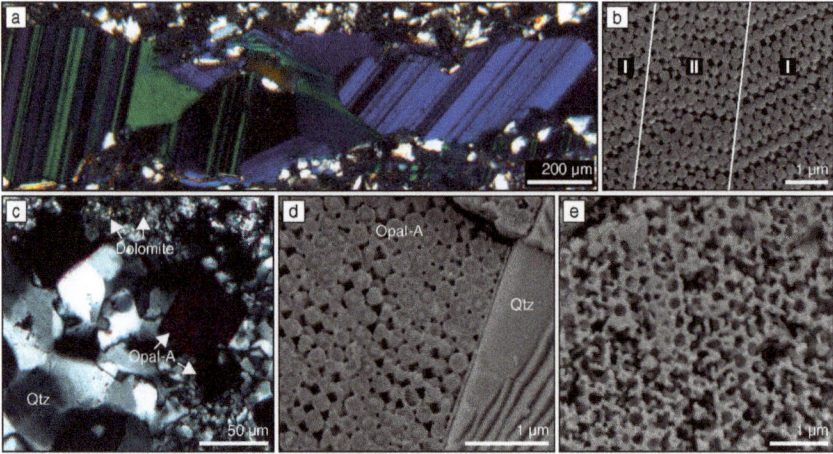

Figure 6. Optical photonic crystal pseudomorphs with precursor preservation from Andamooka. The images (**b**) and (**d**) were captured on HF-etched samples. (**a**) Microphotograph of fracture-filling striated photonic crystals with different colors and orientations between crossed polarizers. (**b**) SE image of the polished surface of a twinned crystal with alternating sphere array orientation in an I-II-I pattern. White lines highlight the interface between sphere stacks with different orientation. (**c**) Microphotograph of opal-A-filled rhombs in a quartz-cemented sample with ooids replaced by a mosaic of interlocked quartz crystals and dolomite. (**d**) SE image of the interface between precious opal-A inside rhombs and quartz on the outside as in (**c**). (**e**) SE image of inverse opal-A in rhombohedral pseudomorphs.

3.4. Polyhedral Particle Shapes and Crystal Bending

Opal-A from all sample localities consists of silica particles with variable size, shape, ordering, internal structure, and degree of cementation. Close-packed particle arrays often consist of grains with a polyhedral shape (Figure 7a). A mild HF etching of these polyhedral spheres reveals a concentric structure composed of subparticles 20–40 nm in diameter. This subparticle structure is common to all opals studied here, irrespective of their fracture-filling or replacive nature, shape, or chemical composition.

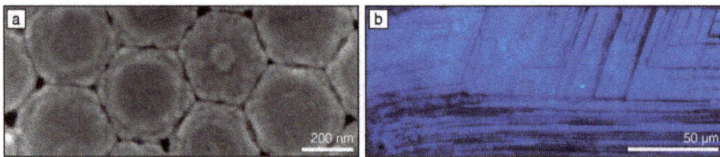

Figure 7. Post-depositional deformation features in precious opal from Andamooka. (**a**) Secondary electron image of HF acid-etched, hexagonal particles with concentric layering and subparticle structure. The pore space is significantly reduced compared to close-packed, ideal spheres. (**b**) Microphotograph of an opal-A photonic crystal between crossed polarizers showing thin black lines indicative of lattice displacement. Black lines on the right side intersect at an angle of ~70°. The lines are continuously bent toward the left side.

Vein-filling common opal frequently contains anhedral photonic regions with uniform color between crossed polarizers. These crystals often display two sets of intersecting thin black lines less than 1 μm wide (Figure 7b). Straight lines form diamond shapes and intersect at an angle of 70.2 ± 1.1°.

However, in most grains, these lines are bent to a variable degree. Locally, the photonic crystals display extensional fractures filled with irregularly arranged spheres.

4. Discussion

Scattering of visible light on the microstructurally diverse opal-A studied here creates color profiles that coincide with micromorphological characteristics, e.g., particle ordering and shape. Several authors have suggested that opal forms from a gel [1,4,5,10,11]. At quiescent conditions in sealed environments, uniform spheres are expected to grow inside this gel and settle gravitationally to form a regular sphere array, i.e., precious opal. However, such a gel matrix strongly immobilizes particles [20], impeding sphere growth, mobility, and arrangement into ordered arrays. Additionally, later particle mobility requires extensive breaking of interparticle bonds [21]. These restrictions equally apply to fracture-filling and replacive opal. In fact, the textural properties of the opals studied here and the aggregative grown single-particle structure point to precipitation and growth of spheres as free units in a sol rather than a gel. After sphere arrangement, dissolved silica from bulk solution or subparticle breakdown precipitates on the available surfaces and acts as a cementing agent [9,22].

4.1. Flow-Induced Ordering

The uniform direction of parabolic fronts in the vein opal indicates that fluid flow directs particles along a pressure gradient toward the fracture tip. The parabolic shape of the opal fronts resembles turbulent non-Newtonian flow velocity profiles [23] and indicates a strong coupling between colloid flow and structural ordering. This observation agrees well with non-Newtonian fluid flow textures preserved in vein opals in previous studies [1,11]. Apparently, colloid flow results in a hierarchy of particle packing, orientation, and deformation.

The uniform diameter of spherical particles in individual vein opal samples indicates that deformation occured when the spheres grew to their final size but were still deformable. The similar chemical composition, microdiffraction and Raman spectroscopic properties, and size of spherical particles point to a pre-existent reservoir of uniform spheres that entered the fracture upon opening. Nonspherical particle morphologies would form within this reservoir only under highly localized, substantially different physicochemical conditions [24]. During colloid ingress into the opening fracture, the pressure gradient forces the particles toward the fracture tip. The increase in particle number within the spatially restricted volume results in a local increase of the fluid's viscosity and jamming of the particles without fusion of the spheres. The flow leads to structuring of the concentrated colloidal suspension and directional ordering of nonspherical particles. The birefringent character of parabolic fronts between crossed polarizers results from the flow-induced uniform orientation of flattened but regularly ordered particles [25].

The texture of Mintabie vein opals illustrates how a flowing colloidal suspension reactivates and deforms opaline structures. It is probable that the concentric roll-ups reflect re-orientation of earlier-formed opal by the current activity of the colloidal suspension. The structure of the reactivated opaline material suggests that the viscosity of this suspension was high enough to induce roll-up structures in cohesive material and keep ripped-up material suspended. Wedge-shaped photonic structures between roll-up structures (Figure 5) contradict a gravitational sphere settling and ordering process, because gravitational settling likely results in horizontally layered photonic structures with a similar orientation. Previous studies have suggested that composite gels with ordered and unordered domains migrate under differential pressure and preserve their structural order [1]. Accordingly, ordered domains should initially possess a complex wedge shape and orientation that fit its final position. However, ordered sphere lattices with a comparable shape are absent from all fracture-filling opal in the present study. This might indicate that the sphere lattice of wedge-shaped photonic regions results from a flow-induced sphere ordering process, analogous to the shear-induced ordering of a disordered colloidal suspension [26,27]. Thus, fluid flow locally transports the spheres into the correct position to create ordered lattices.

4.2. Mineral Replacement

Apart from the textural similarities of vein opals, the highly selective replacement of carbonate and clay minerals, wood, gypsum, fossils, and organic material by opal-A is frequently reported [1,6,12,28]. To date, the replacement of silicates and carbonate fossils by precious opal-A has been attributed to the infilling of cavities, resulting from partial or bulk dissolution, by a silica-rich gel [1,10]. This gel matrix inhibits particle growth and requires considerable bond breaking, analogous to the sphere-forming processes in fractures. Additionally, the pseudomorphs studied here lack evidence for the shrinkage or layering that may indicate gel desiccation or changes in gel composition [10], respectively. These restraints imply that another process induces structural order during pseudomorphic replacement.

The twinned photonic crystals evidence that carbonate replacement is intimately coupled to the precursor crystallographic orientations. Replacement along lattice planes of the precursor crystal ensures the constant physicochemical conditions that are required for uniform sphere sizes, shapes, and arrays [7]. The dissolution reaction creates a sharp localized pH and salinity increase [29] that gradually decreases toward the bulk fluid. This local change lowers the amorphous silica solubility, induces supersaturation and precipitation [21], and confines particle nucleation close to the moving calcite dissolution front. In this interface-coupled dissolution-precipitation process [7,30], the precursor dissolution and silica deposition fronts advance synchronously. During the replacement process, silica nanoparticles continuously nucleate, aggregate, and form close-packed planes parallel to the most stable lattice planes of the crystalline template (Figure 8). The particle size increases at increasing distance from the dissolution front. The nanoporous close-packed sphere geometry (~26% porosity) is permissive and enables mass transport and permanent fluid access to the calcite dissolution front. The textures and random orientation of the crystal pseudomorphs indicate that gravitational settling is negligible in this case. Our observations further show that carbonate precipitation in extensional fractures locally precedes precious opal formation.

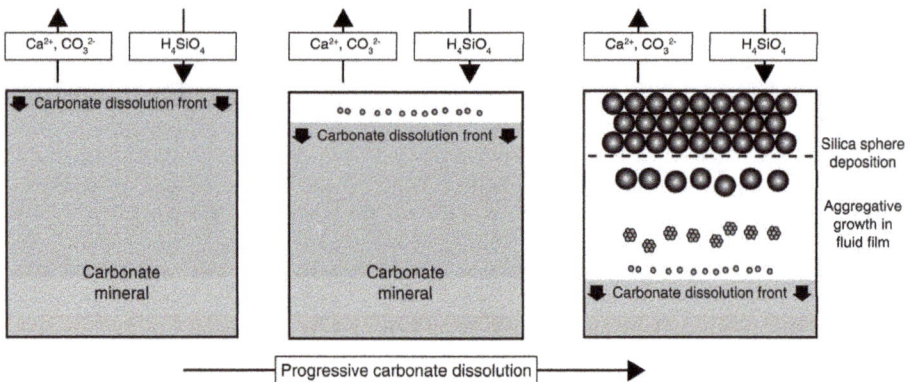

Figure 8. Schematic representation of the volume- and texture-preserving replacement process of a carbonate mineral by amorphous silica nanospheres. Carbonate chemical components are constantly removed into the bulk solution, while dissolved silica (H_4SiO_4) reaches the carbonate surface. Amorphous silica nanoparticles continuously nucleate close to the carbonate surface, aggregate into larger particles in the interfacial fluid film, and self-assemble into close-packed, hexagonal planes. The particle diameter increases at increasing distance from the dissolution front until they reach their final size. The carbonate dissolution and silica deposition fronts advance synchronously during the interface-coupled dissolution-precipitation process. The final spheres form a long-range ordered photonic crystal.

Generally, the microstructure differs between pseudomorphs after silicate minerals (e.g., feldspar) and non-silicates (e.g., carbonate and sulfate minerals). While close-packed sphere arrays replace carbonates and sulfates [2,7], pseudomorphs after feldspar consist of non-uniform spheres in dissolution cavities (Figure 5f). These pseudomorphs are the porous products of the short distance transport and polymerization of dissolved silica. This silicate replacement process does not produce precious opal, due to the small volume fraction of spheres, unless an external source provides additional silica in solution.

Preservation of twin structures in photonic crystal assemblies appears to be a reliable indicator for a replacive interface-coupled dissolution-precipitation mechanism, but the replacement of untwinned crystals is not as straightforward. We observe that randomly oriented rhombohedra consist of ordered uniform spheres. Gravitational sphere settling may fill pseudomorphs but will result in irregular rim structures, horizontal layering, and a late, disordered horizontal region due to limited late-stage particle mobility. Partial infilling of pore space may indicate bulk dissolution of the precursor material. Precious opal in the rhombohedral pseudomorphs studied here lacks all these microstructural features. More likely, close-packed sphere arrays in pseudomorphs predominantly form by an interface-coupled dissolution-precipitation process. This process creates the structural perfection of replacive precious opal compared to those formed by sphere sedimentation.

4.3. Post-Depositional Lattice Deformation

Dislocations and polyhedral particle shapes in ordered sphere arrays indicate post-depositional deformation (Figure 7). At a small shear force, the deformation of soft, ordered spheres into polyhedra decreases pore space and induces a blue-shift of the diffracted light [31]. Therefore, sheared sphere arrays may preserve their long-range order but lose their optical photonic character. At higher shear forces, colloidal crystals can form dislocations or break [32]. Previous studies have shown that dislocations in face-centered cubic (fcc) sphere arrays are visible as permanently extinct lines between crossed polarizers [1,7]. This microscopic observation results from isotropic light scattering on the locally imperfect structure. In the present study, straight, extinct lines intersect at an angle of $70.2 \pm 1.1°$ in untwinned photonic crystals. This intersection angle compares closely with those of ideal fcc {111} planes at 70.53°. Apparently, fcc {111} planes are the preferred slip planes in the natural fcc crystals. Further deformation induces bending and partial disintegration of the crystals. As the wavelength of diffracted light depends on crystal orientation [33], deformation may result in lattice re-orientation and an undulose color of the crystal.

Acknowledgments: Funding for this research was provided by the German Research Council (DFG), grant number MI1205/5-1. We thank Jürgen Ganzel, Berlin, for supplying the samples. Christoph Berthold (Eberhard Karls Universität Tübingen, Tübingen, Germany) provided access to μ-X-ray facilities. Tanja Mohr-Westheide (Naturkundemuseum, Berlin, Germany) provided access to the Raman spectrometer. The authors wish to thank the three anonymous reviewers for their thoughtful comments.

Author Contributions: Moritz Liesegang and Ralf Milke conceived and designed the study. Moritz Liesegang collected and analyzed all data and wrote the paper. All authors discussed the results and commented on the manuscript.

Conflicts of Interest: The authors declare no conflict of interest.

References

1. Rey, P.F. Opalisation of the Great Artesian Basin (central Australia): An Australian story with a Martian twist. *Aust. J. Earth Sci.* **2013**, *60*, 291–314. [CrossRef]
2. Thiry, M.; Milnes, A.R.; Rayot, V.; Simon-Coinçon, S. Interpretation of palaeoweathering features and successive silicifications in the Tertiary regolith of Inland Australia. *J. Geol. Soc.* **2006**, *163*, 723–736. [CrossRef]
3. Williams, L.A.; Crerar, D.A. Silica diagenesis, II, General mechanisms. *J. Sediment. Res.* **1985**, *55*, 312–321. [CrossRef]
4. Darragh, P.; Gaskin, A.; Terrell, B.; Sanders, J. Origin of precious opal. *Nature* **1966**, *209*, 13–16. [CrossRef]

5. Gaillou, E.; Fritsch, E.; Aguilar-Reyes, B.; Rondeau, B.; Post, J.; Barreau, A.; Ostroumov, M. Common gem opal: An investigation of micro- to nano-structure. *Am. Mineral.* **2008**, *93*, 1865–1873. [CrossRef]

6. Liesegang, M.; Milke, R. Australian sedimentary opal-A and its associated minerals: Implications for natural silica sphere formation. *Am. Mineral.* **2014**, *99*, 1488–1499. [CrossRef]

7. Liesegang, M.; Milke, R.; Kranz, C.; Neusser, G. Silica nanoparticle aggregation in calcite replacement reactions. *Sci. Rep.* **2017**, *7*, 14550. [CrossRef] [PubMed]

8. Flörke, O.W.; Graetsch, H.; Röller, K.; Martin, B.; Wirth, R. Nomenclature of micro-and non-crystalline silica minerals, based on structure and microstructure. *Neues Jahrbuch für Mineralogie Abhandlungen* **1991**, *163*, 19–42.

9. Sanders, J.V. Colour of precious opal. *Nature* **1964**, *204*, 1151–1153. [CrossRef]

10. Pewkliang, B.; Pring, A.; Brugger, J. The formation of precious opal: Clues from the opalisation of bone. *Can. Mineral.* **2008**, *46*, 139–149. [CrossRef]

11. Pecover, S.R. Australian Opal Resources: Outback Spectral Fire. *Rocks Miner.* **2007**, *82*, 102–115. [CrossRef]

12. Barnes, L.C.; Townsend, I.J.; Robertson, R.S.; Scott, D.C. *Opal: South Australia's Gemstone (Handbook No. 5)*; Department of Mines and Energy, Geological Survey of South Australia: Parkside, Australia, 1992; pp. 19–26, 37–51. ISBN 978-0730817093.

13. Savitzky, A.; Golay, M.J. Smoothing and differentiation of data by simplified least squares procedures. *Anal. Chem.* **1964**, *36*, 1627–1639. [CrossRef]

14. Jones, J.B.; Segnit, E.R. The nature of opal I. Nomenclature and constituent phases. *J. Geol. Soc. Aust.* **1971**, *18*, 37–41. [CrossRef]

15. McMillan, P. Structural studies of silicate glasses and melts-applications and limitations of Raman spectroscopy. *Am. Mineral.* **1984**, *69*, 622–644.

16. Hartwig, C.M.; Rahn, L.A. Bound hydroxyl in vitreous silica. *J. Chem. Phys.* **1977**, *67*, 4260–4261. [CrossRef]

17. Smallwood, A.G.; Thomas, P.S.; Ray, A.S. Characterisation of sedimentary opals by Fourier transform Raman spectroscopy. *Spectrochim. Acta A* **1997**, *53*, 2341–2345. [CrossRef]

18. Eriksson, P.G.; Simpson, E.L.; Eriksson, K.A.; Bumby, A.J.; Steyn, G.L.; Sarkar, S. Muddy roll-up structures in siliciclastic interdune beds of the c. 1.8 Ga Waterberg Group, South Africa. *Palaios* **2000**, *15*, 177–183. [CrossRef]

19. Carr, S.G.; Olliver, J.G.; Conor, C.H.H.; Scott, D.C. *Andamooka Opal Fields: The Geology of the Precious Stones Field and the Results of the Subsidised Mining Program*; Report of Investigations 51; Department of Mines and Energy, Geological Survey of South Australia: Adelaide, Australia, 1979; pp. 17–18. ISBN 0724354980.

20. Dickinson, E. Structure and rheology of colloidal particle gels: Insight from computer simulation. *Adv. Colloid Interface Sci.* **2013**, *199*, 114–127. [CrossRef] [PubMed]

21. Iler, R.K. *The Chemistry of Silica: Solubility, Polymerization, Colloid and Surface Properties, and Biochemistry*; Wiley: New York, NY, USA, 1979; pp. 222–239. ISBN 978-0471024040.

22. Carcouët, C.C.M.C.; van de Put, M.W.P.; Mezari, B.; Magusin, P.C.M.M.; Laven, J.; Bomans, P.H.H.; Friedrich, H.; Esteves, A.C.C.; Sommerdijk, N.A.J.M.; van Benthem, R.A.T.M.; et al. Nucleation and growth of monodisperse silica nanoparticles. *Nano Lett.* **2014**, *14*, 1433–1438. [CrossRef] [PubMed]

23. Peixinho, J.; Nouar, C.; Desaubry, C.; Théron, B. Laminar transitional and turbulent flow of yield stress fluid in a pipe. *J. Nonnewton. Fluid Mech.* **2005**, *128*, 172–184. [CrossRef]

24. Ding, T.; Long, Y.; Zhong, K.; Song, K.; Yang, G.; Tung, C.H. Modifying the symmetry of colloidal photonic crystals: A way towards complete photonic bandgap. *J. Mater. Chem. C* **2014**, *2*, 4100–4111. [CrossRef]

25. Hong, S.H.; Shen, T.Z.; Song, J.K. Flow-induced alignment of disk-like graphene oxide particles in isotropic and biphasic colloids. *Mol. Cryst. Liq. Cryst.* **2015**, *610*, 68–76. [CrossRef]

26. Wu, Y.L.; Derks, D.; van Blaaderen, A.; Imhof, A. Melting and crystallization of colloidal hard-sphere suspensions under shear. *Proc. Natl. Acad. Sci. USA* **2009**, *106*, 10564–10569. [CrossRef] [PubMed]

27. Vermant, J.; Solomon, M.J. Flow-induced structure in colloidal suspensions. *J. Phys. Condens. Matter* **2005**, *17*, R187–R216. [CrossRef]

28. Scurfield, G.; Segnit, E. Petrifaction of wood by silica minerals. *Sediment. Geol.* **1984**, *39*, 149–167. [CrossRef]

29. Molins, S.; Trebotich, D.; Yang, L.; Ajo-Franklin, J.B.; Ligocki, T.J.; Shen, C.; Steefel, C.I. Pore-scale controls on calcite dissolution rates from flow-through laboratory and numerical experiments. *Environ. Sci. Technol.* **2014**, *48*, 7453–7460. [CrossRef] [PubMed]

Minerals **2018**, *8*, 12

30. Putnis, A.; Putnis, C.V. The mechanism of reequilibration of solids in the presence of a fluid phase. *J. Solid State Chem.* **2007**, *180*, 1783–1786. [CrossRef]
31. Sun, Z.Q.; Chen, X.; Zhang, J.H.; Chen, Z.M.; Zhang, K.; Yan, X.; Wang, Y.F.; Yu, W.Z.; Yang, B. Nonspherical colloidal crystals fabricated by the thermal pressing of colloidal crystal chips. *Langmuir* **2005**, *21*, 8987–8991. [CrossRef] [PubMed]
32. Imhof, A.; van Blaaderen, A.; Dhont, J.K.G. Shear melting of colloidal crystals of charged spheres studied with rheology and polarizing microscopy. *Langmuir* **1994**, *10*, 3477–3484. [CrossRef]
33. Monovoukas, Y.; Gast, A.P. A study of colloidal crystal morphology and orientation via polarizing microscopy. *Langmuir* **1991**, *7*, 460–468. [CrossRef]

minerals

MDPI

Article

The Hydrothermal Breccia of Berglia-Glassberget, Trøndelag, Norway: Snapshot of a Triassic Earthquake

Axel Müller [1,2,*], Morgan Ganerød [3], Michael Wiedenbeck [4], Skule Olaus Svendsen Spjelkavik [5] and Rune Selbekk [†]

1   Natural History Museum, P.O. Box 1172 Blindern, 0318 Oslo, Norway
2   Natural History Museum of London, Cromwell Road, London SW7 5BD, UK
3   Geological Survey of Norway, P.O. Box 6315 Torgard, 7491 Trondheim, Norway; Morgan.Ganerod@ngu.no
4   German Research Center for Geosciences (GFZ), Telegrafenberg, 14473 Potsdam, Germany;
     Michael.Wiedenbeck@gfz-potsdam.de
5   Department of Archaeology and Cultural History, NTNU University Museum, NO-7491 Trondheim,
     Norway; skule.olaus@gmail.com
*   Correspondence: a.b.mueller@nhm.uio.no
†   Deceased 4 December 2017.

Received: 26 February 2018; Accepted: 17 April 2018; Published: 23 April 2018

Abstract: The quartz-K-feldspar-cemented breccia of Berglia-Glassberget in the Lierne municipality in central Norway forms an ellipsoid structure 250 m × 500 m in size. The hydrothermal breccia is barren in terms of economic commodities but famous among mineral collectors for being a large and rich site of crystal quartz of various colours and habits. Despite being a famous collector site, the mineralization is rather unique in respect to its geological setting. It occurs within Late Palaeoproterozoic metarhyolites of the Lower Allochthon of the Norwegian Caledonides regionally isolated from any other contemporaneous hydrothermal or magmatic event. In order to understand better the formation of the Berglia-Glassberget breccia, the chemistry, fluid inclusion petrography and age of the breccia cement were determined. Structural features indicate that the Berglia-Glassberget is a fault-related, fluid-assisted, hydraulic breccia which formed by single pulse stress released by a seismic event. ^{40}Ar-^{39}Ar dating of K-feldspar cement revealed a middle Triassic age (240.3 ± 0.4 Ma) for this event. The influx into the fault zone of an aqueous CO_2-bearing fluid triggered the sudden fault movement. The high percentage of open space in the breccia fractures with cavities up 3 m × 3 m × 4 m in size, fluid inclusion microthermometry, and trace element chemistry of quartz suggests that the breccia was formed at depths between 4 and 0.5 km (1.1 to 0.1 kbar). The origin of the breccia-cementing, CO_2-bearing Na-HCO_3-SO_4 fluid may have been predominantly of metamorphic origin due to decarbonation reactions (T > 200 °C) of limestones of the underlying Olden Nappe. The decarbonation reactions were initiated by deeply derived, hot fluids channelled to sub-surface levels by a major fault zone, implying that the breccia is situated on a deep-seated structure. Regionally, the Berglia-Glassberget occurs at a supposed triple junction of long-lived fault zones belonging to the Møre-Trøndelag, Lærdal-Gjende and the Kollstraumen fault complexes. These fault systems and the associated Berglia-Glassberget earthquake are the expression of rifting and faulting in northern Europe during the middle/late Triassic.

Keywords: quartz; breccia; earthquake; Triassic; Berglia-Glassberget

1. Introduction

Breccias are fragmented rocks which are commonly found in the highest, most fluid-saturated part of the crust, where brittle deformation is dominant e.g., [1,2]. They occur across a wide range of settings: sedimentary breccia, impact breccia, fault breccia (gouge, cataclasite, pseudotachylite), hydrothermal breccia, hydrothermal-magmatic breccia, and purely magmatic breccia. The study of mineralized hydrothermal and hydrothermal-magmatic breccias have been of major interest for ore deposit research due to their potential of hosting economic mineralization, e.g., [3–6] whereas studies of non-mineralized breccias are more scarce. However, understanding the nature and genesis of breccias is important not only economically but also in the context of regional tectonics and earthquake prediction. Brecciated fault zones, for example, preserve a rich historical record of seismic faulting; a record that is yet to be fully studied and understood e.g., [2,7–11].

Breccia bodies exhibit diverse features and are general characterized by: (1) the chemistry, mineralogy and texture of the breccia cement; (2) the nature (lithology, frequency, size, habit, etc.) of the fragments; and (3) the geometry and dimension of the breccia bodies. Genetic aspects like the degree of involvement of magmatic processes and the depth of formation may provide additional information to define hydrothermal and hydrothermal-magmatic breccias, in particular. Classifications of these breccia types commonly include both partly genetic and purely descriptive nomenclature, as discussed by [3,12,13].

In this contribution, the hydrothermal breccia of Berglia-Glassberget in Trøndelag, Norway, is studied. The Berglia-Glassberget breccia is barren in terms of economic commodities but famous among mineral collectors for being a large and rich site of high-quality crystal quartz of various colours and habits found in open cavities [14–16]. The mineralization is rather unique in respect to its geological setting: it occurs within Late Palaeoproterozoic rocks of the Lower Allochthon of the Norwegian Caledonides, regionally isolated from any other contemporaneous hydrothermal or magmatic activities. The breccia formation post-dates the Caledonian deformation and a hydrothermal mineralization of such young age (<390 Ma) has not been described from central Norway. The aims of this study are to better understand the formation of the Berglia-Glassberget breccia in terms of pressure-temperature-composition (P-T-X) conditions, the origin of breccia-cementing fluids, the age of breccia, and the circumstances which have led to the breccia formation. Finally, the results are discussed and evaluated to place the breccia-forming event in a regional context.

2. Regional Geology and Characteristics of the Berglia-Glassberget Breccia

Geographically the quartz-K-feldspar cemented breccia of Berglia-Glassberget is situated in the Lierne municipality in the Trøndelag county of central east Norway (Figure 1). The breccia is hosted by Late Paleoproterozoic mylonitic (very fine-grained), greyish to pinkish metarhyolite of the Formofoss Nappe Complex. The upper greenschist facies Formofoss Nappe Complex represents the upper unit of the Lower Allochthon of the Norwegian Caledonides which overlies the lower greenschist facies Olden Nappe [17,18]. Both nappes form the Grong-Olden Culmination [19] where the Berglia-Glassberget breccia is situated at its eastern edge. The breccia is in the hanging wall of a gentle SE plunging anticline formed by rocks of the Olden Nappe, a few meters above the thrust fault which separates the Olden Nappe and the overlying Formofoss Nappe (Figure 2).

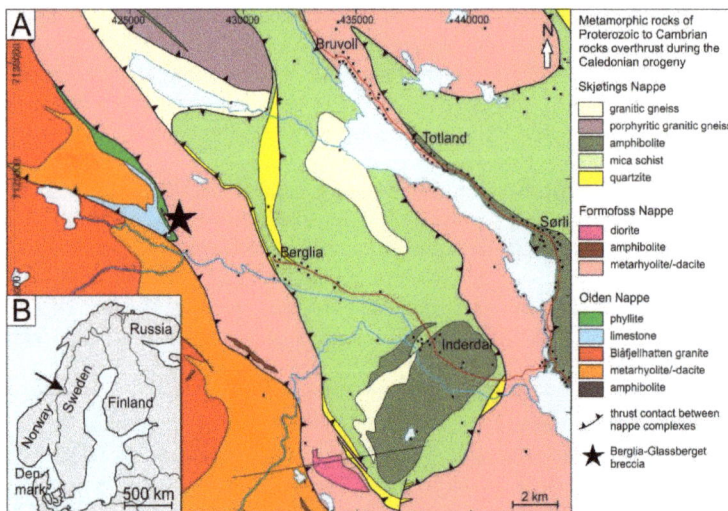

Figure 1. (**A**) Simplified regional geological map according to Fossen et al. [20] with the location of the Berglia-Glassberget breccia. The used coordinate system is Universal Transverse Mercator (UTM) zone 33W. (**B**) Inset showing the Berglia-Glassberget locality in Scandinavia.

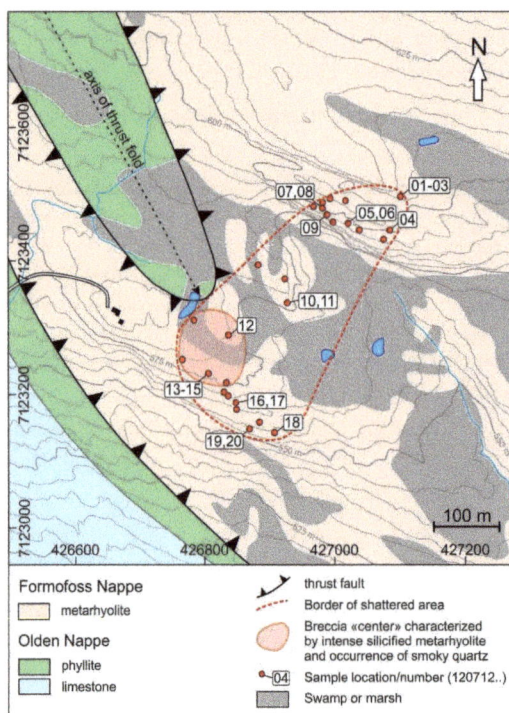

Figure 2. Local geological map of Berglia-Glassberget area according to Fossen et al. [20] with breccia extension and sample locations.

The metarhyolite at Berglia-Glassberget is affiliated to the Transscandinavian Igneous Belt (TIB) [18,21–24], which comprises a giant elongated array of batholiths extending c. 1400 km along the Scandinavian Peninsula from southeasternmost Sweden to Troms in north-western Norway [25,26]. The TIB documents more or less continuous and voluminous magmatic activity between 1850 to 1630 Ma, which developed between the Svecofennian (1920–1790 Ma) [27] and Gothian orogenesis (1640-1520 Ma) [28]. The TIB-related Blåfjellhatten granite constituting the major part of the Olden Nappe west of the Berglia-Glassberget breccia has an emplacement age of 1633.2 ± 2.9 Ma [24] (Figure 1).

The breccia forms a c. 250 × 500 m large, ellipsoid structure (Figure 2) comprising a dense network of randomly orientated, breccia-filled, mainly quartz-cemented and subordinate K-feldspar-cemented fractures (3 cm to 4 m wide). The breccia structure crops out at altitudes ranging from 550 m above sea level (a.s.l.) in the SW to 600 m a.s.l. in the NE. Most of the area is covered with post-glacial soil, woods and swamps. The upper part of the deposit is close to the tree line. The area of most intense brecciation is found in the SW of the structure which is named breccia center in the following. In the center the fragmented metarhyolite is strongly silicified and dark grey to black in colour instead of pinkish grey. The borders of the breccia structure are transitional: the fractures getting thinner and less common with increasing distance from the breccia center.

The randomly orientated fractures hosting hydrothermal breccias are mainly matrix-supported except parts of the breccia center, with 0 to 75 vol % clasts, 25 to 100 vol % matrix, and 0 to 80 vol % open space (cavities). The lithology of the breccia fragments is exclusively metarhyolitic (monomictic) corresponding to the closest wall rock (Figure 3A,B). The size of clasts is highly variable ranging from millimeter-scale to meter-scale. Most clasts are angular, often platy-shaped due to preferentially rock-splitting along foliation planes.

Most of the larger cavities (>0.1 m^3) collapsed shortly after formation and are filled with clay. The cavities contain quartz crystals of varying quality, mostly milky quartz with common crystal sizes of 0.5 to 5 cm. The deposit produced high quantities of collector quality crystal quartz specimen of different colour and habit over a period of about 100 years [14–16,29–35]. The northeastern part of the mineralization has been known by local mineral collectors since its discovery. In the 1980s and 90s collectors as Inge Rolvsen, Egil Skaret and Harald Kvarsvik started to take out quartz crystals for the mineral collector market. In the late 1990s Lars Jørgensen leased the area for systematic collection of specimens. In 2005 large cavities (up to 3 m × 3 m × 4 m in size) with smoky quartz were discovered in the southwestern part of the breccia structure (Figure 3C,D and Figure 4A). Since than there has been a lot of collection activity by amateurs. It is still possible to find good quartz specimens in the area. Before sampling, one of the land owners, Arne Jostein Devik (adevik@online.no), has to be contacted. Despite the intense collection activities there is unfortunately very little documentation and literature about the mineralization [16].

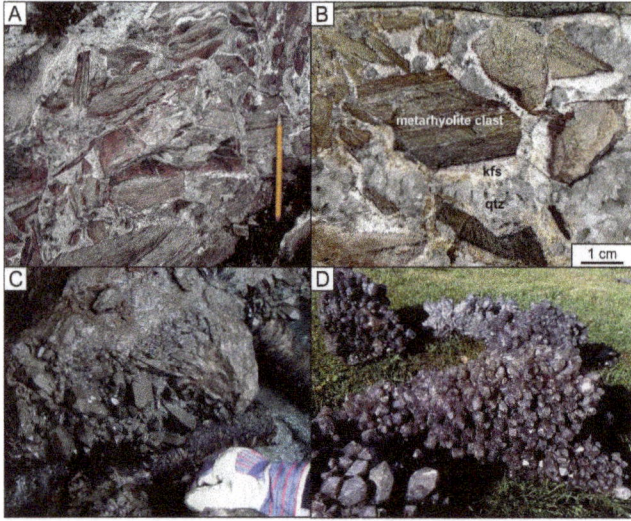

Figure 3. Photographs of the Berglia-Glassberget breccia. (**A**) Outcrop in the center of the Berglia-Glassberget breccia. (**B**) Hand specimen exhibiting typical texture of the Berglia-Glassberget breccia. The white mineral enveloping the metarhyolite clasts is K-feldspar (kfs), which crystallized prior to quartz (qtz). (**C**) Recovering of a large cluster of smoky quartz crystals from the 3 m × 3 m × 4 m cavity discovered in 2005. (**D**) Cleaned crystal clusters from the cavity shown in (**C**). Photograph by Arne Jostein Devik.

Figure 4. Photographs of minerals related to the Berglia-Glassberget breccia. (**A**) Cluster of smoky quartz crystals. The length of the specimen, which was donated by Arne Jostein Devik to the Natural History Museum of Oslo, is 30 cm (NHM collection nr. 41919). Photograph by Øivind Thorensen. (**B**) Small cavity with flesh-coloured adularia crystals overgrown by quartz crystals. Length of specimen 10 cm. (**C**) 14 cm long, smoky quartz crystal donated by Arne Jostein Devik to the Natural History Museum of Oslo (NHM collection nr. 41918). Photograph by Øivind Thorensen. (**D**) Cluster of clear quartz crystals with a Japanese twin in the center donated by Egil Hollund to the Natural History Museum of Oslo (NHM collection nr. 42418). Length of specimen 12 cm. Photograph by Øivind Thorensen.

The volume proportion of flesh-coloured K-feldspar cement (var. adularia) in the breccia veins increases from the breccia center from 5 vol % to up to 100 vol % towards NE and SE of the mineralized area. K-feldspar crystallized prior to quartz and covers the fracture and cavity walls and breccia fragments (Figure 3A,B). In open cavities K-feldspar may form euhedral crystals up to 1 cm in size (Figure 4B).

Most of the cavities contain clear whitish to greyish crystal quartz. Smoky quartz of different shades occurs in the central part of the mineralization (Figure 4C). The largest smoky quartz crystals which have been found were about 30 cm × 10 cm in size and the largest clear quartz crystals are up to 10 cm × 2 cm in size. The largest quartz crystal cluster on a metarhyolite fragment recovered from the large cavity in 2005 weights 1.8 ton and is now part of the collection of the Natural History Museum of Oslo (NHM collection nr. KNR 43853). Scepter quartz and one Japanese twin crystal (Figure 4D) have been found as well as smaller pockets filled with double-termed moron crystals. Calcites of different shapes and colours have been found in some cavities. In addition, albite, galena, rutile, and laumontite have been recorded. Albite is one of last phases crystallized (up to 3 mm in size) and forms crystal lawns on quartz crystal faces. A single find of galena grown on smoky quartz has been described.

According to the exposed textures the process responsible for the formation of the Berglia-Glassberget breccia can be described as a fluid-assisted hydraulic brecciation initiated by a single pulse stress e.g., [36]. The Berglia-Glassberget breccia has not been affected by Caledonian deformation and, thus, post-dates the Caledonian orogenesis. Of particular interest is the fact that there is no other known tectonic, hydrothermal or magmatic activity in the area which can be related to the breccia formation.

3. Sampling and Methods

3.1. Sampling

The sampling was performed in 2012 and covered the entire brecciated area (Figure 2). The samples include breccia-related quartz crystals and K-feldspar, bulk rock samples of the host rock, and samples exhibiting representative breccia textures. The quartz samples were prepared as: (1) double-polished wafers (c. 250 μm) for fluid inclusion microthermometry; and (2) as polished thick sections (c. 300 μm thick) clued on standard glass slides 2.8 cm × 4.8 cm for laser ablation inductively plasma mass spectrometry (LA-ICP-MS) analysis. The bulk compositions of host rocks and K-feldspar were determined at ACME laboratories in Vancouver, Canada [37].

3.2. Scanning Electron Microscope Cathodoluminescence

The quartz crystals were studied with scanning electron microscope cathodoluminescence (SEM-CL) to visualize intra-granular growth zoning and alteration structures and to choose areas for LA-ICP-MS analysis. SEM-CL imaging reveals micro-scale (1 μm to 1 mm) growth zoning, alteration structures and different quartz generations which are not visible with other methods. Grey-scale contrasts visualized by SEM-CL are caused by the heterogeneous distribution of lattice defects (e.g., oxygen and silicon vacancies, broken bonds) and lattice-bound trace elements e.g., [38]. Although the physical background of the quartz CL is not fully understood, the structures revealed by CL give information about crystallisation, deformation and fluid-driven overprint. The used CL detector was a Centaurus BS bialkali-type attached to a LEO 1450VP analytical SEM based at the Geological Survey of Norway in Trondheim, Norway. The applied acceleration voltage and current at the sample surface were 20 kV and 2 nA, respectively. The SEM-CL images were collected from one scan of 43 s photo speed and a processing resolution of 1024 × 768 pixels and 256 grey levels. The brightness and contrast of the collected CL images were improved with PhotoShop software.

3.3. Fluid Inclusion Microthermometry

Fluid inclusion wafers were prepared from 2–3 cm long, euhedral quartz crystals originating from different cavities across the deposit. Microthermometric measurements of 23 fluid inclusions were performed on a calibrated Linkam MDSG 600 heating/freezing stage at NTNU Trondheim, Norway, using the software Linksys 32. All inclusions were rapidly cooled to $-196\ °C$ and held for one minute, then heated at a rate of $30\ °C/min$ and held for one minute to ensure sufficient undercooling. A stepwise heating procedure was performed, during which the temperature was held for one minute at $-50.0\ °C$ and heated at $1\ °C/min$ until ice melting was imminent. The inclusion was then cooled a few degrees to recrystallize the ice in order to observe the final melting temperature more accurately. In cases where Tmice was lower than the holding temperature, the inclusion was frozen again and holding temperature was adjusted.

3.4. Laser Ablation Inductively Coupled Plasma Mass Spectrometry

Concentrations of Li, Be, B, Mn, Ge, Rb, Sr, Na, Al, P, K, Ca, Ti, and Fe were determined by LA-ICP-MS at the Geological Survey of Norway in Trondheim. Five quartz crystals (1 to 3 cm in length) were cut central, parallel to the c-axis and prepared as surface-polished, 300-μm-thick sections mounted on standard glass slides or embedded in Epoxy in 25.4 mm diameter sample mounts. The analyses were undertaken on a double-focusing sector field inductively coupled plasma mass spectrometer, high resolution sector field ICP-MS, model ELEMENT XR from Thermo Scientific at the Geological Survey of Norway in Trondheim, Norway. The instrument is linked to a New Wave Excimer UP193FX ESI laser probe. The 193 nm laser had a repetition rate of 15 Hz, a spot size of 75 μm, and energy fluence about 5 to 6 J/cm^2 on the sample surface. A continuous raster ablation on an area of approximately 150 μm × 300 μm was applied. The approximate depth of ablation was between 10 and 50 μm. An Hitachi CCD video camera, type KP-D20BU, attached to the laser system, was used to observe the laser ablation process and to avoid micro mineral and fluid inclusions. The carrier gas for transport of the ablated material to the ICP-MS was He mixed with Ar. The isotope ^{29}Si was used as the internal standard applying the stoichiometric concentration of Si in SiO_2. External multistandard calibration was performed using three silicate glass reference materials produced by the National Institute of Standards and Technology, USA (NIST SRM 610, 612, 614, and 616). In addition, the applied standards included the NIST SRM 1830 soda-lime float glass (0.1% m/m Al_2O_3), the certified reference material BAM No. 1 amorphous SiO_2 glass from the Federal Institute for Material Research and Testing in Germany, and the Qz-Tu synthetic pure quartz monocrystal provided by Andreas Kronz from the Geowissenschaftliches Zentrum Göttingen (GZG), Germany. Certified, recommended, and proposed values for these reference materials were taken from Jochum et al. [39] and from the certificates of analysis where available. For the calculation of P concentrations, the procedure of Müller et al. [40] was applied. Each measurement comprised 15 scans of each isotope, with the measurement time varying from 0.15 s/scan for K in medium mass resolution mode to 0.024 s/scan of, for example, Li in low mass resolution mode. An Ar blank was run before each reference material and sample measurement to determine the background signal. The background was subtracted from the instrumental response of the reference material/sample before normalization against the internal standard in order to avoid effects of instrumental drift. This was carried out to avoid memory effects between samples. A weighted least squares regression model, including several measurements of the six reference materials, was used to define the calibration curve for each element. Ten sequential measurements on the BAM No.1 SiO_2 quartz glass were used to estimate the limits of detection (LOD) which were based on 3 × standard deviation (3sd) of the 10 measurements.

3.5. $^{40}Ar/^{39}Ar$ Dating of K-Feldspar

The K-feldspar sample (12071216; for origin see Figure 2) was crushed, grounded and subsequently sieved to obtain 180–250 μm fraction. The fraction was washed in acetone and deionized water several times and finally handpicked under a stereomicroscope. Mineral grains with coatings or inclusions were avoided. The sample was packed in aluminum capsules together with the Taylor Creek Rhyolite (TCR) flux monitor standard along with pure (zero age) K_2SO_4 and CaF_2 salts. The sample was irradiated at IFE (Institutt for Energiteknikk, Kjeller, Norway) for c. 140 h with a nominal neutron flux of 1.3×10^{13} n $\times$ (cm$^{-2} \times$ s^{-1}). The correction factors for the production of isotopes from Ca were determined to be $(^{39}Ar/^{37}Ar)_{Ca} = (3.07195 \pm 0.00784) \times 10^{-3}$, $(^{36}Ar/^{37}Ar)_{Ca} = (2.9603 \pm 0.026) \times 10^{-4}$ and $(^{40}Ar/^{39}Ar)_K = (1.3943045 \pm 0.0059335) \times 10^{-1}$ for the production of K (errors quoted at 1sd). The sample was put in a 3.5 mm pit size aluminum sample disk and step heated using a defocused 3.5 mm laser beam with a uniform energy spectrum (Photon Machines Fusions 10.6 at the Geological Survey of Norway, Trondheim). The extracted gases from the sample cell were expanded into a two stage low volume extraction line (c. 300 cm^3), both stages equipped with SAES GP-50 (st101 alloy) getters, the first running hot (c. 350 $°$C) and the second running cold. They were analyzed with an automated Mass Analyzer Products Limited (MAP) 215–50 mass spectrometer in static mode, installed at the Geological Survey of Norway. The peaks and baseline (AMU = 36.2) were determined during peak hopping for 10 cycles (15 integrations per cycle, 30 integrations on mass ^{36}Ar) on the different masses ($^{41-35}$Ar) on a Balzers electron multiplier (SEV 217, analogue mode) and regressed back to zero inlet time. Blanks were analyzed every third measurement. After blank correction, a correction for mass fractionation, ^{37}Ar and ^{39}Ar decay and neutron-induced interference reactions produced in the reactor was applied using in-house software AgeMonster, written by M. Ganerød. It implements the equations of McDougall and Harrison [41] and the newly proposed decay constant for ^{40}K after Renne et al. [42]. A ^{40}Ar/^{36}Ar ratio of 298.56 $\pm$ 0.31 from Lee et al. [43], was used for the atmospheric argon correction and mass discrimination calculation using a power law distribution. We calculated J-values relative to an age of 28.619 $\pm$ 0.036 Ma for the TCR sanidine flux monitor [42]. We define a plateau according to the following requirements: at least three consecutive steps overlapping at the 95% confidence level (1.96σ) using the strict test:

$$abs(age_n - age_{n+1}) < 1.96\sqrt{(\sigma_n^2 + \sigma_{n+1}^2)} \text{ (if errors are quoted at 1}\sigma\text{)}$$

$\geq$50% cumulative ^{39}Ar released, and mean square of weighted deviates (MSWD) less than the two tailed student T critical test statistics for $n - 1$. Weighted mean ages were calculated by weighting on the inverse of the analytical variance.

3.6. Secondary Ion Mass Spectrometry

We conducted $\delta^{18}O$ profiles on the tree quartz crystals 12071206, 12071215 and 1207127; each crystal was embedded in Epoxy in its own individual 25.4 mm diameter sample mount and the given crystal was sectioned by polishing. Surface quality was judged using an optical microscope at high magnification and the total roughness of the sample surfaces were judged to be no more than a couple of micrometers. Thus, along with a separate sample block for calibration materials, there were four such mounts employed for our oxygen isotope determinations. Prior to conducting the $\delta^{18}O$ analyses each crystal was imaged in monochromatic cathode luminescence after which the sample mount was ultrasonically cleaned in high-purity ethanol, was coated with a 35 nm thick, high-purity gold film and was place in a low magnetic susceptibility sample holder and held in place using brass tension springs.

Our Secondary Ion Mass Spectrometry (SIMS) analyses were conducted using the Cameca 1280-HR) instrument in Potsdam using a ~2 nA Gaussian $^{133}Cs^+$ primary beam that was focused to an ~5 μm diameter on the sample's surface. Each analysis was preceded by a 70 s preburn using a 20 μm raster. Low energy electron flooding was used for charge compensation, with the total electron current <1 μA. During data acquisition our primary beam was rastered over a 10 μm $\times$ 10 μm area in order to reduce isotopic drift during the analysis; this rastering was compensated using the dynamic transfer capability of the 1280-HR's extraction optics. Our mass spectrometer was operated in static multi-collection mode with a mass resolution of $M/\Delta M \approx 1900$ at 10% peak height, which is adequate to remove all significant isobaric interferences. The $^{16}O^-$ count rate was measured using our L2' Faraday cup in conjunction with an e+10 Ω resistor, whereas the $^{18}O^-$ count rate was measured using our H2' Faraday cup in conjunction with an e+11 Ω resistor. An analysis consisted of 20 integrations each lasting 4 seconds; hence a single $\delta^{18}O$ analysis, including preburn and automated centring routines, took nearly 3 min. Total sampling mass consumed during data collection was ~260 pg as determined by volume estimates based on white light profilometry. In Figure 5 the 3D surface topographic map of a sputtering crater is shown. All data were collected during a single session lasting 6.3 h, during which we acquired 37 determinations on our quartz samples in addition to 47 calibration runs.

We used the quartz sand reference material NBS28 in order to calibrate our absolute $\delta^{18}O$ values; this material has a published value of $\delta^{18}O_{VSMOW} = 9.57 \pm 0.10$‰ (1sd) [44]. We have used an absolute value for the zero-point on the Vienna Standard Mean Ocean Water (VSMOW) delta-scale set at $^{18}O/^{16}O = 0.00200520$ [45], which we have directly transferred to the zero-point value for the VSMOW scale. A total of 31 determinations were made on NBS28. We also checked for the presence of analytical drift by analysing a piece of NIST SRM 610 silicate glass a total of 16 times; this glass is believed to be homogeneous in its $\delta^{18}O$ composition at the sub-nanogram scale. The piece of polished NIST SRM 610 was embedded in epoxy alongside the NBS28 quartz reference material. No analytical drift could be detected during the course of our run, reflected by an analytical repeatability for the determined $^{18}O^-/^{16}O^-$ ratio of ± 0.095‰ (1sd) on the glass. Our analytical series consisted of multiple analyses on the reference sample mount, followed by a full profile on the 12071206 quartz, followed by multiple analyses again on the reference mount and so on, until all three quartz crystals had been analyzed.

Within run uncertainties based on the standard error from the 20, four second integrations were typically around ± 0.05‰, whereas the analytical repeatability on the NIST SRM 610 glass was ± 0.095‰. An additional source of uncertainty in the δ-values obtained on our three quartz crystals is the assigned uncertainty of ± 0.10‰ (1sd) given on the NBS28 reference sheet (IAEA, 2007). Combining all of these uncertainty sources, we estimate that the total analytical uncertainty on our individual $\delta^{18}O$ determinations to be circa ± 0.14‰ (1sd). The repeatability on the N = 31 determinations of the NBS28 quartz sand was ± 0.32‰ (1sd), which is consistent with the level of heterogeneity reported previously for this material when evaluated at the sub-nanogram sampling scale [46].

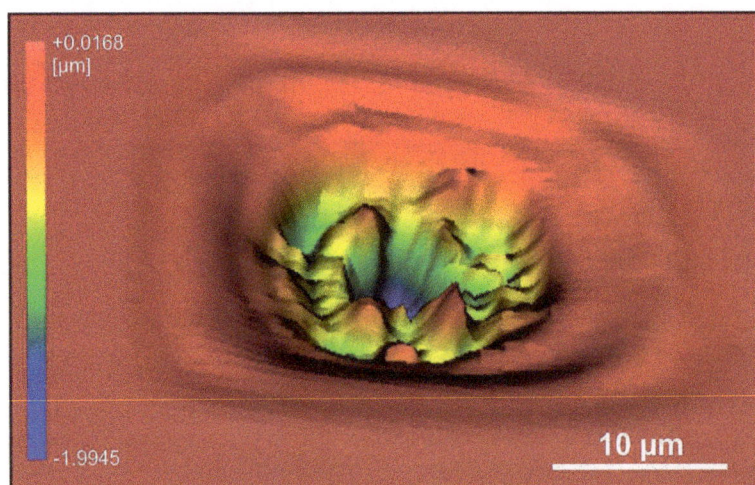

Figure 5. 3D surface topographic map of a sputtering crater created by the described Secondary Ion Mass Spectrometry (SIMS) experiment conditions in hydrothermal quartz of the Berglia-Glassberget breccia.

4. Results

4.1. Chemistry of the Breccia Host Rock

The pinkish grey, very fine-grained metarhyolite, which hosts the breccia, shows a strong mylonitic foliation. Macroscopically feldspars, quartz and biotite are identified. Recrystallized biotite is strongly elongated and stretched parallel to the foliation plane. Five samples of the metarhyolite were selected for major and trace element analysis, which were performed at ACMELabs [37] (Table 1). On the total alkali oxide vs. silica (TAS) diagram (Figure 6A), the unaltered metarhyolites fall in the rhyolite field, but from sampling over a larger area some of the felsic rocks were found to fall in the trachyte field [23]. Compositions overlap with the undeformed Blåfjellhatten granite exposed in the adjacent (2 km to the west) Olden Nappe [24]. Two metarhyolite samples (12071212, 12071213) have very high SiO$_2$ (>93 wt %) due to intense silicification caused by infiltration of breccia-cementing fluids. The occurrence of silicified metarhyolite is limited to the breccia center (Figure 2). On the molar ratio Al$_2$O$_3$/(CaO + Na$_2$O + K$_2$O) (A/CNK) versus Al$_2$O$_3$/(Na$_2$O + K$_2$O) (A/NK) diagram, the unaltered samples are classified as weakly peraluminous I-type granitoids, as in the case of the spatially related Blåfjellhatten granite [24] and the associated felsic volcanic rocks [23] (Figure 6B). The high-K, alkali Berglia-Glassberget metarhyolite has an A-type composition according to Whalen et al. [47] (Figure 6C).

The Nb-Y and Rb-(Y + Nb) diagrams (not shown) of Pearce et al. [48] both show that the Berglia-Glassberget metarhyolite plots in the field for within-plate granites. Within-plate granites are equivalent to the A-type, that is, anorogenic granites e.g., [47]. The relatively high large ion lithophile and high field strength element abundances in the metarhyolite analyses do confirm the A-type characteristics. For further chemical subdivision of the anorogenic granitoids, the diagrams devised by Eby [49] are useful in pointing to the likely source of the magmas. The Rb/Nb vs. Y/Nb plot (not shown), for example, classifies the Berglia-Glassberget metarhyolite in the field (A2) for magmas derived by partial melting of a continental crust that had probably been through a cycle of subduction-zone magmatism [49].

In the Upper continental crust-normalized diagram of selected incompatible elements (Figure 6D) K, Ba and Rb in the silicified samples are extremely depleted together with Th, Nb and Ta, indicating that mainly K-feldspar was leached out by the breccia-forming fluid. These elements are enriched in the unaltered samples compared to average upper crust composition. The adjacent Blåfjellhatten granite has a similar incompatible element signature suggesting a genetic relationship. The Berglia-Glassberget metarhyolite possibly represents the volcanic expression of the Blåfjellhatten granite. In summary, the geochemical data show that the Berglia-Glassberget metarhyolite magma derived from a fairly, but not markedly evolved, crustal source in a continental, extensional and probably anorogenic setting which is typically for TIB granites.

Table 1. Bulk analyses of metarhyolite, the host rock of the Berglia-Glassberget breccia. Analyses were performed at ACMELabs [37].

Sample nr.	12071201	12071210	12071209	12071213	12071212
Rock type	metarhyolite	metarhyolite	metarhyolite	silicified metarhyolite	silicified metarhyolite
Major elements (wt %)					
SiO_2	70.78	72.75	75.63	93.1	97.29
Al_2O_3	15.42	13.95	13.07	4.29	1.39
Fe_2O_3	1.13	1.38	1.1	0.23	0.08
MgO	0.18	0.22	0.22	0.08	0.03
CaO	0.08	0.49	0.13	0.02	0.01
Na_2O	3.27	3.84	3.22	0.03	<0.01
K_2O	7.92	6.13	5.83	1.29	0.28
TiO_2	0.28	0.34	0.26	0.18	0.16
P_2O_5	0.04	0.04	0.02	0.03	0.02
MnO	0.01	0.04	0.02	<0.01	<0.01
LOI	0.8	0.7	0.4	0.7	0.7
Sum	99.88	99.88	99.93	99.97	99.98
Trace elements (μg/g)					
Ba	395	353	124	37	9
Be	1	3	3	<1	<1
Cs	2.9	1.8	3.1	0.7	0.1
Ga	12.4	14.7	14.4	4.8	1.7
Hf	7.9	8.6	7.2	3.4	3.8
Mo	<0.1	0.2	<0.1	<0.1	<0.1
Pb	5.2	13.3	3.3	2.9	1.1
Nb	21.1	21.1	22.6	4.9	3.9
Rb	333.4	226.9	244.7	70.6	18.3
Sn	2	3	2	<1	<1
Sr	48.6	61.7	44.4	28	30.1
Ta	1.3	1.3	1.5	0.4	0.3
Th	16.3	17.4	14.9	6.8	6.9
U	4	3.9	4	1.6	0.8
W	2.4	0.9	0.9	0.9	1.7
Zr	268.7	294.2	221.4	118	123.7
Y	28.6	30.6	18.4	10.2	10.2
La	15.7	32.3	7.8	11.3	17.3
Ce	57.4	76.3	68.3	35.9	34.1
Pr	3.23	7.27	1.79	1.97	2.94
Nd	11.1	25.5	6.3	6.2	9.7
Sm	2.03	4.35	1.36	1.04	1.43
Eu	0.2	0.48	0.09	0.16	0.18
Gd	2.47	3.9	1.57	1.01	1.25
Tb	0.51	0.65	0.36	0.23	0.24
Dy	4.18	4.49	2.56	1.72	1.56
Ho	0.92	1.12	0.68	0.38	0.33
Er	3.31	3.27	2.49	1.19	1.1
Tm	0.54	0.55	0.43	0.2	0.19
Yb	3.54	3.81	3.3	1.15	1.34
Lu	0.6	0.61	0.55	0.19	0.22

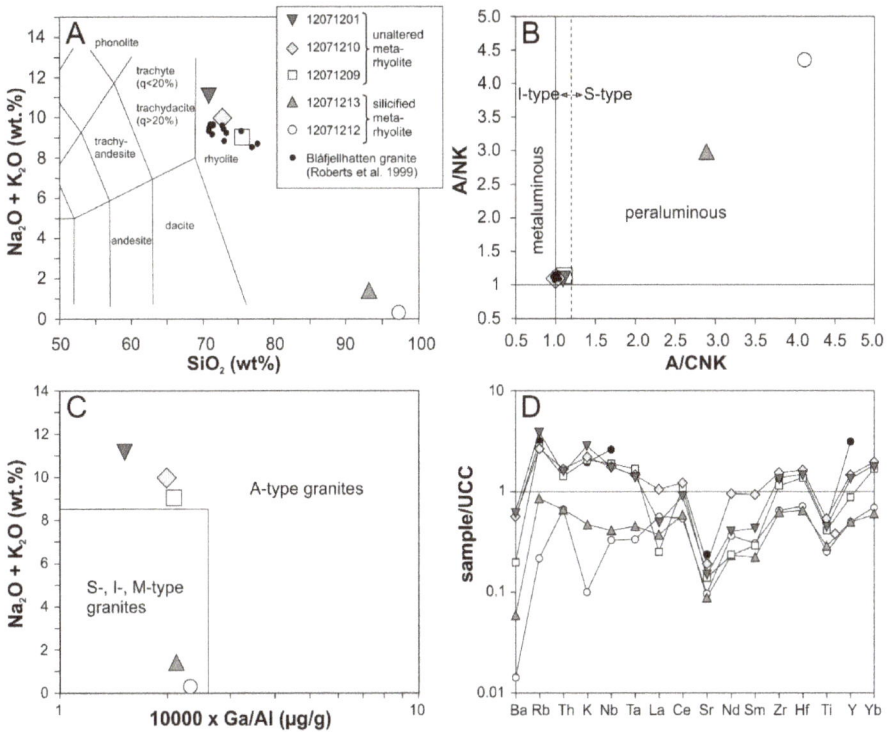

Figure 6. Whole rock chemistry of breccia host rocks. (**A**) Total alkali oxide vs. silica (TAS) classification diagram for volcanic rocks. (**B**) A/CNK versus A/NK (A/NK = molar ratio of $Al_2O_3/(Na_2O + K_2O)$; A/CNK = molar ratio of $Al_2O_3/(CaO + Na_2O + K_2O)$ discrimination plot according to Shand [50]. (**C**) discrimination plot of A-type and other granites according to Whalen et al. [47]. (**D**) Upper continental crust (UCC)-normalized diagram of selected incompatible elements. UCC values are from Rudnick and Gao [51].

4.2. Chemistry and Age of the K-Feldspar Cement

The average composition of the K-feldspar cement of the Berglia-Glassberget breccia is given in Table 2. The K-feldspar has relative pure orthoclase composition $An_{1.6}Ab_{0.7}Or_{97.7}$ with high Rb (792 µg/g), Ba (1937 µg/g) and Th (77 µg/g) and relative low Sr (70 µg/g). The trace element concentrations are generally higher compared to the chemistry of low-temperature K-feldspar from other hydrothermal vein deposits [52,53]. The composition is, however, similar to K-feldspar found in moderately evolved granitic rocks e.g., [54].

The main results of the $^{40}Ar/^{39}Ar$ dating of the K-feldspar cement and the age spectrum and inverse isochron plots are displayed in Table 3 and Figure 7. The raw degassing experiments, corrected for blanks, can be found in the Supplementary Materials Table S1. The degas pattern shows climbing apparent ages in the first part of the heating experiment and stabilize into a plateau from steps 19–25 (Figure 7A). From those steps an inverse weighted mean plateau age (WMPA) of 240.3 ± 0.4 Ma (MSWD = 0.589, P = 0.739) is calculated. The same steps yield an inverse isochron age (Figure 7B) of 240.2 ± 3.9 (MSWD = 0.765, P = 0.575) with a trapped $^{40}Ar/^{36}Ar$ ratio of modern atmospheric value [43], thus, an excess ^{40}Ar component in the spectrum can be ruled out. Given the similarities between the apparent ages from the WMPA and the inverse isochron determinations, we interpret the WMPA as the best age estimate of these K-Feldspars.

Table 2. Composition of K-feldspar cement. Analyses were performed at ACMELabs [37].

Major Elements (wt %)		Trace Elements (µg/g)			
SiO_2	62.34	Ba	1937	Y	34.7
TiO_2	0.69	Cs	0.9	La	94.1
Al_2O_3	17.88	Ga	11	Ce	190.1
Fe_2O_3	0.6	Hf	23	Pr	18.07
MnO	<0.01	Nb	28.1	Nd	57.6
MgO	0.03	Pb	25.4	Sm	8.68
CaO	0.31	Rb	792	Eu	0.95
Na_2O	0.08	Sr	70.2	Gd	6.24
K_2O	16.11	Ta	2.0	Tb	0.86
P_2O_5	0.24	Th	77	Dy	5.39
LOI	1.3	U	7.9	Ho	1.08
Sum	99.58	V	46	Er	4.01
An	1.6	W	22.1	Tm	0.68
Ab	0.7	Zn	13	Yb	5.59
Or	97.7	Zr	814.2	Lu	0.97

Minerals **2018**, *8*, 175

Table 3. Results of ^{40}Ar/^{39}Ar K-feldspar dating.

| Sample nr. | | UTM 33W | | | | Spectrum | | | | | Inverse Isochron | | |
	Material	E	N	Steps(n)	%^{39}Ar	Age ± 1.96σ	MSWD(P)	TGA ± 1.96σ	K/Ca ± 1.96σ	Age ± 1.96σ	MSWD (P)	Trapped ^{40}Ar/^{36}Ar	Spread (%)
12071206	K-Feldspar	426849	7123189	19–25(7)	51.26	**240.26 ± 0.36**	0.59(0.74)	238.38 ± 0.39	72.82 ± 2.74	240.19 ± 3.87	0.76(0.58)	298.71 ± 9.02	5.6

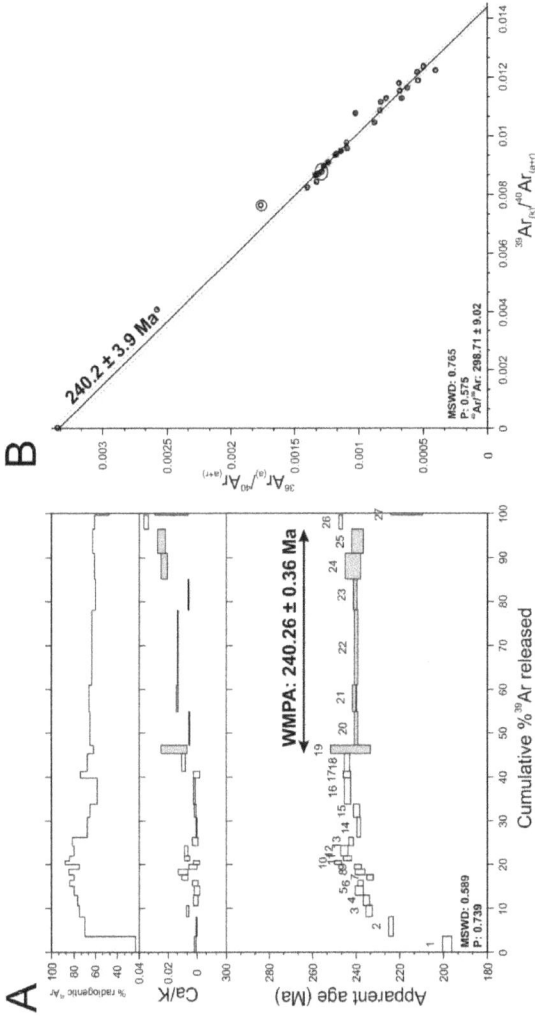

Figure 7. ^{40}Ar/^{39}Ar degassing spectra and inverse isochron results. See text for denotations. Uncertainties are reported at the 1.96sd level (95% confidence).

93

4.3. SEM-CL and Chemistry of the Quartz Cement

Applying SEM-CL imaging quartz crystals of the breccia cement show complex intra-granular growth structures (Figure 8). All investigated crystals have a relative homogeneous, bright luminescent core, with only a few oscillatory zones of low contrast. Oscillatory growth zones with less intense CL become more and more prominent towards the grain margin. The outermost growth zones are almost non-luminescent and appear black in SEM-CL images. Sector zoning is typically developed (Figure 8E). Sector zoning refers to a compositional difference between coeval growth sectors in a crystal and results from differences in fluid-crystal element partitioning between nonequivalent faces of the crystal. The growth and sector zoning, generally considered as primary structures, are superimposed by dull to non-luminescent healed microfractures associating patchy domains of recrystallized quartz. The patchy domains are abundant in sample 12071217 occupying about 1/3 of the crystal volume (Figure 8D). These superimposing textures are considered as secondary quartz and formed after quartz crystallization.

Trace elements of five quartz crystals (1 to 3 cm in length) originating from five different cavities were analyzed by LA-ICP-MS. Six analyses were performed on each crystal (Table 4, Figure 8). Sample 12071215 (smoky quartz) originates from a cavity in the center of the brecciated area, samples 12071217 and -18 from the SE and samples 12071204 and -06 from NE of the breccia structure (Figure 2). Concentrations of Be, B, Ge, Rb, Na, P, K, Ca, Ti, and Fe are mostly below the LOD and are not provided (LOD: Be = 0.5, B – 1.4, Ge – 0.07, Rb – 0.1, Na – 7.9, P – 5.0, K = 20.4, Ca = 6.3, Ti = 0.5, Fe = 0.5). Aluminium, Li, Mn, and Sr are the only elements which have concentrations consistently higher than the LODs (Table 4).

Aluminium concentrations show an extreme data spread ranging from <4 to 2471 µg/g. The data spread within one crystal can be similar as the concentration range of the entire data set (sample 12071218). Aluminium concentrations >1000 µg/g have been described previously from hydrothermal quartz but seem not so common considering published data [55–59]. In addition to Al the studied quartz is strongly enriched in Li. Concentrations vary from <1 µg/g to 319 µg/g; highest Li is observed in sample 12071218. The Li enrichment is stronger than, for example, in quartz of Li-enriched granites of the Land's End pluton in Cornwall, UK [60] and Li-rich pegmatites of the Tres Arroyos granite-pegmatite suite, Spain [61] (Figure 9A). In the Al versus Li plot of Figure 9A the data show a weak positive correlation due to the substitution $Si^{4+} \leftrightarrow Al^{3+} + Li^+$, where Al replaces Si in the tetrahedral site and Li^+ enter interstitial lattice position e.g., [62]. Higher Li values plot close to the 2:1 Al:Li atomic ratio line indicating that about half of the Al^{3+} defects are charged compensated by Li^+; alternative charge compensator are H^+, Na^+, or K^+. Non-luminescent, secondary quartz, which is most prominent in sample 12071217, is strongly depleted in both Al and Li, whereas growth zones with bright CL have high concentrations of both elements. The correlation of high Al and Li with bright CL has been described previously from hydrothermal quartz by Ramseyer and Mullis [63] and Jourdan et al. [59].

Manganese concentrations vary between <0.2 to 0.6 µg/g and do not show correlations with Al or Li. Strontium concentrations range from <0.03 to 0.18 µg/g. Germanium is mostly below the LOD of <0.07 µg/g except for sample 12071215 where Ge values rise up to 15 µg/g.

Titanium concentrations are consistently below the LOD of 0.5 µg/g. For illustration, Ti concentrations ($\leq$0.5 µg/g; hatched area in Figure 9B) plotted in the logarithmic Ti and Al diagram according to Rusk [58] for comparison with data of c. 30 porphyry-type, orogenic Au, and epithermal deposits (colour-shaded fields in Figure 9B). The diagram is used to fingerprint the type of ore deposit based on the trace element composition of quartz. Data from Berglia-Glassberget overlap with epithermal deposits which confirm the hydrothermal nature of the Berglia-Glassberget breccia.

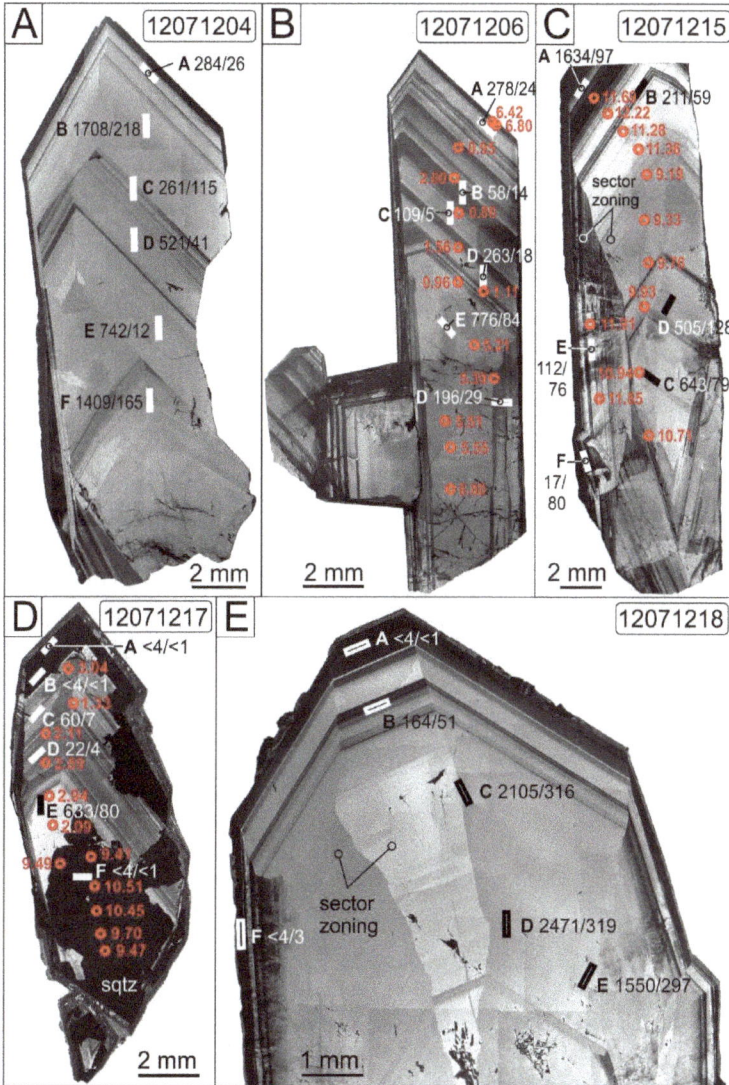

Figure 8. SEM-CL images of quartz crystals of the Berglia-Glassberget mineralization. The white and black bars indicate the location of ICP-MS ablation rasters with the corresponding Al/Li concentrations. The red circles mark the location of SIMS measurements with corresponding $\delta^{18}O_{VSMOW}$ values in ‰ (in red). (**A**) Sample 12071204, (**B**) Sample 12071206, (**C**) Sample 12071215, (**D**) Sample 12071217, (**E**) Sample 12071218. Sqtz—secondary quartz.

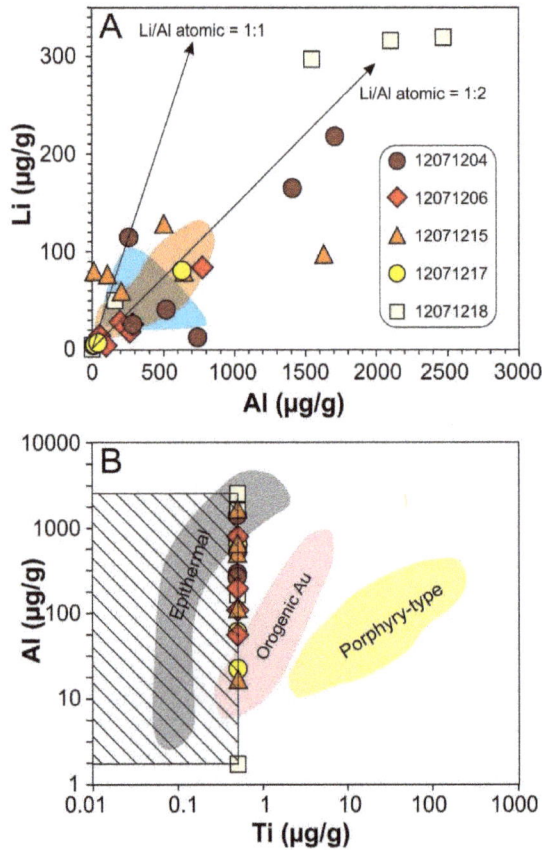

Figure 9. Trace element concentrations of quartz crystals forming the matrix of the Berglia-Glassberget. Concentrations were determined with LA-ICP-MS. (**A**) Al versus Li plot. Arrows indicate atomic ratios 1:1 and 2:1. The orange-shaded area corresponds to values of quartz from the Li-rich Tres Arroyos granite-pegmatite suite (Spain) [61] and the blue-shaded area those from the Li-bearing Land's End pluton in Cornwall (UK) [60]. (**B**) Logarithmic Ti versus Al plot of quartz of the Berglia-Glassberget breccia compared with data from Rusk [58] including c. 30 porphyry-type (Cu-Mo-Au) deposits, orogenic Au deposits, and epithermal deposits. The hatched field corresponds to the Berglia-Glassberget dataset because the Ti contents are below the detection limit of 0.5 µg/g. For symbol explanation see Figure 9A.

Table 4. Concentrations of trace elements (µg/g) of quartz crystals determined by LA-ICP-MS.

	CL Intensity	Li	Mn	Ge	Sr	Al	P	Ca
LOD		0.8	0.2	0.07	0.03	4.2	5.0	6.3
12071204-A	moderate	26.0	0.2	2.83	0.06	284.3	<5.0	<6.3
12071204-B	high	217.9	0.2	0.08	0.12	1707.6	<5.0	<6.3
12071204-C	moderate	115.3	0.6	0.07	0.14	260.6	<5.0	11.5
12071204-D	high	41.1	0.3	<0.07	<0.03	521.4	5.3	<6.3
12071204-E	high	12.5	0.3	0.08	<0.03	742.2	10.0	13.9
12071204-F	high	165.2	0.3	0.07	<0.03	1408.9	<5.0	<6.3
12071206-A	high	24.3	<0.2	<0.07	0.09	278.1	6.7	<6.3

Table 4. *Cont.*

	CL Intensity	Li	Mn	Ge	Sr	Al	P	Ca
12071206-B	moderate	14.1	0.4	<0.07	0.09	55.7	7.2	<6.3
12071206-C	low	4.8	0.3	<0.07	0.05	109.0	8.2	<6.3
12071206-D	moderate	18.1	0.2	<0.07	0.04	262.6	5.4	7.0
12071206-E	high	83.9	0.4	0.08	<0.03	775.5	8.1	45.4
12071206-F	moderate	28.7	0.4	<0.07	0.14	196.3	7.0	<6.3
12071215-A	low	96.8	<0.2	15.03	0.15	1634.3	<5.0	<6.3
12071215-B	high	59.3	0.3	3.45	0.07	210.6	<5.0	<6.3
12071215-C	high	79.1	0.3	<0.07	0.07	643.3	5.2	13.7
12071215-D	high	128.1	0.5	<0.07	0.18	505.1	<5.0	<6.3
12071215-E	low	76.5	0.3	2.29	0.10	112.1	<5.0	22.6
12071215-F	low	80.0	0.4	2.71	0.10	16.7	<5.0	<6.3
12071217-A	very low	<0.8	0.4	0.29	0.04	<4.2	<5.0	21.4
12071217-B	very low	<0.8	0.4	<0.07	0.06	<4.2	<5.0	<6.3
12071217-C	low	6.8	0.4	<0.07	<0.03	59.8	5.9	29.8
12071217-D	low	3.7	0.4	<0.07	0.05	22.4	<5.0	<6.3
12071217-E	high	80.0	0.3	0.09	0.03	633.2	<5.0	<6.3
12071217-F	very low	<0.8	0.3	0.08	0.03	<4.2	<5.0	11.4
12071218-A	very low	<0.8	0.6	<0.07	0.04	<4.2	<5.0	<6.3
12071218-B	low	50.6	0.2	0.17	0.05	164.0	<5.0	<6.3
12071218-C	very high	316.2	0.3	0.14	0.07	2104.8	7.5	<6.3
12071218-D	very high	319.3	0.4	0.15	0.15	2471.3	<5.0	<6.3
12071218-E	very high	297.2	0.4	0.11	0.11	1549.7	<5.0	<6.3
12071218-F	low	3.2	0.4	<0.07	0.03	<4.2	<5.0	<6.3

4.4. Petrography and Microthermometry of Fluid Inclusions

About 150 inclusions in quartz crystals from 5 different cavities of the Berglia-Glassberget breccia were examined. Microthermometric measurements were performed on representative 23 fluid inclusions based on the observation that only three genetically related types of pseudosecondary fluid inclusion were identified (Table 5; secondary inclusions are not considered):

Type I: Inclusions occur in small three-dimensional clusters or in short trails preferentially in the core of euhedral crystals. For these inclusions a pseudosecondary origin is suggested [64]. The aqueous fluid of the inclusions contains two-phase bubbles (liquid and gaseous CO_2) occupying 25–40 vol % of the inclusion at room temperature (Figure 10A). The presence of CO_2 was confirmed by Raman spectrometry performed at German Research Center of Geosciences in Potsdam (GFZ), Germany (Rainer Thomas, personal communication). Type I inclusions contain several types of relative large solids (5–25 μm) most of them have been identified as nahcolite and alkali sulphates by Raman spectroscopy. The group shows a narrow range of ice melting temperatures, from −0.4 to −0.6 °C, and liquid homogenization, from 242 to 248 °C. First melting was observed at −24 to −22 °C, implying some K^+ in addition to Na^+ [65]. Salinities are low, 0.7 to 1.0 NaCl wt % equivalent. Assuming a pressure during breccia formation of 1.1 to 0.1 kbar the trapping temperature of these fluid inclusions would be between 247 and 329 °C (utilizing the HokieFlincs_H2O-NaCl spreadsheet by Steele-MacInnis et al. [66]).

Type II: Inclusions occur in short trails or in lineations ending at intra-granular growth zones. For these inclusions an early pseudosecondary origin is suggested [64]. These inclusions contain smaller bubbles than type I inclusions and one or two solids at room temperature (Figure 10B). The solids were identified with Raman spectrometry as nahcolite and sulphates (Rainer Thomas, personal communication). Th and Tmice vary between 203 to 214 °C and −29 to −24 °C, respectively. The ice melt temperatures range from −4.6 to −0.4 °C and calculated salinities (according to Steele-MacInnis et al. [66]) vary from 0.4 to 7.3 NaCl wt % equivalent. One inclusion displayed clathrate formation with a melting temperature of 2.2 °C.

Type III: Inclusions contain aqueous liquid and vapor, and occasionally needle-like solids identified as nahcolite (Figure 10C). They occur in short trails ending at intra-crystal growth zones.

These inclusions are also interpreted to be pseudosecondary [64]. Type III has Th from 136 to 189 °C. Tmice shows a wide range from −8.9 and 19.2 °C due to clathrate formation in some inclusions. The ice melt temperatures for non-clathrate-forming inclusions range from −8.9 to −0.2 °C and those for clathrate-forming inclusions 2.2 to 19.2 °C. Salinities of non-clathrate-forming inclusion vary from 0.4 to 12.7 NaCl wt % equivalent. Two Tm of clathrates exceed the maximum temperature of 10 °C in the H_2O-CO_2 system, and these may reflect clathrate metastability or the presence of other gases e.g., CH_4 [67].

Summarizing, the breccia was cemented by K-feldspar and quartz which crystallized from an aqueous CO_2-bearing Na-HCO_3-SO_4 fluid with very minor Cl. However, primary fluid inclusion could not be identified. The crystallization temperature consistently decreases, whereas the salinity consistently increases from type I to type III inclusions.

Table 5. Summary of petrographic and microthermometric data of primary and pseudosecondary fluid inclusions in quartz crystals from the Berglia-Glassberget breccia. Te—temperature of first melting, Tmice—temperature of final ice melting, Th—temperature of homogenization.

	Type I	Type II	Type III
inclusion type	as groups, pseudosecondary	as groups, pseudosecondary	in trails, pseudosecondary
inclusion petrography	multi-phase, isometric	multi-phase, isometric	two- and multi-phase, isometric to longish
size (µm)	35 to 60	15 to 25	25 to 55
degree of fill	0.60 to 0.75	0.80 to 0.95	0.70 to 0.95
Te (°C)	−24 to −22	−29 to −24	−41 to −21
Tmice (°C)	−0.4 to −0.6	−4.6 to 2.2	−8.9 to 19.2
Th (°C)	242 to 248	203 to 214	136 to 189
number of analyzed inclusions	4	4	15

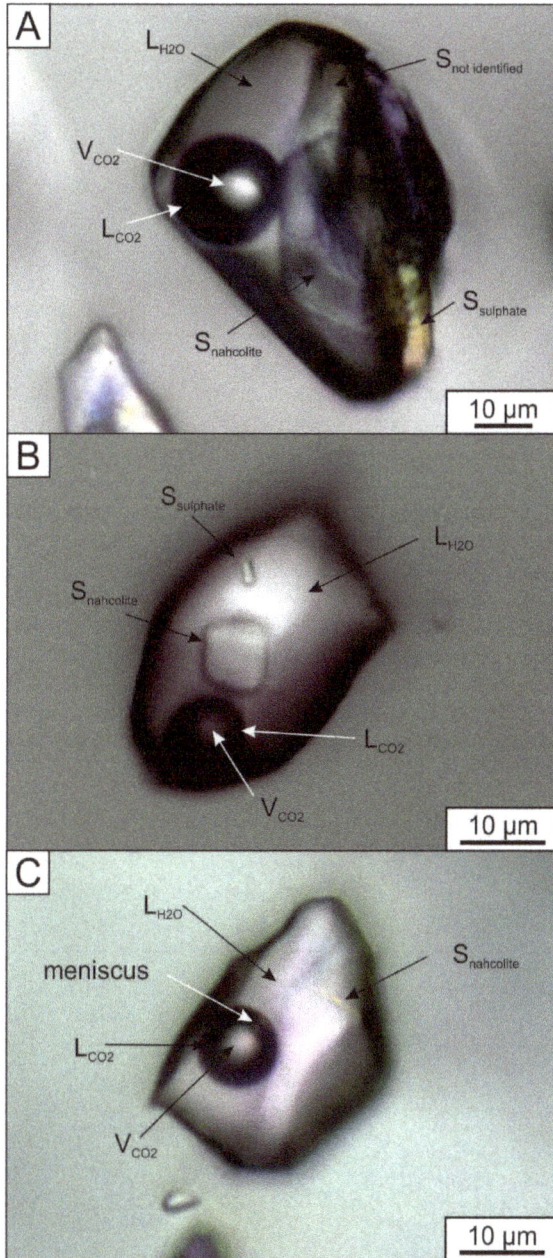

Figure 10. Representative fluid inclusions in quartz crystals of the breccia of Berglia-Glassberget. (**A**) Aqueous Type I inclusion containing a CO_2 bubble and several large solids including nahcolite and a sulphate, (**B**) Aqueous Type II inclusion with a CO_2 bubble and two solids, nahcolite and a sulphate, (**C**) Aqueous Type III inclusion with a CO_2 bubble and a nahcolite crystal. L—liquid phase, V—volatile phase, S—Solid.

4.5. Oxygen Isotope Geochemistry of Quartz

The results of SIMS oxygen isotope measurements are provided in Table 6, and the full SIMS data set including the results of the calibration runs is given in the electronic Supplementary Materials Table S1. Profiles of $\delta^{18}O$ values across three selected quartz crystals (12071206, 12071215, 12071217) are plotted in Figure 11A–C and the locations of measurement spots are indicated in Figure 8. Figure 11D–F illustrate the $\delta^{18}O$ variation in form of histograms.

The measured $\delta^{18}O$ values display a large variation from 0.9 to 12.2‰ $\delta^{18}O$. Highest $\delta^{18}O$ values are observed across the entire 12071215 crystal from the breccia center (mean 10.8 ± 1.1‰; n = 12) and in the late secondary, non-luminescent quartz, overprinting crystal 12071217 (mean 9.8 ± 0.5‰; n = 6). Average isotope ratios of crystals form the breccia margin are much lower: 3.8 ± 2.4‰ $\delta^{18}O$ for crystal 12071206 and 2.4 ± 0.7‰ $\delta^{18}O$ for 12071217 (excluding $\delta^{18}O$ values of secondary formed quartz). Thus, the $\delta^{18}O$ values in quartz decrease considerably from the center towards the margin of the breccia. In contrast to the relative consistent isotope ratios of primary precipitated quartz of crystals 12071215 and 12071217 (not considering secondary quartz in crystal 12071217), the $\delta^{18}O$ values in 12071206 display two composition steps (Figure 11C). The first step (in growth direction) from 5.6 ± 0.4‰ to 1.3 ± 0.7‰ corresponds to the transition from the homogeneous crystal core to the crystal margin rich in oscillatory growth zoning as documented in the CL images (Figure 8B). The second step is marked by the outermost, non-luminescent growth zone in which the isotope ratio increases abruptly to 6.6 ± 0.3‰.

In addition to the documented large-scale $\delta^{18}O$ variations controlled by the quartz-forming fluid, there is a subordinate systematic difference between different, adjacent crystal growth faces. Different growth faces are visualized in CL in the form of sector zoning (Figure 8). Examples of this variation are displayed in the bottom of crystal 12071215 (Figure 8C), where $\delta^{18}O$ values in one zone are 10.5 ± 0.5‰ (n = 3) and in the adjacent zone 11.9 ± 0.1‰ $\delta^{18}O$ (n = 2), and in the top of crystal 12071217 (Figure 8D), where one sector zone has 1.3‰ and the neighboring zone 2.6 ± 0.7‰ $\delta^{18}O$ (n = 2). The observed $\delta^{18}O$ differences in adjacent crystal faces are between 0.6 to 2.1‰, which are in the same range as $\delta^{18}O$ differences described by Onasch and Vennemann [68] and Jourdan et al. [59] from hydrothermal quartz. No correlation was found between $\delta^{18}O$ values and quartz trace element contents in the three quartz crystals that sere analyzed.

Table 6. Results of oxygen isotope microanalyses by SIMS on selected hydrothermal quartz crystals (samples 12071206, 12071215, 12071217).

12071215—Breccia Center		12071217—Breccia Margin		12071206—Breccia Margin	
Analysis nr.	$\delta^{18}O$ (‰ ± 0.1)	Analysis nr.	$\delta^{18}O$ (‰ ± 0.1)	Analysis nr.	$\delta^{18}O$ (‰ ± 0.1)
01	11.7	01	3.0	01	6.4
02	12.2	02	1.3	02	6.8
03	11.3	03	2.1	03	1.0
04	11.4	04	2.9	04	2.6
05	9.2	05	2.9	05	0.9
06	9.3	06	2.1	06	1.6
07	9.8	07	9.4	07	1.0
08	9.9	08	9.5	08	1.1
09	11.9	09	10.5	09	5.2
10	10.9	10	10.4	10	5.4
11	11.8	11	9.7	11	5.5
12	10.7	12	9.5	12	5.5
				13	6.2

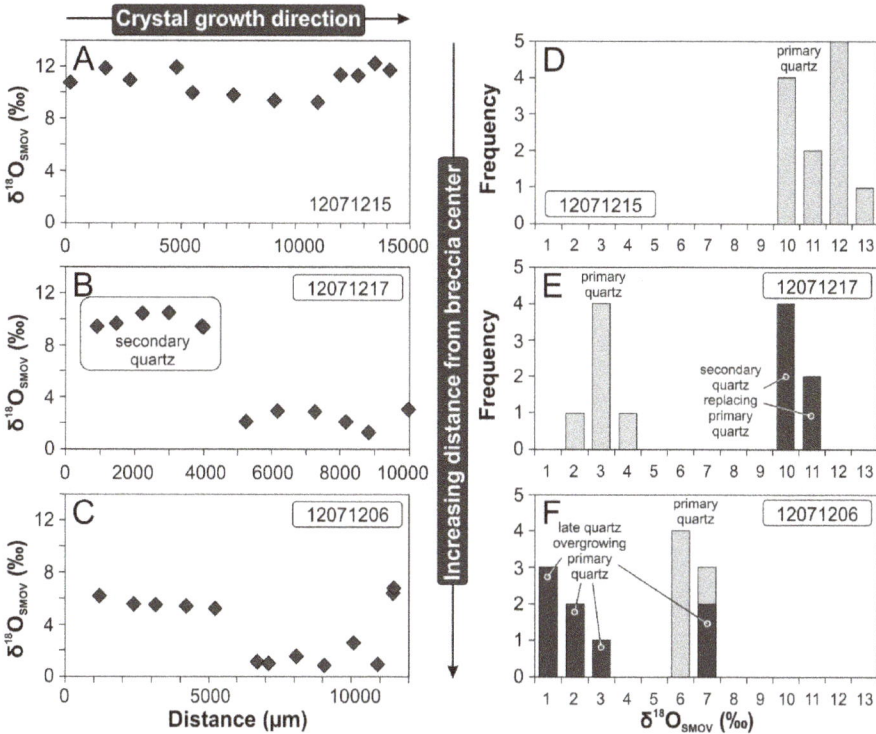

Figure 11. (**A–C**) Variation profiles of $\delta^{18}O$ values across three quartz crystals (see also Figure 8 for location of the SIMS measurements within the crystals). (**D–F**) Histograms of measured $\delta^{18}O$ values shown in the profiles (**A–C**). Grey bars are $\delta^{18}O$ values related to major quartz crystal growth (primary quartz). Black bars represent quartz $\delta^{18}O$ values affected by late- to post-crystallization processes.

5. Discussion

5.1. Processes Responsible for the Formation of the Berglia-Glassberget Breccia

The structural characteristics described above classify the Berglia-Glassberget mineralization genetically as fault-related, fluid-assisted hydraulic breccia formed by a single pulse stress typically for upper crust levels e.g., [1,36]. Such fault rocks represent implosion breccias, formed by the 'sudden creation of void space and fluid pressure differentials at dilational fault jogs during earthquake rupture propagation' [2]. This implosion forms commonly fitted-fabric to chaotic fault breccias, cemented by hydrothermal fluids e.g., [36,69–74]. The implosion hypothesis envisages that faulting-induced voids are transient and filled coseismically by a dilated mass of rock fragments. The dilation breccias probably formed during the seismic event of strike-slip displacement on an irregular fault.

The influx of fluids into fault zones can trigger short-term weakening mechanism that facilitate fault movement and earthquake nucleation by reducing the shear stress or frictional resistance to slip e.g., [75]. Crustal fluids can be trapped by low-permeability sealed fault zones or stratigraphic barriers. The development of fluid overpressures at the base of the fault zone can help to facilitate fault slip, which was likely the cause for the Berglia-Glassberget breccia formation. Once the barrier is ruptured by an earthquake, permeability increases and fluids are redistributed from high to low pressure areas. In this model, a fault is considered to act as a valve [76,77].

The seismic energy released by brittle failure lead to rapid, seconds-long, fragmentation and dilation of at least 30 million m^3 of metarhyolitic rock at Berglia-Glassberget. Hydraulic fracturing was mainly responsible for the rock fragmentation. The breccia was resealed soon after the seismic event by quartz and K-feldspar cement. The very high Al and Li concentrations in the quartz cement suggest disequilibrium growth, meaning that the quartz crystallized very fast, maybe in hours or days, so that impurities were incorporated as in large quantities as defect clusters during rapid crystal growth. The resealing strengthened the fractured rock, so that rebrecciation of the same volume is uncommon, except where repeated dilational strains are focused at fault bends or jogs [11,78]. The Berglia-Glassberget does not show evidence for rebrecciation or persistent refracturing and has been formed by a single seismic event. The single event of brecciation is evident from the occurrence of only one quartz cement generation as documented by similar trace element chemistry, CL structures and fluid inclusion assemblage. However, the nature of the regional fault structure remains unclear because of the overburden covering large areas of the studied locality (see also discussion below).

The proportion of open space (cavities) in brecciated fractures is up to 80 vol % in the breccia center. This high percentage of open space indicates that the hydrostatic pressure was much greater than the lithostatic pressure. In fact, the lithostatic pressure was so low that large cavities up to 3 m × 3 m × 4 m in size remained stable after the seismic event. The characteristics described above suggest that the brecciation took place at a maximum depth of about 4 km [1]. However, the post-late-Triassic erosion in the study area did not exceed 2 km as indicated by regional fission track analyses e.g., [79]; thus the formation depth of the Berglia-Glassberget breccia was presumably much less than 4 km.

5.2. Characteristics and Origin of Breccia-Forming Fluids

The observed fluid inclusion assemblage characterizes the breccia-forming fluid as aqueous CO_2-bearing Na-HCO_3-SO_4 fluid with very minor Cl. The minimum temperature of the breccia-cementing fluid was in the range of 247 and 329 °C considering type I inclusions as being representative for the fluid pumped by the seismic event into the brecciation level and assuming a formation depth of 0.5 to 4 km (0.01 to 0.11 GPa). The crystallization of paragenetic, almost Na-free, low-temperature K-feldspar (var. adularia) and low Ti concentrations (<0.5 µg/g) in the quartz indicate formation temperatures of circa 350 °C or less [80–82]. Type II and type III inclusions record gradually decreasing crystallization temperatures and increasing salinity of a single-source fluid during progressing cement crystallization. The fluid leached K, Ba, Rb, Th, Nb and Ta from the metarhyolite in the SW part of shattered area (breccia center). The dissolved K, Ba, Rb and Th were preferentially transported to the periphery of the breccia structure where they precipitated mainly as breccia-cementing adularia. The element transport is recorded by the increasing K-feldspar/quartz cement ratio in the breccia faults towards the margin of the breccia structure. Such a leaching explains the composition of the K-feldspar cement; its composition is inherited from the leached, rock-forming K-feldspar of the metarhyolite.

The oxygen isotope ratios of breccia-cementing quartz display a large variation (0.9 to 12.2‰ $\delta^{18}O$) which can generally result from: (1) variations in temperature; (2) variations in the $\delta^{18}O$ of the parent fluid; (3) disequilibrium effects; and (4) a combination of those three. Disequilibrium partitioning of oxygen isotopes (cause 3) causes $\delta^{18}O$ variation in the range of 0.6 to 2.1‰ and, thus, has only a subordinate effect considering the overall variation of about 12‰. The high and relative consistent $\delta^{18}O$ values of crystal 12071215 from the breccia center (10.8 ± 1.1‰) presumably reflect the initial oxygen isotope ratio of the breccia-cementing fluid. Such high $\delta^{18}O$ values could be caused by any of these reservoirs: granitic rocks, metamorphic rocks or sedimentary rocks e.g., [83]. In any case, meteoric water can be excluded as the primary source of this oxygen. The average $\delta^{18}O$ values in quartz decrease from the center towards the margin of the breccia, which is interpreted as dilution of the breccia-cementing fluid in heavy ^{18}O during progressing fluid migration and accompanied quartz precipitation. A minor decrease in fluid temperature might also have contributed to the $\delta^{18}O$ decrease. The sudden drop in $\delta^{18}O$ in the overgrowth of crystal 12071206 is explained by the influx of "local" meteoric water during final quartz growth. However, this seems to be a special feature of this particular cavity because similar drops are not developed in the other two investigated crystals. The authors have no plausible explanation of the high $\delta^{18}O$ values (9.8 ± 0.5‰) observed in the secondary quartz replacing primary quartz of crystal 12071217 and in the outermost growth zone of crystal 12071206 (6.6 ± 0.3‰).

In general, the origin of CO_2 recorded in fluid inclusions of hydrothermal mineralization is still a matter of debate. Fluid inclusion and breccia characteristics indicate a deep-seated source for CO_2 that could be of mantle, magmatic or metamorphic in origin. The source of CO_2 in hydrothermal mineralization in metamorphic terranes has usually been considered as a relic of mantle and/or lower crustal (magmatic and/or metamorphic) fluids e.g., [69,84,85] that might have been channelled along through-going faults toward Earth's surface [86] and/or transported to higher levels in the crust by magmas [87,88]. Devolatilisation of supracrustal sequences during prograde metamorphism is the widely accepted hypothesis for the origin of CO_2 for a range of hydrothermal deposit types [69,89]. Alternatively, CO_2-dominated fluids may exsolve from felsic magmas formed at depths greater than 5 km in the crust; they are a typical feature of intrusion-related deposits [88]. Furthermore, these fluids can also be associated with granulite facies metamorphism and with charnockitic magmatism [90,91].

In the case of the Berglia-Glassberget breccia the CO_2 has most likely a metamorphic origin due to decarbonation reactions (T > 200 °C) of limestones in a thin autochthonous succession, the Bjørndalen Formation, lying unconformably upon the granitic and felsic volcanic rocks of the Olden Nappe. The Berglia-Glassberget breccia is situated immediately above a thrust fold anticline formed by the limestone-bearing autochthonous sequence (Figure 2). Thus, decarbonation reactions of this limestone might be the possible source of the CO_2. This implies that low-grade metamorphism or hydrothermal activity initiated thermo-metamorphic reactions, involving silicate and carbonate rocks, took place in the middle Triassic. However, there was definitely no low-grade metamorphism in the area in the middle Triassic. The most plausible explanation is that CO_2 was produced from limestone of the Bjørndalen Formation by hydrothermal reactions initiated by deeply derived, hot fluids channelled to sub-surface levels by a major fault zone. The high SO_4 content probably relates to oxidation of sulfides of wall rocks through which the fluids was channelled.

5.3. The Formation of the Berglia-Glassberget Breccia in the Regional Context

Tectonic, hydrothermal or magmatic activities of Middle Triassic age have not been recorded in the vicinity of the Berglia-Glassberget mineralization which could be directly related to the breccia formation. In respect to Caledonian structures, the Berglia-Glassberget breccia is at the eastern edge of the Grong-Olden Culmination, which is a Caledonian nappe structure [19]. The breccia is situated in the hanging wall, above the fold axis of a gently SW plunging anticline of the Olden Nappe, a few meters above the thrust fault (Figure 2). This the thrust fault separates the Olden Nappe from the overlying Formofoss Nappe Complex. The close position of the breccia above the thrust fault may have had an effect on the location of breccia formation. However, these flat-dipping Caledonian structures are not known to have been reactivated during Triassic time. The eastern limb of the thrust fold and the foliation of the metarhyolite dip 30 to 35° NE but the breccia structures are discordant (generally sub-vertical) and transect these Caledonian structures.

On a global scale, the middle/late Triassic boundary (230 ± 5 Ma) marks the incipient dispersal of Pangea by the onset of continental rifting [92]. In NW Europe including southern Norway the Triassic was a period of major rifting and faulting [93–98], involving many long-lived fault zones, such as the Great Glen Fault [99] and the Møre Trøndelag Fault Complex (MTFC) [79,100–105] (Figure 12). The MTFC is the northernmost of several important regional structures identified in southern Norway and is widely agreed to have played an important role in the development of the Norwegian margin e.g., [106,107]. The MTFC is a large-scale SW-NE-striking tectonic zone which extends onshore for about 350 km from Ålesund in the SW to Snåsa and further towards the NE. It is a long-lived tectonic feature possibly with Precambrian roots [103,108,109] and principally comprised by the Hitra-Snåsa and Verran faults (Figure 12). The faults became multiply reactivated during the late Paleozoic, Mesozoic and late Cretaceous [79,104]. Fission-track dating of apatite, zircon and titanite by Grønlie et al. [79] along the NE part of the Verran fault revealed several Triassic reactivations overlapping in time with the formation of the Berglia-Glassberget breccia. According to Redfield et al. [104,105] the major Verran fault peters out about 30 km W of the breccia. However, a dense network of SW-NE striking structures transecting the Grong-Olden Culmination [19,20] document MTFC-related faults in the E and NE extension of the major Verran fault, immediately adjacent to the Berglia-Glassberget breccia.

Considering the regional context of these long-lived fault zones, the Berglia-Glassberget breccia is situated at the NE end of the Verran fault, probably forming a triple junction with the N-S striking faults related to the northern end of the Lærdal-Gjende fault system (coming from S, e.g., [110]) and the SE end of the Kollstraumen detachment (coming from NW; e.g., [111]) (Figure 12). Based in this geotectonic positon, the Berglia-Glassberget breccia can be considered as a seismic expression related to the Triassic reactivation of the MTFC. However, the specific nature of the major fault along which the seismic rupture occurred cannot be defined due to the heavy overburden, but might be a solved by future studies.

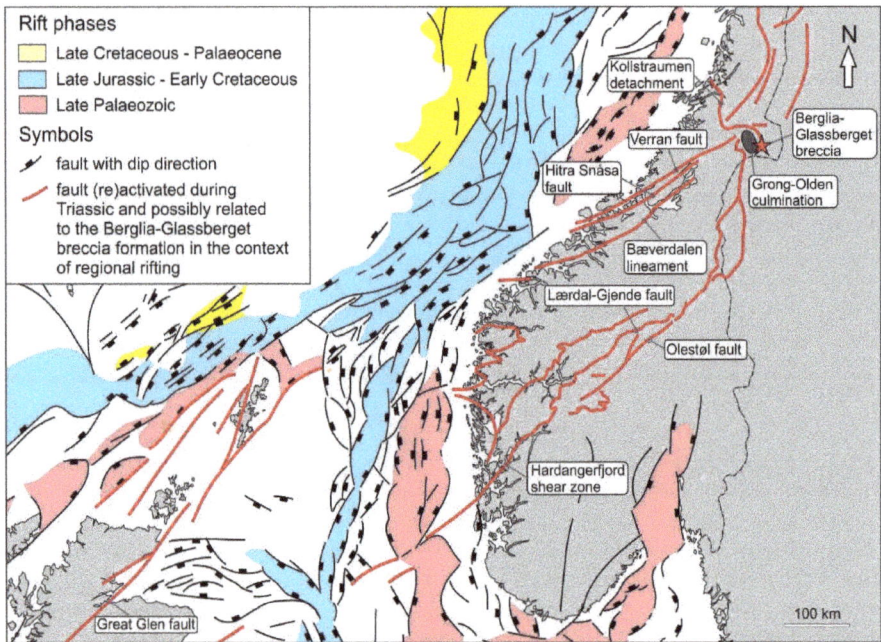

Figure 12. Geotectonic setting and relationship to the Møre Trøndelag Fault Complex (MTFC) comprising the Hitra-Snåsa and Verran faults. Main post-Caledonian structural elements of southern Norway, the Norwegian continental shelf and adjacent areas related to different rift phases affecting the NE Atlantic region (from Faleide et al. [112,113] updated with onshore elements from Redfield et al. [104,105]).

6. Summary

The results of the study can be summarized as follows:

- The Berglia-Glassberget mineralization is a fluid-assisted, hydraulic breccia (250 m × 500 m in lateral dimension) which formed by single-pulse stress released by a seismic event during middle Triassic (240.3 ± 0.4 Ma).

- The influx of aqueous CO_2-bearing Na-HCO_3-SO_4 fluids into the fault zone triggered a short-lived weakening mechanism that facilitated fault movement by reducing the shear stress.

- The intruding fluid, pumped by the seismic event to higher levels, leached K, Ba, Rb, Th, Nb and Ta metarhyolitic host rock and simultaneously silicified the host rock and its breccia fragments in the SW part of the shattered area (breccia "center"). The dissolved K, Ba, Rb and Th were precipitated mainly as breccia-cementing, low-temperature K-feldspar (var. adularia) followed by quartz.

- A high percentage of open space in the breccia fractures up to 80 vol % with cavities up 3 m $\times$ 3 m $\times$ 4 m in size, fluid inclusion microthermometry, and trace element chemistry of quartz suggest that the breccia was formed at depths between 4 and 0.5 km (1.1 to 0.1 kbar). The minimum temperature of the breccia cementing fluid was probably in the range of 247 and 329 °C. However, truly primary fluid inclusions could not be identified. The open space within the breccia body enabled the crystallization of large quantities of quartz crystals, which makes the locality attractive for mineral collectors.

- The origin of the CO_2-bearing, breccia-cementing fluids may be of predominantly metamorphic origin due to decarbonation reactions (T > 200 °C) of limestones in an autochthonous sequence above felsic magmatic rocks of the underlying Olden Nappe. The decarbonation reactions were possibly initiated by deeply derived, hot fluids channelled to sub-surface levels by a major fault zone.

- In the regional context, the Berglia-Glassberget breccia is interpreted to be situated at a triple junction of long-lived fault zones belonging to the Møre-Trøndelag, Lærdal-Gjende and the Kollstraumen fault complexes. These fault systems are the expression of major rifting and faulting in northern Europe during middle/late Triassic. By this means the Berglia-Glassberget breccia contributes to a better understanding of the extensional tectonics of the Norwegian mainland during that period.

Supplementary Materials: The complete SIMS data set including the results of the calibration runs is available online at http://www.mdpi.com/2075-163X/8/5/175/s1..

Acknowledgments: We are very grateful to Arne Solli and David Roberts for fruitful discussions, Øyvind Thorensen who took photographs of mineral specimen, and Arne Jostein Devik who provided information and material about mineral collection activities at Berglia-Glassberget. Rainer Thomas is thanked for performing Raman spectroscopy on fluid inclusions. Frédéric Couffignal conducted the SIMS analyses. We are very thankful to the constructive reviews by two anonymous reviewers.

Author Contributions: Axel Müller conceived and designed the experiments; Morgan Ganerød, Michael Wiedenbeck, Axel Müller, and Skule Olaus Svendsen Spjelkavik performed the experiments; Rune Selbekk provided the sample material and existing data and information about the Berglia-Glassberget breccia; Axel Müller, Morgan Ganerød and Michael Wiedenbeck analysed the data; Axel Müller, Morgan Ganerød and Michael Wiedenbeck wrote the paper.

Conflicts of Interest: The authors declare no conflict of interest.

References

1. Sibson, R.H. Fault rocks and fault mechanisms. *J. Geol. Soc.* **1977**, *133*, 191–213. [CrossRef]
2. Sibson, R.H. Brecciation processes in fault zones: Inferences from earthquake rupturing. *Pure Appl. Geophys.* **1986**, *124*, 159–174. [CrossRef]
3. Sillitoe, R.H. Ore-related breccias in volcanoplutonic arcs. *Econ. Geol.* **1985**, *80*, 1467–1514. [CrossRef]
4. Taylor, R.G.; Pollard, P.J. *Mineralized Breccia Systems. Method of Recognition and Interpretation*; Contributions of the Economic Geology Research Unit; Economic Geology Research Unit (EGRU): Townsville, Australia, 1993; Volume 46, p. 31.
5. Fournier, R.O. Hydrothermal processes related to movement of fluid from plastic into brittle rock in the magmatic-epithermal environment. *Econ. Geol.* **1999**, *94*, 1193–1211. [CrossRef]
6. Landtwing, M.R.; Dillenbeck, E.D.; Leake, M.H.; Heinrich, C.A. Evolution of the breccia-hosted porphyry Cu-Mo-Au deposit at Agua Rica, Argentina: Progressive unroofing of a magmatic hydrothermal system. *Econ. Geol.* **2002**, *97*, 1273–1292. [CrossRef]
7. Sibson, R.H. Earthquake faulting as a structural process. *J. Struct. Geol.* **1989**, *11*, 1–14. [CrossRef]

8. Roberts, G.P. Displacement localization and palaeo-seismicity of the Rencurel Thrust Zone, French Sub-Alpine Chains. *J. Struct. Geol.* **1994**, *16*, 633–646. [CrossRef]

9. Cowan, D.S. Do faults preserve a record of seismic slip? A field geologist's opinion. *J. Struct. Geol.* **1999**, *21*, 995–1001. [CrossRef]

10. Micklethwaite, S.; Cox, S.F. Fault-segment rupture, aftershock-zone fluid flow, and mineralization. *Geology* **2004**, *32*, 813–816. [CrossRef]

11. Woodcock, N.H.; Dickson, J.A.D.; Tarasewicz, J.P.T. Transient fracture permeability and reseal hardening in fault zones: Evidence from dilation breccia textures. In *Fractured Reservoirs*; Lonergan, L., Jolly, R.J.H., Rawnsley, K., Sanderson, D.J., Eds.; Special Publications; Geological Society: London, UK, 2007; Volume 270, pp. 43–53.

12. Baker, E.M.; Kirwin, D.J.; Taylor, R.G. *Hydrothermal Breccia Pipes*; Contributions of the Economic Geology Research Unit; Geology Department, James Cook University: Townsville, Australia, 1986; Volume 12, p. 45.

13. Lawless, J.V.; White, P.J. Ore-related breccias: A revised genetic classification, with particular reference to epithermal deposits. In *Proceedings of the 12th New Zealand Geothermal Workshop*; Harvey, C.C., Browne, P.R.L., Freestone, D.H., Scott, G.L., Eds.; Geothermal Institute, University of Auckland: Auckland, New Zealand, 1990; pp. 197–202.

14. Ewensson, T. Jakten på den Svarte Krystallen. *Stein* **2000**, *27*, 17–18. (In Norwegian)

15. Nordrum, F.S. Mineralogische Neuigkeiten aus Norwegen. *Miner. Welt* **2002**, *13*, 56–59. (In German)

16. Jørgensen, L. Berglia-Glassberget kvartsforekomst, Sørli i Lierne. *Nor. Bergverksmus. Skr.* **2003**, *25*, 39–40. (In Norwegian)

17. Asklund, B. *Hauptzüge der Tektonik und Stratigraphie der Mittleren Kaledoniden in Schweden*; C-417; Sveriges Geologiska Undersökning: Uppsala, Sweden, 1938; pp. 1–99. (In German)

18. Johansson, L. Basement and Cover Relationships in the Vestranden-Grong-Olden Region, Central Scandinavian Caledonides: Petrology, Age Relationships, Structures and Regional Corrrelations. Ph.D. Thesis, Lund University, Lund, Sweden, 1986.

19. Roberts, D. Tectonostratigraphy within the area of 1:50,000 map-sheet 'Grong', Nord-Trøndelag, Central Norway. *Geol. Fören. Stockh. Förh.* **1989**, *111*, 404–407. [CrossRef]

20. Fossen, H.; Nissen, A.L.; Roberts, D. *Bergrunnskart Blåfjellhatten 1923 III, M 1:50,000*; Norges Geologiske Undersøkelse: Trondheim, Norway, 2013.

21. Klingspor, I.; Tröeng, B. Rb-Sr and K-Ar age determinations of the Proterozoic Olden granite, Central Caledonides, Jämtland, Sweden. *Geol. Fören. Stockh. Förh.* **1980**, *102*, 515–522. [CrossRef]

22. Roberts, D. *Geologisk kart over Norge. Berggrunnsgeologisk kart Grong, M 1:250,000*; Norges Geologiske Undersøkelse: Trondheim, Norway, 1997.

23. Roberts, D. Geochemistry of Paleoproterozoic porphyritic felsic volcanites from the Olden and Tømmerås windows, Central Norway. *Geol. Fören. Stockh. Förh.* **1997**, *119*, 141–148.

24. Roberts, D.; Nissen, A.L.; Walker, N. U-Pb zircon age and geochemistry of the Blåfjellhatten granite, Grong-Olden Culmination, Central Norway. *Nor. Geol. Tidsskr.* **1999**, *79*, 161–168. [CrossRef]

25. Gorbatschev, R. Precambrian basement of the Caledonides. In *The Caledonide Orogen—Scandinavia and Related Areas*; Gee, D.G., Sturt, B.A., Eds.; John Wiley & Sons: Chichester, UK, 1985; pp. 197–212.

26. Högdahl, K.; Andersson, U.B.; Eklund, O. *The Transscandinavian Igneous Belt (TIB) in Sweden: A Review of Its Character and Evolution*; Special Paper; Geological Survey of Finland: Espoo, 2004; Volume 37, p. 125.

27. Lahtinen, R.; Garde, A.A.; Melezhik, V.A. Paleoproterozoic evolution of Fennoscandia and Greenland. *Episodes* **2008**, *31*, 20–28.

28. Bingen, B.; Andersson, J.; Söderlun, U.; Möller, C. The Mesoproterozoic in the Nordic countries. *Episodes* **2008**, *31*, 29–34.

29. Nordrum, F.S. Noen funn av mineraler i Norge 2000–2001, part II. *Stein* **2001**, *28*, 16–24. (In Norwegian)

30. Nordrum, F.S. Nyfunn av mineraler i Norge 2001–2002. *Stein* **2002**, *29*, 4–10. (In Norwegian)

31. Nordrum, F.S. Nyfunn av mineraler i Norge 2002–2003. *Nor. Bergverksmus. Skr.* **2003**, *25*, 82–89.

32. Nordrum, F.S. Nyfunn av mineraler i Norge 2002–2003. *Stein* **2003**, *30*, 4–10. (In Norwegian)

33. Nordrum, F.S. Nyfunn av mineraler i Norge 2004–2005. *Nor. Bergverksmus. Skr.* **2005**, *30*, 117–124. (In Norwegian)

34. Nordrum, F.S. Nyfunn av mineraler i Norge 2006–2007. *Stein* **2007**, *34*, 14–26. (In Norwegian)

35. Nordrum, F.S. Nyfunn av mineraler i Norge 2007–2008. *Stein* **2008**, *35*, 8–20. (In Norwegian)

36. Jébrak, M. Hydrothermal breccias in vein-type ore deposits: A review of mechanisms, morphology and size distribution. *Ore Geol. Rev.* **1997**, *12*, 111–134. [CrossRef]

37. ACMELabs, 2017. Bureau Veritas Mineral Laboratories. Available online: http://acmelab.com/ (accessed on 27 July 2017).

38. Götze, J.; Plötze, M.; Habermann, D. Origin, spectral characteristics and practical applications of the cathodoluminescence (CL) of quartz—A review. *Mineral. Petrol.* **2001**, *71*, 225–250. [CrossRef]

39. Jochum, K.P.; Weis, U.; Stoll, B.; Kuzmin, D.; Yang, Q.; Raczek, I.; Jacob, D.E.; Stracke, A.; Birbaum, K.; Frick, D.A.; et al. Determination of reference values for NIST SRM 610-617 glasses following ISO guidelines. *Geostand. Geoanal. Res.* **2011**, *35*, 397–429. [CrossRef]

40. Müller, A.; Wiedenbeck, M.; Flem, B.; Schiellerup, H. Refinement of phosphorus determination in quartz by LA-ICP-MS through defining new reference material values. *Geostand. Geoanal. Res.* **2008**, *32*, 361–376. [CrossRef]

41. McDougall, I.; Harrison, T.M. *Geochronology and Thermochronology by the $^{40}Ar/^{39}Ar$ Method*; Oxford University Press: New York, NY, USA, 1999; p. 269.

42. Renne, P.R.; Mundil, R.; Balco, G.; Min, K.W.; Ludwig, K.R. Joint determination of K-40 decay constants and $^{40}Ar/^{40}K$ for the Fish Canyon sanidine standard, and improved accuracy for $^{40}Ar/^{39}Ar$ geochronology. *Geochim. Cosmochim. Acta* **2010**, *74*, 5349–5367. [CrossRef]

43. Lee, J.Y.; Marti, K.; Severinghaus, J.P.; Kawamura, K.; Yoo, H.S.; Lee, J.B.; Kim, J.S. A redetermination of the isotopic abundances of atmospheric Ar. *Geochim. Cosmochim. Acta* **2006**, *70*, 4507–4512. [CrossRef]

44. International Atomic Energy Agency. Reference Sheet for Reference Materials NBS 28 and NBS 30. 2007. Available online: https://nucleus.iaea.org/rpst/Documents/NBS28_NBS30.pdf (accessed on 20 February 2018).

45. Baertschi, P. Absolute ^{18}O content of standard mean ocean water. *Earth Planet. Sci. Lett.* **1976**, *31*, 341–344. [CrossRef]

46. Ramsey, M.H.; Wiedenbeck, M. Quantifying isotopic heterogeneity of candidate Reference Materials at the picogram sampling scale. *Geostand. Geoanal. Res.* **2018**, *42*, 5–24. [CrossRef]

47. Whalen, J.B.; Currie, K.L.; Chappell, B.W. A-type granites: Geochemical characteristics, discrimination and petrogenesis. *Contrib. Mineral. Petrol.* **1987**, *95*, 407–419. [CrossRef]

48. Pearce, J.A.; Harris, N.B.W.; Tindle, A.G. Trace element discrimination diagrams for the tectonic interpretation of granitic rocks. *J. Petrol.* **1984**, *25*, 956–983. [CrossRef]

49. Eby, G.N. Chemical subdivision of the A-type granitoids: Petrogenic and tectonic implications. *Geology* **1992**, *20*, 641–644. [CrossRef]

50. Shand, S.J. *Eruptive Rocks. Their Genesis, Composition, Classification, and Their Relation to Ore-Deposits with a Chapter on Meteorite*; John Wiley & Sons: New York, NY, USA, 1943; p. 350.

51. Rudnick, R.L.; Gao, S. Composition of the continental crust. In *Treatise on Geochemistry*; Holland, H.D., Turekian, K.K., Eds.; Elsevier: Amsterdam, The Netherlands, 2004; Volume 3, pp. 1–64.

52. Teertstra, D.K.; Hawthorne, F.C.; Černý, P. Identification of normal and anomalous compositions of minerals by electron-microprobe analysis; K-rich feldspar as a case study. *Can. Mineral.* **1998**, *36*, 87–95.

53. Correcher, V.; García-Guinea, J. On the luminescence properties of adularia feldspar. *J. Lumin.* **2001**, *93*, 303–312. [CrossRef]

54. Sanchez-Munoz, L.; Müller, A.; Andrés, S.L.; Martin, R.F.; Modreski, P.J.; De Moura, O.J.M. The P-Fe diagram for K-feldspars: A preliminary approach in the discrimination of pegmatites. *Lithos* **2017**, *272–273*, 116–127. [CrossRef]

55. Perny, B.; Eberhardt, P.; Ramseyer, K.; Mullis, J.; Pankrath, R. Microdistribution of Al, Li and Na in alpha quartz: Possible causes and correlation with short-lived cathodoluminescence. *Am. Mineral.* **1992**, *77*, 534–544.

56. Rusk, B.G.; Reed, M.H.; Dilles, J.H.; Kent, A.J.R. Intensity of quartz cathodoluminescence and trace-element content in quartz from the porphyry copper deposit at Butte, Montana. *Am. Mineral.* **2006**, *91*, 1300–1312. [CrossRef]

57. Rusk, B.G.; Lowers, H.; Reed, M.H. Trace elements in hydrothermal quartz; relationships to cathodoluminescent textures and insights into hydrothermal processes. *Geology* **2008**, *36*, 547–550. [CrossRef]

58. Rusk, B. Cathodoluminescent textures and trace elements in hydrothermal quartz. In *Quartz: Deposits, Mineralogy and Analytics*; Götze, J., Möckel, R., Eds.; Springer: Heidelberg, Germany; New York, NY, USA, 2012; pp. 307–329.

59. Jourdan, A.-L.; Vennemann, T.W.; Mullis, J.; Ramseyer, K.; Spiers, C.J. Evidence of growth and sector zoning in hydrothermal quartz from Alpine veins. *Eur. J. Mineral.* **2009**, *21*, 219–231. [CrossRef]

60. Drivenes, K.; Larsen, B.R.; Müller, A.; Sørensen, B.E. Crystallization and uplift path of late-Variscan granites evidenced by quartz chemistry and fluid inclusions: An example from the Land's End granite, SW England. *Lithos* **2016**, *252–253*, 57–75. [CrossRef]

61. Garate-Olave, I.; Müller, A.; Roda-Robles, E.; Gil-Crespo, P.P.; Pesquera, A. Extreme fractionation in a granite-pegmatite system documented by quartz chemistry: The case study of Tres Arroyos (Central Iberian Zone, Spain). *Lithos* **2017**, *286–287*, 162–174. [CrossRef]

62. Dennen, W.H. Stoichiometric substitution in natural quartz. *Geochim. Cosmochim. Acta* **1966**, *30*, 1235–1241. [CrossRef]

63. Ramseyer, K.; Mullis, J. Factors influencing short-lived blue cathodoluminescence of a-quartz. *Am. Mineral.* **1990**, *75*, 791–800.

64. Roedder, E. *Fluid Inclusions*; Reviews in Mineralogy; Mineralogical Society of America, Book Crafters, Inc.: Chelsea, MI, USA, 1984; Volume 12, p. 644.

65. Bakker, R.J. Package FLUIDS. Part 4: Thermodynamic modelling and purely empirical equations for H_2O–NaCl–KCl solutions. *Mineral. Petrol.* **2012**, *105*, 1–29. [CrossRef]

66. Steele-MacInnis, M.; Lecumberri-Sanchez, P.; Bodnar, R.J. HokieFlincs_H_2O–NaCl: A Microsoft Excel spreadsheet for interpreting microthermometric data from fluid inclusions based on the PVTX properties of H_2O–NaCl. *Comput. Geosci.* **2012**, *49*, 334–337. [CrossRef]

67. Diamond, L.W. Introduction to gas-bearing aqueous fluid inclusions. *Short Course Ser. Mineral. Assoc. Can.* **2003**, *32*, 101–158.

68. Onasch, C.M.; Vennemann, T.W. Disequilibrium partitioning of oxygen isotopes associated with sector zoning in quartz. *Geology* **1995**, *23*, 1103–1106. [CrossRef]

69. Phillips, G.N.; Powell, R. Link between gold provinces. *Econ. Geol.* **1993**, *88*, 1084–1098. [CrossRef]

70. Sibson, R.H. Earthquake rupturing as a mineralizing agent in hydrothermal systems. *Geology* **1987**, *15*, 701–704. [CrossRef]

71. Pavlis, T.L.; Serpa, L.F.; Keener, C. Role of seismogenic processes in fault-rock development: An example from Death Valley, California. *Geology* **1993**, *21*, 267–270. [CrossRef]

72. Williams, C.L.; Thompson, T.B.; Powell, J.L.; Dunbar, W.W. Goldbearing breccias of the Rain Mine, Carlin trend, Nevada. *Econ. Geol.* **2000**, *95*, 391–404. [CrossRef]

73. Clark, C.; James, P. Hydrothermal brecciation due to fluid pressure fluctuations: Examples from the Olary Domain, South Australia. *Tectonophysics* **2003**, *366*, 187–206. [CrossRef]

74. Labaume, P.; Carrio-Schaffhauser, E.; Gamond, J.-F.; Renard, F. Deformation mechanisms and fluid-driven mass transfers in the recent fault zones of the Corinth Rift (Greece). *C. R. Geosci.* **2004**, *336*, 375–383. [CrossRef]

75. Collettini, C.; Cardellini, C.; Chiodini, G.; De Paola, N.; Holdsworth, R.E.; Smith, S.A.F. Fault weakening due to CO_2 degassing in the Northern Apennines: Short- and long-term processes. *Geol. Soc. Lond. Spec. Publ.* **2008**, *299*, 175–194. [CrossRef]

76. Sibson, R.H. Fluid flow accompanying faulting: Field evidence and models. In *Earthquake Prediction: An International Review*; Simpson, D.W., Richards, P.G., Eds.; Maurice Ewing Series 4; American Geophysical Union: Washington, DC, USA, 1981; pp. 593–603.

77. Sibson, R.H. Implications of fault-valve behavior for rupture nucleation and recurrence. *Tectonophysics* **1992**, *211*, 283–293. [CrossRef]

78. Tarasewicz, J.P.T.; Woodcock, N.H.; Dickson, J.A.D. Carbonate dilation breccias: Examples from the damage zone to the Dent Fault, northwest England. *Geol. Soc. Am. Bull.* **2005**, *117*, 736–745. [CrossRef]

79. Grønlie, A.; Naeser, C.W.; Naeser, N.D.; Mitchell, J.G.; Sturt, B.A.; Ineson, P. Fission track and K/Ar dating of tectonic activity in a transect across the Møre Trøndelag Fault Zone, Central Norway. *Nor. Geol. Tidsskr.* **1994**, *74*, 24–34.

80. Henley, R.W.; Ellis, A.J. Geothermal systems ancient and modern: A geochemical review. *Earth Sci. Rev.* **1983**, *19*, 1–50. [CrossRef]

81. Henley, R.W. Epithermal gold deposits in volcanic terranes. In *Gold Metallogeny and Exploration*; Foster, R.P., Ed.; Springer: Boston, MA, USA, 1991; pp. 133–164.

82. Huang, R.; Audétat, A. The titanium-in-quartz (TitaniQ) thermobarometer: A critical examination and re-calibration. *Geochim. Cosmochim. Acta* **2012**, *84*, 75–89. [CrossRef]

83. Hoefs, J. *Stable Isotope Geochemistry*; Springer: Berlin/Heidelberg, Germany; New York, NY, USA, 1997.

84. Klemd, R.; Hirdes, W. Origin of an unusual fluid composition in Early Proterozoic Palaeoplacer and lode-gold deposits in Birimian greenstone terranes of West Africa. *S. Afr. J. Geol.* **1997**, *100*, 405–414.

85. Schmidt-Mumm, A.; Oberthür, T.; Vetter, U.; Blenkinsop, T.G. High CO_2 content of fluid inclusions in gold mineralisations in the Ashanti Belt, Ghana: A new category of ore forming fluids? *Miner. Depos.* **1997**, *32*, 107–118. [CrossRef]

86. Sibson, R.H. A brittle failure mode plot defining conditions for high-flux flow. *Econ. Geol.* **2000**, *95*, 41–48. [CrossRef]

87. Lowenstern, B. Carbon dioxide in magmas and implications for hydrothermal systems. *Miner. Depos.* **2001**, *36*, 490–502. [CrossRef]

88. Baker, T. Emplacement depth and carbon dioxide-rich fluid inclusions in intrusion-related gold deposits. *Econ. Geol.* **2002**, *97*, 1111–1117. [CrossRef]

89. Kerrick, D.M.; Caldera, K. Metamorphic CO_2 degassing from orogenic belts. *Chem. Geol.* **1998**, *145*, 213–232. [CrossRef]

90. Santosh, M.; Jackson, D.H.; Harris, N.B.W.; Mattey, D.P. Carbonic fluid inclusions in South Indian granulites: Evidence for entrapment during charnockite formation. *Contrib. Mineral. Petrol.* **1991**, *108*, 318–330. [CrossRef]

91. Wilmart, E.; Clocchiatti, R.; Duchesne, J.C.; Touret, J.L.R. Fluid inclusions in charnockites from the Bjerkreim–Sokndal massif (Rogaland, southwestern Norway): Fluid origin and in situ evolution. *Contrib. Mineral. Petrol.* **1991**, *108*, 453–462.

92. Veevers, J.J. Middle/Late Triassic (230 ± 5 Ma) singularity in the stratigraphic and magmatic history of the Pangean heat anomaly. *Geology* **1989**, *17*, 784–787. [CrossRef]

93. Ziegler, P.A. Evolution of sedimentary basins in North-West Europe. In *Petroleum Geology of the Continental Shelf of North-West Europe*; Institute of Petroleum: London, UK, 1981; pp. 3–39.

94. Roberts, A.M.; Yielding, G.; Kusznir, N.J.; Walker, I.M.; Dorn-Lopez, D. Mesozoic extension in the North Sea: Constraints from flexural backstripping, forward modelling and fault populations. In *Petroleum Geology of Northern Europe*; Parker, J.R., Ed.; Geological Society: London, UK, 1993; pp. 1123–1136.

95. Roberts, A.M.; Yielding, G.; Kusznir, N.J.; Walker, I.M.; Dorn-Lopez, D. Quantitative analysis of Triassic extension in the northern Viking Graben. *J. Geol. Soc.* **1995**, *152*, 15–26. [CrossRef]

96. Færseth, R.B. Interaction of Permo-Triassic and Jurassic extensional fault-blocks during the development of the northern North Sea. *J. Geol. Soc.* **1996**, *153*, 931–944. [CrossRef]

97. Eide, E.; Torsvik, T.H.; Andersen, T.B. Absolute dating of brittle fault movements: Late Permian and late Jurassic extensional fault breccias in western Norway. *Terra Nova* **1997**, *9*, 135–139. [CrossRef]

98. Torsvik, T.H.; Andersen, T.B.; Eide, E.A.; Walderhaug, H.J. The age and tectonic significance of dolerite dykes in western Norway. *J. Geol. Soc.* **1997**, *154*, 961–973. [CrossRef]

99. Frostick, L.; Reid, I.; Jarvis, J.; Eardley, H. Triassic sediments of the inner Moray Firth, Scotland: Early rift deposits. *J. Geol. Soc.* **1988**, *145*, 235–248. [CrossRef]

100. Gabrielsen, R.H.; Ramberg, B. Fracture patterns in Norway from LandSAT imagery: results and potential use. In *Proceedings of the Norwegian Sea Symposium Tromsø*; NSS/23; Norwegian Petroleum Society: Stavanger, Norway, 1979; pp. 1–28.

101. Doré, A.G.; Gage, M.S. Crustal alignments and sedimentary domains in the evolution of the North Sea, North East Atlantic Margin and Barents Shelf. In *Petroleum Geology of North West Europe*; Brooks, K., Glennie, K., Eds.; Graham and Trotman: London, UK, 1987; pp. 1131–1148.

102. Grønlie, A.; Torsvik, T. On the origin and age of hydrothermal thorium enriched carbonate veins and breccias in the Møre Trøndelag Fault Zone, central Norway. *Nor. Geol. Tidsskr.* **1989**, *69*, 1–19.

103. Grønlie, A.; Roberts, D. Resurgent strike-slip duplex development along the Hitra-Snåsa and Verran faults, Møre-Trøndelag Fault Zone, Central Norway. *J. Struct. Geol.* **1989**, *11*, 295–305. [CrossRef]

104. Redfield, T.F.; Torsvik, T.H.; Andriessen, P.A.M.; Gabrielsen, R.H. Mesozoic and Cenozoic tectonics of the Møre Trøndelag Fault Complex, central Norway: Constraints from new apatite fission track data. *Phys. Chem. Earth* **2004**, *29*, 673–682. [CrossRef]

105. Redfield, T.F.; Braathen, A.; Gabrielsen, R.H.; Osmundsen, P.T.; Torsvik, T.H.; Andriessen, P.A.M. Late Mesozoic to Early Cenozoic components of vertical separation across the Møre-Trøndelag Fault Complex, Norway. *Tectonophysics* **2005**, *395*, 233–249. [CrossRef]

106. Gabrielsen, R.H.; Odinsen, T.; Grunnaleite, I. Structuring of the Northern Viking Graben and the Møre Basin; the influence of basement structural grain and the particular role of the Møre Trøndelag Fault Complex. *Mar. Petrol. Geol.* **1999**, *16*, 443–465. [CrossRef]

107. Braathen, A. Kinematics of post-Caledonian polyphase brittle faulting in the Sunnfjord region, western Norway. *Tectonophysics* **1999**, *302*, 99–121. [CrossRef]

108. Aanstad, K.; Gabrielsen, R.H.; Hagevang, T.; Ramberg, I.B.; Torvanger, O. Correlation of offshore and onshore structural features between 62N and 68N, Norway. In *Proceedings of the Norwegian Symposium on Exploration*; NSE/11; Norwegian Petroleum Society: Stavanger, Norway, 1981; pp. 1–25.

109. Seranne, M. Late Paleozoic kinematics of the Møre Trøndelag Fault Zone and adjacent areas, central Norway. *Nor. Geol. Tidskr.* **1992**, *72*, 141–158.

110. Andersen, T.B.; Torsvik, T.H.; Eide, E.; Osmundsen, P.T.; Faleide, J. Permian and Mesozoic extensional faulting within the Caledonides of Central South Norway. *J. Geol. Soc.* **1999**, *156*, 1073–1080. [CrossRef]

111. Nordgulen, Ø.; Braathen, A.; Corfu, F.; Osmundsen, P.T.; Husmo, T. Polyphase kinematics and geochronology of the Kollstraumen detachment. *Nor. J. Geol.* **2002**, *82*, 299–316.

112. Faleide, J.I.; Tsikalas, F.; Breivik, A.J.; Mjelde, R.; Ritzmann, O.; Engen, Ø.; Wilson, J.; Eldholm, O. Structure and evolution of the continental margin off Norway and the Barents Sea. *Episodes* **2008**, *31*, 82–91.

113. Faleide, J.I.; Bjørlukke, K.; Gabrielsen, R.H.O. Geology of the Norwegian Continental Shelf. In *Petroleum Geoscience: From Sedimentary Environments to Rock Physics*; Bjørlukke, K., Ed.; Springer: Berlin/Heidelberg, Germany, 2010; Chapter 22; pp. 467–499.

Article

Amethyst Occurrences in Tertiary Volcanic Rocks of Greece: Mineralogical, Fluid Inclusion and Oxygen Isotope Constraints on Their Genesis

Panagiotis Voudouris [1,*]**, Vasilios Melfos** [2]**, Constantinos Mavrogonatos** [1]**, Alexandre Tarantola** [3]**, Jens Götze** [4]**, Dimitrios Alfieris** [5]**, Victoria Maneta** [6] **and Ioannis Psimis** [7]

1   Faculty of Geology and Geoenvironment, National and Kapodistrian University of Athens, 15784 Athens, Greece; kmavrogon@geol.uoa.gr
2   Faculty of Geology, Aristotle University of Thessaloniki, 54124 Thessaloniki, Greece; melfosv@geo.auth.gr
3   GeoRessources, Faculté des Sciences et Technologies, UMR Université de Lorraine, 54506 Nancy, France; alexandre.tarantola@univ-lorraine.fr
4   Institute of Mineralogy, TU Bergakademie Freiberg, 09599 Freiberg, Germany; jens.goetze@mineral.tu-freiberg.de
5   6 Kairi str., 15126 Athens, Greece; dimitrisal@hotmail.com
6   Department of Earth Sciences, St. Francis Xavier University, 5005 Chapel Square, Antigonish, NS B2G 2W5, Canada; victoria.maneta@gmail.com
7   Maccaferri Hellas Ltd., 13674 Athens, Greece; giannis.psimis@gmail.com
*   Correspondence: voudouris@geol.uoa.gr; Tel.: +30-21-0727-4129

Received: 19 June 2018; Accepted: 25 July 2018; Published: 28 July 2018

Abstract: Epithermally altered volcanic rocks in Greece host amethyst-bearing veins in association with various silicates, carbonates, oxides and sulfides. Host rocks are Oligocene to Pleistocene calc-alkaline to shoshonitic lavas and pyroclastics of intermediate to acidic composition. The veins are integral parts of high to intermediate sulfidation epithermal mineralized centers in northern Greece (e.g., Kassiteres–Sapes, Kirki, Kornofolia/Soufli, Lesvos Island) and on Milos Island. Colloform–crustiform banding with alternations of amethyst, chalcedony and/or carbonates is a common characteristic of the studied amethyst-bearing veins. Hydrothermal alteration around the quartz veins includes sericitic, K-feldspar (adularia), propylitic and zeolitic types. Precipitation of amethyst took place from near-neutral to alkaline fluids, as indicated by the presence of various amounts of gangue adularia, calcite, zeolites, chlorite and smectite. Fluid inclusion data suggest that the studied amethyst was formed by hydrothermal fluids with relatively low temperatures (~200–250 °C) and low to moderate salinity (1–8 wt % NaCl equiv). A fluid cooling gradually from the external to the inner parts of the veins, possibly with subsequent boiling in an open system, is considered for the amethysts of Silver Hill in Sapes and Kassiteres. Amethysts from Kornofolia, Megala Therma, Kalogries and Chondro Vouno were formed by mixing of moderately saline hydrothermal fluids with low-salinity fluids at relatively lower temperatures indicating the presence of dilution processes and probably boiling in an open system. Stable isotope data point to mixing between magmatic and marine (and/or meteoric) waters and are consistent with the oxidizing conditions required for amethyst formation.

Keywords: amethyst; volcanic rock; hydrothermal alteration; gemstones

---

1. Introduction

Amethyst is a quartz variety often used in jewellery and occurs in varying shades of violet colors [1,2]. Its name comes from the ancient Greek words "a" (not) and "methystos" (intoxicated), a reference to the belief that the stone protected its owner from drunkenness. The colors of amethyst range from bluish violet to purple-violet and red-violet and its origin has been controversial for a long time [2–6]. According to Cox [3] as cited also in Fritsch and Rossman [4] the purple color in amethyst is due to $O^{2-} \rightarrow Fe^{4+}$ inter valence charge transfer, which absorbs light in the middle of the visible region. The Fe^{4+} ion, at the Fe^{4+} site (e.g., FeO_4) is formed from Fe^{3+} by the action of ionizing radiation and is important for the coloration of amethyst [2]. However, is remains controversial if the Fe^{4+} site is substitutional, e.g., [3], or interstitial [5].

Brazil is the main amethyst producer today, counting more than ten very important deposits [1,6]. Other amethyst producing countries are Mexico, Uruguay, Canada, South Korea, Russia, Zambia, Sri Lanka and India, but the best varieties of amethysts have been extracted from the central Ural mountains [1,7,8].

Amethyst is formed in igneous, metamorphic and sedimentary rocks as well as in hydrothermal veins, metasomatic and hot spring deposits [1,9,10]. Well-formed crystals occur as filling druses in various igneous rocks, such as granites, or volcanic rocks, mainly lavas. In metamorphic rocks, amethyst is a relatively common mineral in the so-called alpine-type fissures.

Greece includes several areas with major and minor occurrences of amethyst deposits [11–15]. These deposits are associated with magmatic-hydrothermal (e.g., skarns, volcanic rocks) and metamorphic environments of various ages. Amethyst of best quality has been found in well-formed crystals within extensional alpinotype fissures in metamorphic core complexes, which are related to tectonic exhumation of the Rhodope- and Attico-Cycladic massifs [11]. Gem quality amethyst crystals are also found in the skarn of Serifos Island, where amethyst occurs as scepter on prase quartz crystals. However, the above occurrences are of minor importance compared to the occurrences hosted in volcanic rocks, which can also be considered as potential deposits for possible future exploitation.

This study focuses on five amethyst deposits which are related to the Tertiary volcanic rocks of Greece. Three amethyst-bearing areas are located in Northern Greece (Kassiteres–Sapes, Kirki, Kornofolia), while the remaining two occur on Lesvos and Milos Islands in the Aegean Sea. Host rocks for the studied amethyst occurrences are lavas and pyroclastics of calc-alkaline to shoshonitic composition and Oligocene to Pleistocene age. This work summarizes earlier work and presents new geological, mineralogical, microthermometric and oxygen isotope data, which aim to a better understanding of the conditions of amethyst formation in the studied deposits.

2. Meterials and Methods

Thirty-five thin and ten thin-and-polished sections of amethyst-bearing veins and host rocks were studied by optical and a JEOL JSM 5600 scanning electron microscope equipped with back-scattered imaging capabilities, respectively, at the Department of Mineralogy and Petrology at the University of Athens (Greece). Quantitative analyses were carried out at the University of Athens, Department of Geology, using a JEOL JSM 5600 scanning electron microscope equipped with automated OXFORD ISIS 300 energy dispersive analysis system. Analytical conditions were 20 kV accelerating voltage, 0.5 nA beam current, <2 μm beam diameter and 60 s count times. The X-ray lines used were AlKα, SiKα, BaLα, CaKα, CeLα, KKα, FeKα, NaKα, TiKα, PKα, CrKα, MnKα, MgKα, and SrLα. The reference substances used were orthoclase, albite, and wollastonite (for K, Na, Si and Ca), pyrite for Fe, and synthetic Ti, Cr, Mn, MgO and Al_2O_3 (for Ti, Cr, Mn, Mg and Al).

Fluid inclusion spatial relationships and phase changes during heating/freezing runs within the inclusions were microscopically observed in a total of 20 doubly polished thin sections from Silver Hill of Sapes, Kassiteres, Kornofolia of Soufli, Kirki, Megala Therma of Lesvos Island and Chondro Vouno and Kalogries of Milos Island. Routine heating and freezing runs were performed with a LINKAM THM-600/TMS 90 stage coupled to a Leitz SM-LUX-POL microscope at the Department of Mineralogy, Petrology and Economic Geology at Aristotle University of Thessaloniki (Greece). Part of the microthermometric studies were carried out at the Institute of Mineralogy-Petrology of Hamburg University (Germany), using a CHAIXMECA heating and freezing stage. Calibration of the stages was achieved using organic reference substances with known melting points and ice (H_2O). The precision of the measurements was ±0.2 °C during low-temperature measurements and ±1 °C during high-temperature measurements. The SoWat program [16] was used to process fluid inclusion data based on equations of Bodnar [17] in the system H_2O-NaCl.

Stable isotope analyses were performed at the Stable Isotope and Atmospheric Laboratories, Department of Geology, Royal Holloway, University of London (UK). The oxygen isotope composition of quartz was obtained using a CO_2 laser fluorination system similar to that described by Mattey [18]. Each mineral separate or standard is weighed at 1.7 mg ± 10%. These were loaded into the 16-holes of a nickel sample tray, which was inserted into the reaction chamber and then evacuated. The oxygen was released by a 30W Synrad CO_2 laser in the presence of BrF_5 reagent. The yield of oxygen was measured as a calibrated pressure based on the estimated or known oxygen content of the mineral being analyzed. Low yields result in low $\delta^{18}O$ values for all mineral phases, so accurate yield calculations are essential. Yields of >90% are required for most minerals to give satisfactory $\delta^{18}O$ values. The oxygen gas was measured using a VG Isotech (now GV Instruments, Wythenshawe, UK) Optima dual inlet isotope ratio mass spectrometer (IRMS).

All values are reported relative to Vienna Standard Mean Ocean Water (V-SMOW). The data are calibrated to a quartz standard (Q BLC) with a known $\delta^{18}O$ value of +8.8‰ V-SMOW from previous measurements at the University of Paris-6 (France). This has been further calibrated for the RHUL laser line by comparison with NBS-28 quartz. Each 16-hole tray contained up to 12 sample unknowns and 4 of the Q BLC standard. For each quartz run a small constant daily correction, normally less than 0.3‰, was applied to the data based on the average value for the standard. Overall, the precision of the RHUL system based on standard and sample replicates is better than ±0.1‰.

Sixteen fresh rock samples from the volcanic rocks hosting amethyst were selected for whole-rock geochemical analysis. Major elements were analyzed on lithium tetraborate glass beads by X-ray fluorescence (XRF) using a Philips PW 1410 spectrometer at the Institute of Mineralogy and Petrology at Hamburg University (Germany). Detection limits for trace elements are 10 ppm for Ba, Cr, Cu, Nb, Ni, Pb, Rb, Sr, V, Y, Zn, Zr, Th, 20 ppm for La, Nd, and 25 ppm for Ce. The precision for major elements is better than ±0.4% for SiO_2, ±0.13% for Al_2O_3 and Fe_2O_3, ±0.22% for MgO, ±0.10% for CaO, ±0.33% for Na_2O, and 0.02%–0.04% for K_2O, TiO_2, P_2O_5 and SO_3.

3. Geological Setting

3.1. Regional Geology

The Hellenide orogen formed as a result of the collision between the African and Eurasian plates above the north-dipping Hellenic subduction zone from the Late Jurassic to the present [19]. From north to south, it consists of three continental blocks (Rhodopes, Pelagonia, and Adria-External Hellenides) and two oceanic domains (Vardar and Pindos Suture Zones) [19,20], (Figure 1a). In the Aegean region, continuous subduction of both oceanic and continental lithosphere beneath the Eurasian plate since the Early Cretaceous resulted in a series of magmatic arcs from the north (Rhodope massif) to the south (Active South Aegean Volcanic Arc) [19,21], (Figure 1a). The progressive southward migration of the magmatic centres from the Oligocene magmatic belts in the Rhodope massif, through the Miocene Aegean Islands (e.g., Limnos and Lesvos Islands) to the active South Aegean Volcanic Islands (e.g., Milos) has been attributed to slab retreat in a back-arc setting [22]. In the Rhodope Massif, Late Cretaceous-Tertiary exhumation of metamorphic core complexes along detachment faults and extensional collapse was accompanied by voluminous Late Eocene to Early Miocene calc-alkaline, high-K alkaline and shoshonitic magmatism [23–26], (Figure 1a–c). It is suggested that this magmatism in the Rhodope Massif was caused by convective removal of the lithospheric mantle (lithospheric delamination) and subsequent upwelling of the asthenosphere [24,25,27]. The Tertiary magmatism in the Rhodope Massif shows a decreasing influence of crustal contamination with time and an increasing input from the mantle until the eruption of purely asthenospheric magmas [25]. On Lesvos Island, thick successions of Early Miocene calc-alkaline to shoshonitic lavas and pyroclastics of basic to acidic composition occupy large parts of the island [27], (Figure 1d). Geochemical data suggest that the volcanic rocks on Lesvos Island (similarly to those from the Rhodope area) were derived from sub-continental lithospheric mantle and/or the lower crust, with a minimal contribution from the upper crust [27].

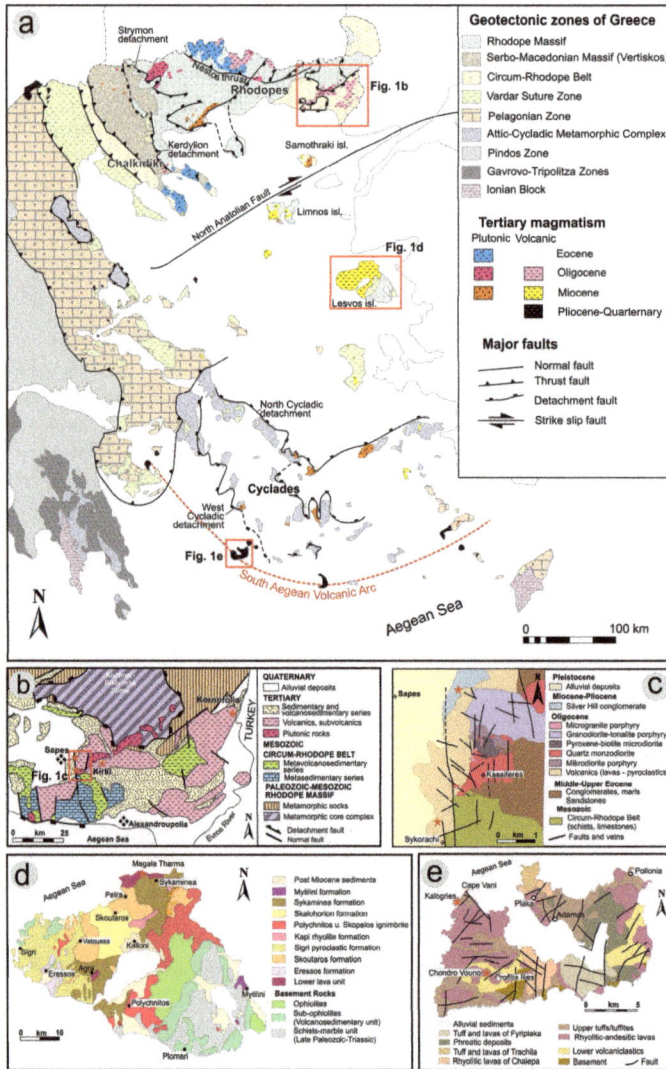

Figure 1. (**a**) Simplified geological map of Greece showing the main tectonic zones and the distribution of Cenozoic igneous rocks (modified after Melfos and Voudouris [26], and references therein); (**b**) Geological map of Western Thrace showing locations of amethyst deposits at Kornofolia, Sapes and Kirki areas (modified after Melfos and Voudouris [26]); (**c**) Geological map showing the location of amethyst in the Kassiteres–Sapes area (modified after Ottens and Voudouris [28]); (**d**) Generalized geological map of Lesvos Island and the location of amethyst at Megala Therma (modified after Innocenti et al. [23]); (**e**) Generalized geologic map of Milos Island depicting the location of amethyst at Kalogries and Chondro Vouno (modified after Fytikas et al. [29]).

Finally, on Milos Island, calc-alkaline volcanic arc activity spans a period from ~3.5 Ma to the present and originated from several emergent eruptive centers characterized by both explosive and effusive activity [29–31], Figure 1e). The magmatic rocks hosting amethyst mineralization are metaluminous to slightly peraluminous andesites, trachyandesites to dacites with calc-alkaline, high-K calc-alkaline to shoshonitic (e.g., Lesvos Island) character (Table 1; Figure 2). All amethyst-bearing areas host epithermal style of mineralization. They are characterized by intense hydrothermal alteration of the volcanic rocks, including adularization, sericitization, as well as propylitic, and zeolitic alteration, which are closely related to the formation of amethyst-bearing quartz veins.

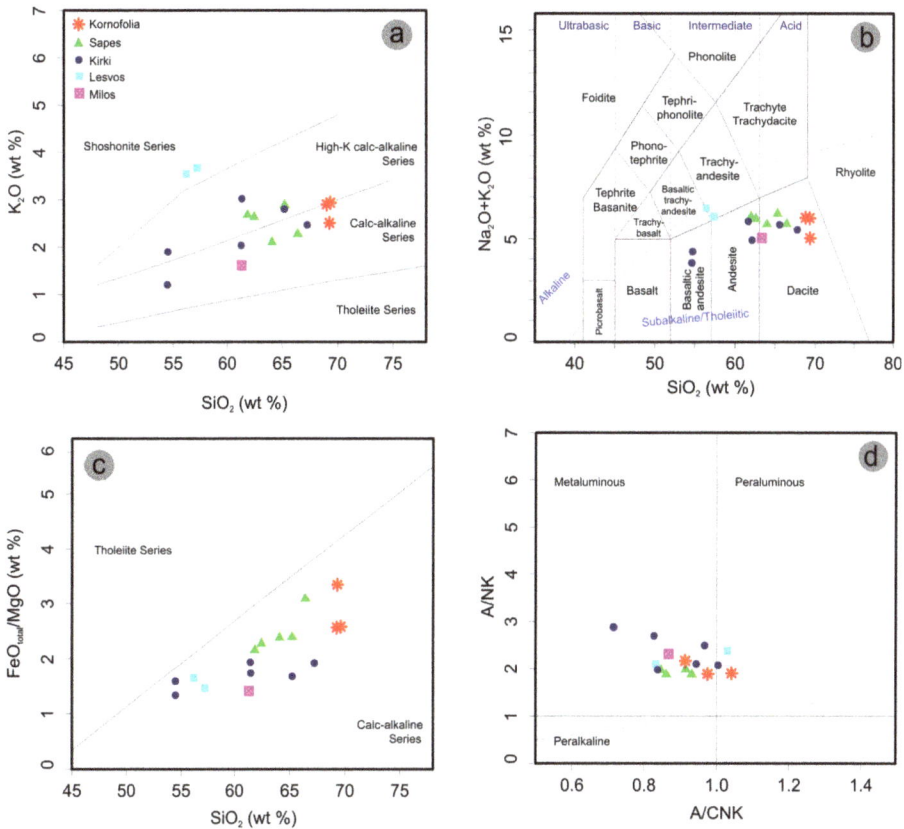

Figure 2. Classification of the studied volcanic rocks hosting amethyst mineralization in Greece. (a) SiO_2 vs. K_2O diagram [32]; (b) Total alkali vs. silica diagram. Rock fields are from Le Bas et al. [33], whereas the boundary between the alkaline and subalkaline fields is from Irvine and Baragar [34]; (c) SiO_2 vs. FeO_t/MgO plot from Miyashiro [35]; (d) A/CNK vs. A/NK plot from Shand [36] discriminating metaluminous, peraluminous and peralkaline compositions. (A/NK = molar ratio of $Al_2O_3/(Na_2O + K_2O)$; A/CNK = molar ratio of $Al_2O_3/(CaO + Na_2O + K_2O)$).

Table 1. Chemical composition of volcanic rocks hosting amethyst at Kornofolia (1–3), Kirki (4–8), Kassiteres–Sapes (9–14), Lesvos (15–16) and Milos (17, from Liakopoulos [37]). Values for major oxides and trace elements are in wt % and ppm, respectively.

	1	2	3	4	5	6	7	8	9	10	11	12	13	14	15	16	17
SiO_2	69.34	68.99	69.26	66.40	61.85	64.09	62.47	65.23	54.45	54.49	61.46	61.41	67.23	65.19	56.20	57.21	61.96
TiO_2	0.46	0.46	0.50	0.56	0.59	0.56	0.54	0.78	0.75	0.73	0.62	0.48	0.48	0.51	0.85	1.01	0.66
Al_2O_3	15.54	15.50	15.02	16.94	16.36	16.42	16.18	16.19	16.78	15.98	17.31	16.01	16.01	16.25	18.12	18.93	16.57
Fe_2O_3	3.23	3.57	3.43	6.14	5.80	5.90	4.91	9.42	8.90	6.23	5.84	3.48	3.48	4.11	6.60	6.26	6.03
MnO	0.05	0.06	0.06	0.12	0.08	0.08	0.07	0.09	0.23	0.13	0.10	0.08	0.08	0.11	0.12	0.11	0.10
MgO	0.87	1.25	1.19	2.56	2.19	2.33	1.85	5.42	6.01	3.21	2.68	1.63	1.63	2.20	3.61	3.87	2.65
CaO	4.18	3.52	5.22	5.48	5.29	6.07	4.80	9.37	7.73	6.04	6.01	4.64	4.64	5.20	7.20	5.74	6.63
Na_2O	3.12	3.12	2.56	3.43	3.63	3.34	3.35	2.63	2.51	2.91	2.92	2.91	2.91	2.85	2.94	2.42	3.33
K_2O	2.94	2.89	2.51	2.68	2.11	2.64	2.89	1.18	1.93	3.00	2.02	2.51	2.51	2.82	3.55	3.67	1.61
P_2O_5	0.13	0.14	0.15	0.11	0.10	0.10	0.11	0.14	0.22	0.14	0.13	0.13	0.13	0.12	0.42	0.48	-
Total	99.86	99.50	99.99	99.87	99.24	99.91	99.93	99.67	99.54	99.83	99.04	99.90	99.10	99.36	99.61	99.79	99.87
Ba	726	732	462	560	663	526	627	679	376	881	809	507	1175	705	1466	1077	383
Ce	61	65	43	74	52	58	62	43	-	32	38	87	55	64	139	120	-
Cr	6	5	8	14	12	12	12	10	89	135	40	20	10	19	16	34	-
Ga	14	17	17	16	18	14	18	14	19	14	15	10	-	14	22	21	-
La	28	39	26	58	23	35	39	27	-	16	29	38	13	46	88	57	-
Nb	6	5	2	7	5	6	6	7	4	5	8	6	8	8	-	2	-
Nd	28	27	20	25	24	23	24	17	9	14	15	4	16	25	55	50	-
Pb	23	30	52	25	20	19	22	21	-	27	24	20	30	17	28	24	14
Rb	86	75	50	61	55	49	56	76	34	35	202	75	128	88	131	132	34
Sc	11	10	14	22	11	24	21	16	-	-	-	-	-	-	16	24	-
Sr	273	281	345	264	321	318	296	281	393	309	320	297	353	341	942	748	215
Th	8	12	3	8	8	10	4	9	5	7	11	13	18	13	15	19	-
U	5	9	12	2	1	5	5	5	-	-	-	0	-	-	1	8	-
V	87	95	96	153	141	152	138	133	237	221	160	121	91	111	160	179	118
Y	22	25	22	25	32	26	22	24	27	25	27	28	20	22	24	25	-
Zr	122	113	112	144	145	144	137	135	89	84	147	115	113	128	208	202	-

3.2. Local Geology

3.2.1. Northeastern Greece (Sapes, Kirki and Kornofolia)

In northeastern Greece, amethyst mainly occurs in three epithermally altered volcanic environments of Oligocene age, namely at Kornofolia, Kassiteres–Sapes and Kirki areas [11,12,38,39], (Figure 1b,c). At Kassiteres, deep-violet amethyst occurs within colloform and crustiform banded epithermal quartz-chalcedony veins (up to 1 m thick), crosscutting mostly sericitic and K-feldspar (adularia) altered andesitic/dacitic lavas and pyroclastics (Figures 1c and 3a,b). The amethyst in the veins is often accompanied by gangue adularia (Figure 4a).

Figure 3. Field photographs and hand specimens of volcanic-hosted amethyst in Greece. (**a**) Boulder of dark violet amethyst alternating with chalcedony from Kassiteres, Sapes area; (**b**) Amethyst crystals from Kassiteres area; (**c**) Alternations of amethyst with chalcedony from Silver Hill, Sapes area; (**d**) Amethyst crystals overgrown on chalcedony within a cavity of dacite, Kornofolia area; (**e**) Amethyst crystals from Kornofolia area; (**f**) Veinlets of amethyst and calcite crosscutting zeolite-altered lavas in Kirki area; (**g**) Prismatic amethyst from Megala Therma, Lesvos Island; (**h**) Amethyst crystals associated with barite from Chondro Vouno, Milos Island; (**i**) Short prismatic amethyst crystals from Kalogries, Milos Island.

In the northern part of the area, massive amethyst-chalcedony boulders (up to 2 m × 2 m) occur within the "Silver Hill" conglomerate formation, which is probably a phreatomagmatic maar-diatreme breccia [40], (Figures 1c and 3c). In both localities, amethyst may also form hexagonal prismatic crystals (up to 3 cm in length), sometimes developing atop a lower part composed of smoky quartz. Prismatic amethyst crystals are rare. The presence of silicified wood within the tuffs hosting amethyst veins at Kassiteres–Sapes area, and the observation that the amethyst veins crosscut the fossilized wood [41], indicate probably a very shallow environment for amethyst deposition.

In Kornofolia area, amethyst-bearing veins are hosted within zeolite-altered to fresh dacitic lavas (Figures 1b and 3d). The quartz veins are composed of an external chalcedony layer, followed by deposition of coarse-grained quartz towards the vein centre. Open spaces are filled by short prismatic amethyst crystals up to 3 cm in length. In some veins amethyst is missing and vugs are filled by botryoidal or stalactitic pinkish chalcedony. Smectite, zeolites and calcite are accessory minerals both within the veins and in the wall rocks (Figure 4b). Similar to Kassiteres–Sapes area, in the broad Kornofolia area, the amethyst-chalcedony veins crosscut fossiliferous limestone reefs and silicified wood hosted in volcano-sedimentary layers, thus suggesting a very shallow submarine to transitional environment [41], and accordingly a very shallow depth of amethyst formation, probably close to the seafloor.

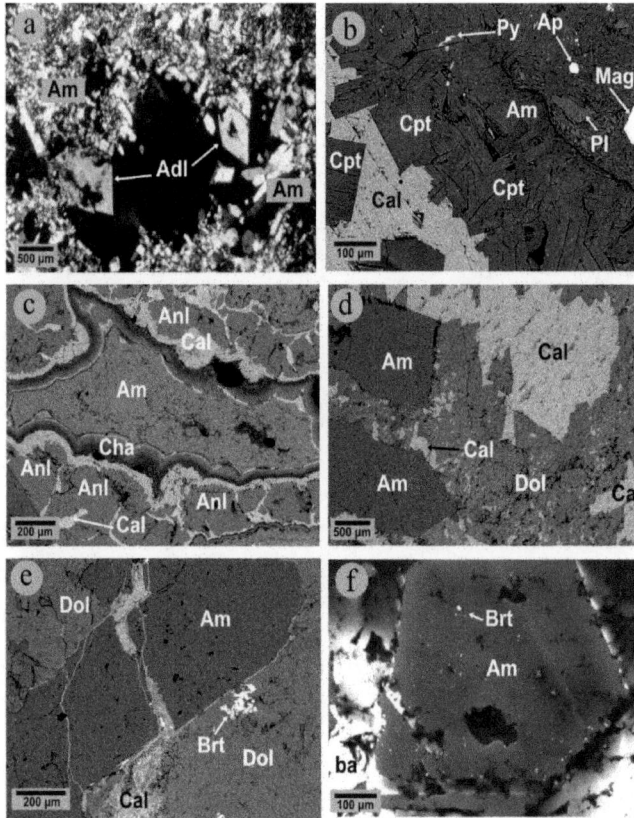

Figure 4. Microphotographs showing mineralogical assemblages of amethyst-bearing veins (**a**) Adularia (Adl) in association with amethyst (Am) from Kassiteres area (transmitted light, plane polarized); (**b**) Clinoptilolite-Ca (Cpt), including pyrite (Py), and calcite (Cal) that overgrow amethystine quartz (Am) fill cavities of dacitic lavas characterized by plagioclase (Pl) and disseminated crystals of apatite (Ap) and magnetite (Mag), Kornofolia area (SEM-BSE image); (**c**) Cavity filled with analcime (Anl), then calcite (Cal), which is rimmed by chalcedony (Cha), with final deposition of amethyst (Am), Kirki area (SEM-BSE image); (**d,e**) Amethyst (Am) is rimmed or crosscut by calcite (Cal), dolomite (Dol) and barite (Brt), Lesvos Island (SEM-BSE images); (**f**) Barite (Brt) included in amethyst (Am) from Milos Island (SEM-BSE image).

Unaltered to zeolite-altered lavas of dacitic composition are the host rocks of chalcedony-amethyst bearing NNW-trending veins at Kirki area, in a locality resembling that of Kornofolia (Figures 1b and 3f). Open spaces within the veins are filled by up to 1 cm long amethyst crystals and/or clear quartz. Amethyst in the veins is associated with the zeolites laumontite, heulandite, and analcime, and with platy calcite (Figure 4c).

3.2.2. Lesvos Island

In the northern part of Lesvos Island, the Megala Therma locality consists of NNE- and NW-trending amethyst-bearing epithermal veins crosscutting propylitic or sericitic altered andesitic lavas of the Lower lava unit (Figure 1d). The veins are banded and brecciated. They are composed of amethystine quartz, carbonates, chlorite and fluorite. Idiomorphic amethyst crystals up to 10 cm in length mostly occur in the centers of the veins. The amethyst crystals resemble those from Mexico with typical Muzo-type habit [11,13,15]. Sceptre and window forms are very common (Figure 3g). The Megala Therma amethyst was formed in the flanks of the Stypsi stratovolcano and probably under shallow depth. This conclusion may be derived from the fact that similar amethyst veins crosscut the central parts of the Stypsi caldera (some km to the south of Megala Therma), and about 150 m beneath a subhorizontal shallow-level advanced argillic lithocap, corresponding to the paleowater table approximately at the time of amethyst formation [42].

3.2.3. Milos Island

The Island of Milos represents a Plio-Pleistocene volcanic edifice with shallow submarine to subaerial, dacitic to rhyolitic subvolcanics, lavas and pyroclastics hosting epithermal-style precious- and base metal mineralization related to quartz-chalcedony veins [31]. Amethyst is present in two localities, namely the Chondo Vouno and the Kalogries area (Figure 1e). In both areas, amethyst is found within banded quartz-chalcedony-barite veins crosscutting sericitic and adularia altered rhyolitic subvolcanic bodies (Chondro Vouno) and propylitic lavas of dacitic composition (Kalogries). The veins display typical epithermal features (colloform-crustiform banding) and reach a thickness of up to 2 m and a length of several tens of meters [31,43]. Chalcedonic quartz is deposited on both sides of the vein walls and is followed by the formation of amethyst crystals (reaching up to 3 cm in length at Chondro Vouno) in the center of the veins (Figure 3h,i). Amethyst in Kalogries area (Figure 3i) is overgrown by late botryoidal aragonite. A very shallow submarine to subaerial environment of amethyst formation at Kalogries is evidenced by the presence of fossilized vertabrates within volcanoclastic material overlying the lavas, which are crosscut by the quartz-chalcedony veins [44]. We suggest that amethyst at Kalogries formed at, or just below the seafloor level, at a water depth not exceeding 100 m.

4. Mineralogy and Mineral Chemistry

Amethyst in the veins is accompanied by various mineralogical associations, suggesting specific conditions of crystallization. Quartz (var. amethyst) and chalcedony are by far the most abundant minerals in the veins, while other vein minerals include carbonates, barite, zeolites, chlorite, adularia and in minor amounts pyrite, smectite, goethite and lepidocrocite (Figure 4). Microanalyses are presented in Table 2 and the mineral-chemical data are plotted in terms of binary and ternary diagrams (Figure 5).

Adularia is present as vein and wallrock alteration mineral in Kassiteres and Silver Hill at Sapes, as well as at the Chondro Vouno amethyst deposits. In the K-feldspar alteration zones, adularia usually replaces plagioclase, primary clinopyroxene and amphibole of the volcanic host rocks. In the veins, adularia forms idiomorphic crystals (Figure 4a), overgrows amethyst crystals or can be present as inclusions in amethystine quartz. Microanalyses revealed a stoichiometric composition for the adularia from Sapes area, with Ba up to 0.7 wt %, substituting for K. A small percentage of Na_2O (up to 0.32 wt %) also substitutes for K (Table 2).

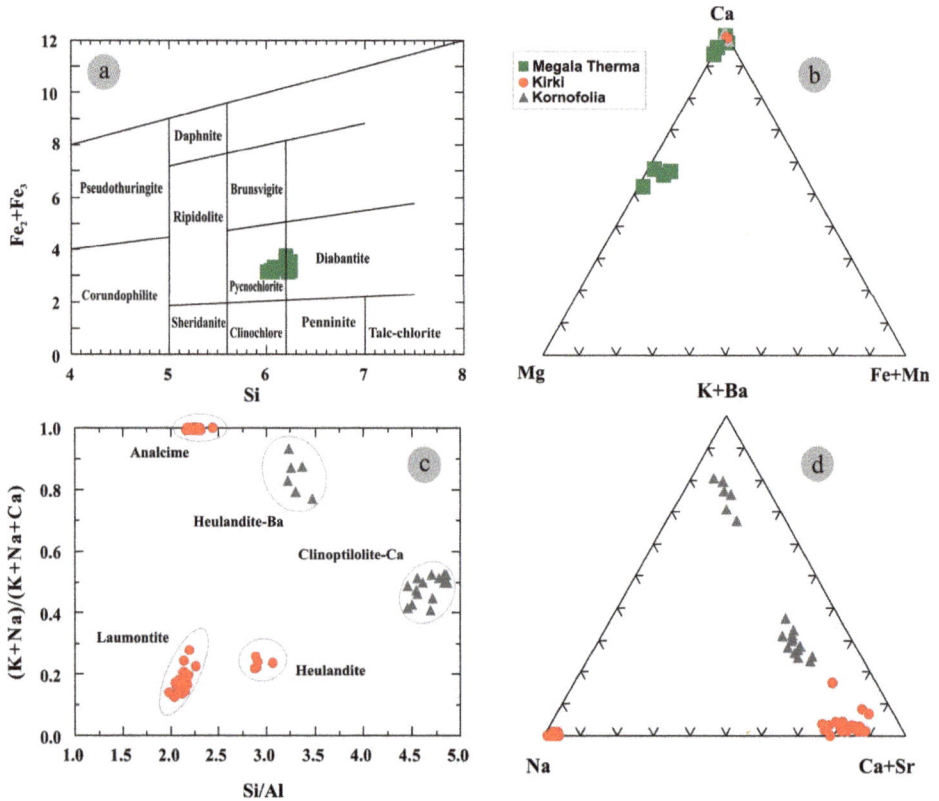

Figure 5. (a) Binary Fe_{total}-Si classification diagram of chlorite from Lesvos amethyst deposit (after Hey [45]); (**b**) Ternary Ca-Mg-Mn plot for calcite and dolomite associated with amethyst from Kirki, Kornofolia and Lesvos Island; (**c**) Binary plot ((K + Na)/(Ca + Na + K)) versus (Si/Al) of zeolites accompanying amethyst; (**d**) Ternary Ba-K-Na plot of zeolites related with amethyst.

Chlorite occurs in minor amounts in the amethyst-bearing veins at Megala Therma in association with carbonates and barite. It is a common mineral in the propylitic altered lavas hosting the amethyst, where, in conjunction with carbonates and sericite, it replaces pyroxene and hornblende phenocrysts. Microanalyses of vein chlorite are presented in Table 2. The Lesvos chlorites are classified as pycnochlorite and diabantite (Figure 5a).

Calcite and dolomite accompany amethyst at Megala Therma, Lesvos Island, Kornofolia and Kirki (Figure 4b–e). In the last two localities, calcite pre- and/or postdates amethyst in the veins (Figure 4b,c). At Megala Therma, the amethyst crystals are overgrown and crosscut first by dolomite and then by calcite and barite (Figure 4d,e). Microanalyses indicated up to 18.1 wt % MgO and 1.93 wt % FeO substituting for Ca in the dolomite (Table 2). Minor amounts of Mn (MnO up to 0.8 wt %) and Sr (SrO up to 0.2 wt %), are detected in calcite from Kornofolia and Lesvos, respectively (Table 2, Figure 5b).

Barite accompanies amethyst in the veins of Kalogries and Chondro Vouno, where it pre- or postdates amethyst deposition. At Kalogries, it can also be found as inclusions in amethyst and chalcedony (Figure 3f). The Milos barite contains up to 3.41 wt % SrO, substituting for Ba in the structure. It also occurs in minor amounts at Megala Therma and in the Kassiteres and Silver Hill amethyst occurrences.

Table 2. Representative chemical microanalyses of silicates and carbonates associated with amethyst: 1–3: adularia (Kassiteres); 4, 5: chlorite (Megala Therma, Lesvos); 6, 7: calcite, dolomite (Lesvos); 8: calcite (Kornofolia); 9: calcite (Kirki); 10, 11: clinoptilolite-Ca (Kornofolia); 12: heulandite-Ba (Kornofolia); 14, 15: analcime (Kirki); 16: laumontite (Kirki); bd = below detection limit, na = not analyzed, Ctn = cations.

	1	2	3	4	5	6	7	8	9	10	11	12	13	14	15	16
SiO_2	66.15	65.63	65.77	30.41	31.50	na	na	na	na	66.36	67.56	53.27	53.48	56.38	55.12	51.36
Al_2O_3	17.92	18.19	18.30	17.60	19.65	na	na	na	na	11.69	12.31	13.37	13.97	20.85	20.51	21.28
FeO	0.03	bd	0.01	20.48	19.81	0.08	1.93	0.12	0.36	0.11	0.10	1.01	0.39	bd	0.02	bd
MgO	na	na	na	17.15	17.83	0.16	18.06	bd	0.08	bd	0.08	0.29	0.02	0.08	bd	bd
MnO	na	na	na	0.37	0.39	bd	bd	0.82	0.45	bd	0.09	0.05	bd	0.04	bd	bd
CaO	bd	0.01	bd	0.71	0.45	50.03	29.53	49.00	49.24	3.95	4.38	0.55	0.50	0.04	0.11	10.37
Na_2O	0.04	0.08	0.26	0.63	0.65	na	na	na	na	0.90	0.68	0.66	0.90	10.97	11.52	0.59
K_2O	16.13	16.54	15.99	0.04	0.19	na	na	na	na	1.97	1.64	1.88	4.18	0.05	bd	0.89
SrO	na	na	na	0.86	0.11	bd	0.18	0.01	bd	2.33	1.59	0.79	1.87	0.43	0.58	1.10
BaO	0.03	0.20	0.55	na	na	na	na	bd	bd	0.52	0.29	15.97	14.66	bd	bd	0.14
Total	100.33	100.65	100.88	87.64	90.60	50.27	49.7	49.95	50.13	87.84	88.72	87.87	90.56	88.80	87.86	85.73
Atoms	8(O)	8(O)	8(O)	28(O)	28(O)	1(Ctn)	2(Ctn)	1(Ctn)	1(Ctn)	72(O)	72(O)	72(O)	72(O)	96(O)	96(O)	48(O)
Si	3.03	3.02	3.01	6.23	6.20	-	-	-	-	29.61	29.52	27.27	26.82	33.72	33.60	16.14
Al	0.97	0.98	0.99	4.24	4.55	-	-	-	-	6.12	6.30	8.10	8.28	14.76	14.76	7.86
Fe^{3+}	0.00	0.00	0.00	0.24	0.32	0.00	0.05	0.00	0.01	0.00	0.00	0.45	0.18	0.00	0.00	0.00
Fe^{2+}	0.00	0.00	0.00	3.24	2.90	0.00	0.00	0.00	0.00	0.00	0.00	0.00	0.00	0.00	0.00	0.00
Mg	-	-	-	5.22	5.22	0.01	0.90	0.00	0.00	0.00	0.09	0.18	0.00	0.00	0.00	0.00
Mn	-	-	-	0.07	0.07	0.00	0.00	0.01	0.01	0.00	0.00	0.00	0.00	0.00	0.00	0.00
Ca	0.00	0.00	0.00	0.11	0.11	0.99	1.05	0.99	0.98	1.89	2.07	0.27	0.27	0.00	0.12	3.48
Na	0.00	0.01	0.02	0.25	0.25	-	-	-	-	0.81	0.54	0.63	0.90	12.72	13.56	0.36
K	0.94	0.97	0.94	0.04	0.04	-	-	-	-	1.08	0.90	1.26	2.70	0.00	0.00	0.36
Sr	-	-	-	0.11	0.00	0.00	0.00	-	-	0.63	0.36	0.27	0.54	0.12	0.24	0.18
Ba	0.00	0.00	0.01	-	-	-	-	-	-	0.09	0.09	3.24	2.88	0.00	0.00	0.00

Clinoptilolite-Ca and heulandite-Ba at Kornofolia, and analcime, laumontite and heulandite at Kirki are closely related with amethyst in the veins (Figure 4b–d). At Kornofolia, a chalcedony layer overgrows zeolitized wallrocks followed by deposition of amethyst, then by heulandite-Ba included in clinoptilolite-Ca [46,47] and finally calcite (Figure 4b). Deposition in the Kirki veins started with alternations of thin layers of quartz, smectite, laumontite and heulandite, then analcime and calcite, followed by chalcedony, and finally by amethyst, which forms short prismatic crystals in the centre of the veins (Figure 4c). Microanalyses of clinoptilolite indicated an almost stoichiometric composition, with relatively elevated contents of Ca and K corresponding up to 4.6 and 2.4 wt %, respectively (Table 2). The analysed heulandite-Ba from Kornofolia revealed variable cation contents with K (K_2O up to 4.2 wt %) substituting for Na and relatively high Ba contents (BaO up to 15.97 wt %), corresponding to 3.24 apfu. Microanalyses of analcime from Kirki revealed stoichiometric compositions with a stable $(K + Na)/(K + Na + Ca)$ ratio close to 1, whereas the Si/Al ratio shows a small variance between 2.0 and 2.4 (Table 2, Figure 5c,d).

Pyrite and smectite at Kornofolia and Kirki predate amethyst in the veins, whereas goethite and lepidocrocite are included in amethyst at Chondro Vouno [48].

5. Fluid Inclusions

5.1. Morphology and Types of Fluid Inclusions

Fluid inclusions were studied in amethysts from all localities apart from Kirki where the samples did not contain any inclusions. The samples contain clear, sometimes strained, granular euhedral amethystine quartz crystals. These crystals were formed filling open spaces in colloform and crustiform banded epithermal veins of the volcanic rocks. Chalcedony often forms an external selvage along the veins. Fluid inclusions in amethyst occur mainly close to the external rather than the internal part of the crystals in the veins where they are more scattered.

Fluid inclusions were studied in the amethysts on the basis of the criteria introduced by Roedder [49], Van den Kerkhof and Hein [50], and Goldstein and Reynolds [51], such as shape similarity, size diversity, constant liquid to vapour ratios and the occurrence along growth zones. All studied fluid inclusions are two-phase aqueous liquid–vapour (Figure 6a) with almost constant liquid–vapour ratios (~10 volume % vapor in Silver Hill, Megala Therma, Chondro Vouno and Kalogries, and ~20 volume % vapor in Kassiteres and Kornofolia). During microthermometry analysis, they homogenized to the liquid upon heating.

Post-entrapment phenomena, such as necking down or leakage of fluid inclusions, were observed in most localities. Liquid- or vapour-only fluid inclusions in Soufli, Lesvos and Milos arranged along healed fractures possibly show a significant post-entrapment modification due to leakage. Necking and leakage phenomena were observed in the amethysts of Kassiteres, Soufli, Lesvos and Chondro Vouno of Milos producing a significant amount of vapour (Figure 6b,c). These inclusions are characterized by variable liquid to vapour ratio.

All inclusions affected by post-entrapment modifications are frequently found along healed micro-cracks of amethyst grains (Figure 6c) due to mechanical intracrystalline strain, as described by Audétat and Günther [52], Tarantola et al. [53,54], Diamond and Tarantola [55], and Stünitz et al. [56]. Volume changes of the inclusions are probably attributed to a "stretching" process which may lead to multiple micro-fractures that induce leakage of fluid inclusions after trapping and expanding of the vapour phase [57,58] (Figure 6b). These inclusions were avoided in this study. Furthermore, post-entrapment "stretching" processes destroyed the fluid inclusions in amethysts from Kirki; therefore, no workable inclusions were found in the amethysts from this site. The impact of post-entrapment modifications on amethyst inclusions was minimized by selecting only regularly-shaped inclusions with constant liquid to vapour ratios.

Most fluid inclusions are arranged in clusters or along trails. The trails underline either crystal growth zones (Figure 6d) demonstrating a primary origin of the inclusions, or occur within the grains along healed intragranular micro-cracks as pseudosecondary inclusions (Figure 6e) as a result of fracturing during growth of the crystals. The studied fluid inclusions have a maximum length of 75 μm but mostly ranges between 6 and 20 μm. The largest inclusions occur in Megala Therma amethysts. Isolated inclusions with negative-crystal shape are rare.

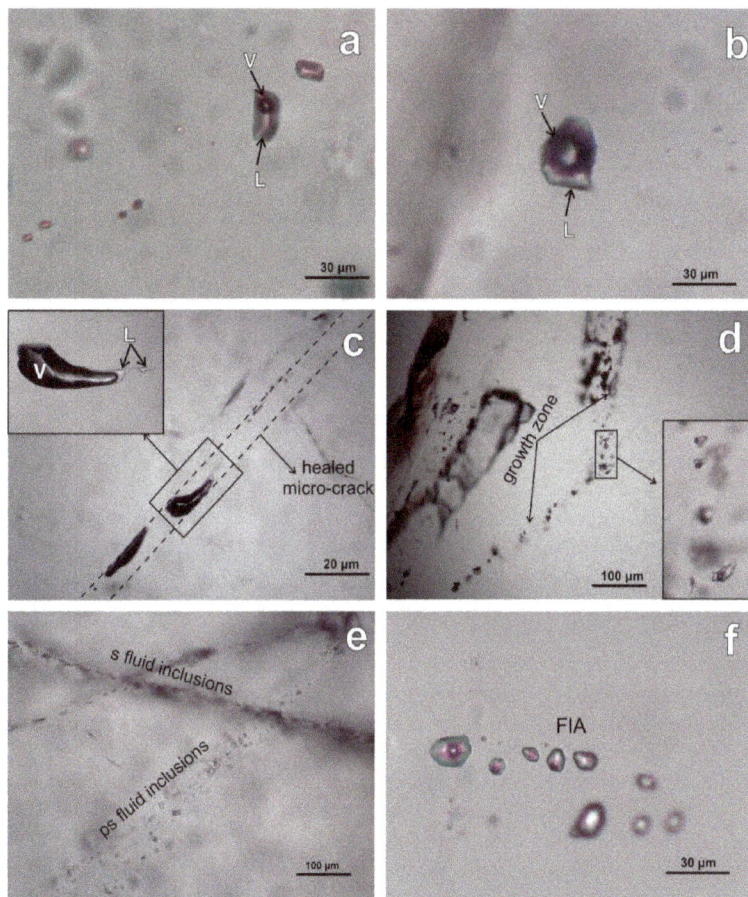

Figure 6. Photomicrographs of fluid inclusion populations from the studied amethysts under plane polarized light. (**a**) Primary two-phase liquid-rich fluid inclusion (Kornofolia of Sapes, sample AS1); (**b**) vapor-rich fluid inclusion with leakage phenomena due to "stretching" processes (Megala Therma of Lesvos, sample LS1); (**c**) deformed and stressed vapor-rich fluid inclusions along healed micro-cracks of amethyst (Megala Therma of Lesvos, sample LS1); (**d**) primary FIA suitable for microthermometry along a growth zone and a close-up photo showing two-phase fluid inclusions (Megala Therma of Lesvos, sample LS3); (**e**) pseudosecondary (ps) two-phase liquid-rich inclusions crosscut by later secondary (s) inclusions (Kalogries of Milos, sample ML2); (**f**) a FIA which includes a group of fluid inclusions formed under the same temperature and pressure (Megala Therma of Lesvos, sample LS1). **L** liquid phase, **V** vapour phase, **FIA** Fluid inclusion assemblage.

Fluid inclusions were evaluated based on fluid inclusion assemblages (FIAs; Figure 6f). A FIA defines a group of coeval and cogenetic fluid inclusions that trapped a fluid of the same chemical composition [51]. This implies that they were formed under the same temperature and pressure conditions. Amethysts contain three well defined types of FIAs with primary, pseudosecondary and secondary inclusions. The primary origin of FIAs in the studied amethysts is identified by the parallel groups of fluid inclusions along the growth zones and crystal faces (Figure 6d).

In addition to the primary FIAs, pseudosecondary fluid inclusion assemblages were also used for microthermometry as they represent fluids which were incorporated before amethyst growth was complete [59]. They were trapped in intragranular fractures caused by internal stresses in the host crystal due to rapid growth or by external rock stresses which are mechanically transferred to the host crystal. Pseudosecondary FIAs are arranged along trails which are crosscut by secondary fluid inclusions, interpreted to be older in age (Figure 6e).

Secondary inclusions are distributed along trails and planes in healed microfractures and are interpreted as late, post-dating amethyst crystallization. Besides their relation with host-crystals, pseudosecondary and secondary inclusions are difficult to distinguish. In the studied samples the secondary inclusions did not have constant volumes thus resulting in highly variable microthermometric data, especially T_h. This was interpreted as probably due to the post-entrapment refilling of the fluid inclusions with a single phase in the area of intersection, as it was described by Goldstein [59]. These inclusions are interpreted as secondary and as a consequence they do not represent the original fluid trapped during amethyst formation and were not considered in the further discussion.

5.2. Microthermometry

Microthermometric results of the studied fluid inclusions are summarized in Table 3. The primary inclusions trapped along growth zones and the pseudosecondary trails show similar homogenization temperatures (T_h) and salinities. Freezing of the fluid inclusions hosted within the studied amethysts revealed apparent eutectic temperatures (T_e) between −22.5 and −21.0 °C, indicating that the dissolved salt is dominated by NaCl with very likely the presence of KCl [60].

The temperatures of final ice melting ($T_{m(ice)}$ = −1.2 to −0.5 °C) in amethyst from Silver Hill at Sapes indicate salinities of 0.9 to 2.1 wt % NaCl equiv [17,61]. Homogenization temperatures range from 188 to 246 °C (Figure 7a). A total of 8 primary or pseudosecondary FIAs were analyzed. Two-phase fluid inclusions within single FIAs homogenized within ranges of 2 to 19 °C. This consistency confirms that the inclusions are part of a FIA which trapped a single homogeneous fluid not affected by subsequent geological events [51]. These temperatures consistently decrease with decreasing age of the inclusions and host minerals. Older amethysts close to the vein walls contain FIAs that homogenized between 201 and 230 °C while younger amethysts contain inclusions that homogenized from 189 to 205 °C (Table 3). Fluid inclusions with higher or lower homogenization temperatures (T_h > 230 °C or <189 °C) than the majority of the data are attributed to post-entrapment phenomena rather than heterogeneous entrapment [55,62,63]. Similar salinity and T_h in fluid inclusions in amethysts from Silver Hill were reported by Melfos [12].

Fluid inclusions in amethysts from Kassiteres exhibit low salinities (0.5−3.4 wt % NaCl equiv) based upon the $T_{m(ice)}$, which varies from −2.0 to −0.3 °C. Homogenization temperatures range between 211 and 275 °C (Figure 7b) and show a progressive T_h decrease from hotter fluids (T_h = 238−275 °C) in the amethyst crystals close to the vein walls to cooler fluids (T_h = 211−239 °C) at the inner parts of the veins (Table 3).

The FIAs of amethysts from Kornofolia of Soufli exhibit $T_{m(ice)}$ of −2.8 to −1.2 °C, which corresponds to a fluid salinity between 2.1 and 4.7 wt % NaCl equiv. The inclusions homogenized at a large variety of temperatures from 168 to 297 °C (Figure 7c). Based on five primary two-phase fluid inclusion assemblages, the homogenization temperatures are clustered between 191 and 223 °C, showing that the assemblages are true FIAs and, therefore, this range probably correspond to a homogeneous fluid. Fluid inclusions which homogenized at higher or lower temperatures (>223 °C or <191 °C) are considered to have stretched or leaked after entrapment. Melfos [12] reported similar salinity but higher homogenization temperatures, up to 378 °C, which may be attributed to post-entrapment processes.

In Megala Therma of Lesvos Island the salinity of the fluid inclusions in amethyst varies from 3.1 to 4.8 wt % NaCl equiv ($T_{m(ice)}$ = −2.9 to −1.8 °C). Homogenization temperatures range between 211 and 250 °C, with a maximum at 230 °C (Figure 7d). Primary two-phase fluid inclusions in most FIAs show two T_h clusters: 235–246 °C (4 FIAs) and 219–235 °C (5 FIAs) (Table 3). These amethysts appear to have trapped more than one episode of fluid flow and possibly correspond to different crystal formation stages. The earlier stage with higher T_h corresponds to the basal parts of the amethyst, closer to the vein selvage, whereas the later stage with slightly lower T_h characterizes the outer parts of the crystals towards the vein centre. A few inclusions with high temperatures of homogenization (>246 °C) reported in this study are interpreted as due to post-entrapment changes.

Table 3. Homogenization temperatures and salinity of fluid inclusions in volcanogenic amethysts from Greece. Salinities were calculated based on equations of Bodnar [17] in the system H_2O-NaCl. (FI = fluid inclusion; FIA = fluid inclusions assemblage; L = liquid phase; V = vapour phase; equiv = equivalent; *n* = number of microthermometric analyses in each FIA; n.a. = not analysed).

Sample	Location	FI Types	FIA	T_h (°C)	$T_{m(ice)}$ (°C)	Salinity (wt % NaCl equiv)
				High T_h (°C)		
		L + V → L	FIA 1	201–219 (*n* = 5)	−0.9 to −0.6 (*n* = 5)	1.1 to 1.6
		L + V → L	FIA 2	205–212 (*n* = 4)	n.a.	-
		L + V → L	FIA 3	205–215 (*n* = 3)	−1.0 (*n* = 1)	1.7
ASH, KP627	Silver Hill, Sapes	L + V → L	FIA 4	207–218 (*n* = 5)	n.a.	-
		L + V → L	FIA 5	218–230 (*n* = 4)	−1.1 to −0.7 (*n* = 5)	1.2 to 1.9
				Low Th (°C)		
		L + V → L	FIA 6	189–190 (*n* = 5)	−1.2 (*n* = 3)	2.1
		L + V → L	FIA 7	189–200 (*n* = 3)	−1.1 to −0.5 (*n* = 3)	0.9 to 1.9
		L + V → L	FIA 8	190–203 (*n* = 5)	n.a.	-
				High T_h (°C)		
		L + V → L	FIA 1	238–243 (*n* = 3)	−1.6 to −1.3 (*n* = 2)	2.2 to 2.7
		L + V → L	FIA 2	244–263 (*n* = 13)	−1.1 to −0.8 (*n* = 11)	1.4 to 1.9
		L + V → L	FIA 3	252–258 (*n* = 9)	−1.1 to −0.7 (*n* = 6)	1.2 to 1.9
288, 347, 900, 494, 886,	Kassiteres	L + V → L	FIA 4	267–275 (*n* = 9)	−0.9 to −0.7 (*n* = 6)	1.2 to 1.6
888, 495, 812, 363, 818				**Low T_h (°C)**		
		L + V → L	FIA 5	211–227 (*n* = 5)	−1.2 to −0.7 (*n* = 5)	1.2 to 2.1
		L + V → L	FIA 6	218–227 (*n* = 6)	−0.6 (*n* = 3)	1.1
		L + V → L	FIA 7	218–234 (*n* = 7)	−1.8 to −1.7 (*n* = 5)	2.9 to 3.1
		L + V → L	FIA 8	225–234 (*n* = 5)	−1.7 to −0.3 (*n* = 3)	0.5 to 2.9
		L + V → L	FIA 9	229–239 (*n* = 6)	−1.2 to −0.5 (*n* = 5)	0.9 to 2.1
		L + V → L	FIA 1	191–205 (*n* = 5)	−2.7 to −1.2 (*n* = 5)	2.1 to 4.5
		L + V → L	FIA 2	191–212 (*n* = 6)	n.a.	-
SF2, AS1	Kornofolia, Soufli	L + V → L	FIA 3	194–219 (*n* = 8)	−2.4 to −1.4 (*n* = 8)	2.4 to 4.0
		L + V → L	FIA 4	206–220 (*n* = 6)	−2.4 to −1.9 (*n* = 3)	3.2 to 4.0
		L + V → L	FIA 5	211–223 (*n* = 5)	n.a.	-
				High T_h (°C)		
		L + V → L	FIA 1	235–245 (*n* = 7)	−2.8 to −2.7 (*n* = 5)	4.5 to 4.7
		L + V → L	FIA 2	236–246 (*n* = 9)	−2.6 to −2.5 (*n* = 2)	4.2 to 4.4
		L + V → L	FIA 3	236–246 (*n* = 5)	−2.7 to −2.5 (*n* = 3)	4.2 to 4.5
	Megala Therma,	L + V → L	FIA 4	238–241 (*n* = 3)	n.a.	-
LS1, LS3	Lesvos isl.			**Low T_h (°C)**		
		L + V → L	FIA 5	219–229 (*n* = 6)	−2.9 to −2.2 (*n* = 6)	3.7 to 4.8
		L + V → L	FIA 6	219–235 (*n* = 8)	−2.4 to −2.1 (*n* = 4)	3.6 to 4.0
		L + V → L	FIA 7	219–230 (n= 6)	−2.9 to −2.0 (*n* = 6)	3.4 to 4.8
		L + V → L	FIA 8	226–234 (*n* = 5)	n.a.	-
		L + V → L	FIA 9	227–232 (*n* = 9)	−2.9 to −2.1 (*n* = 9)	3.6 to 4.8
				High T_h (°C)		
		L + V → L	FIA 1	204–211 (*n* = 4)	−5.0 to −4.2 (*n* = 4)	6.7 to 7.9
		L + V → L	FIA 2	205–219 (*n* = 5)	−5.0 to −4.1 (*n* = 2)	6.6 to 7.9
	Chondro	L + V → L	FIA 3	205–221 (*n* = 5)	−4.7 to −4.0 (*n* = 4)	6.5 to 7.5
M1	Vouno,	L + V → L	FIA 4	209–212 (*n* = 3)	−5.0 to −4.2 (*n* = 3)	6.7 to 7.9
	Milos isl.			**Low T_h (°C)**		
		L + V → L	FIA 5	193–198 (*n* = 6)	−4.7 to −3.9	6.3 to 7.5
		L + V → L	FIA 6	195–201 (*n* = 4)	−4.9 to −4.7 (*n* = 2)	7.5 to 7.7
		L + V → L	FIA 7	196–197 (*n* = 3)	n.a.	-
		L + V → L	FIA 8	196–203 (*n* = 8)	n.a.	-
		L + V → L	FIA 1	190–201 (*n* = 5)	−3.1 to −2.9 (*n* = 3)	4.8 to 5.1
		L + V → L	FIA 2	193–197 (*n* = 3)	n.a.	-
M2, ML2	Kalogries,	L + V → L	FIA 3	193–202 (*n* = 6)	−2.9 to −2.4 (*n* = 2)	4.0 to 4.8
	Milos isl.	L + V → L	FIA 4	193–204 (*n* = 3)	−2.5 (*n* = 1)	4.2
		L + V → L	FIA 5	193–206 (*n* = 5)	−2.9 to −2.5 (*n* = 4)	4.2 to 4.8
		L + V → L	FIA 6	196–207 (*n* = 4)	−3.4 to −2.5 (*n* = 4)	4.2 to 5.6

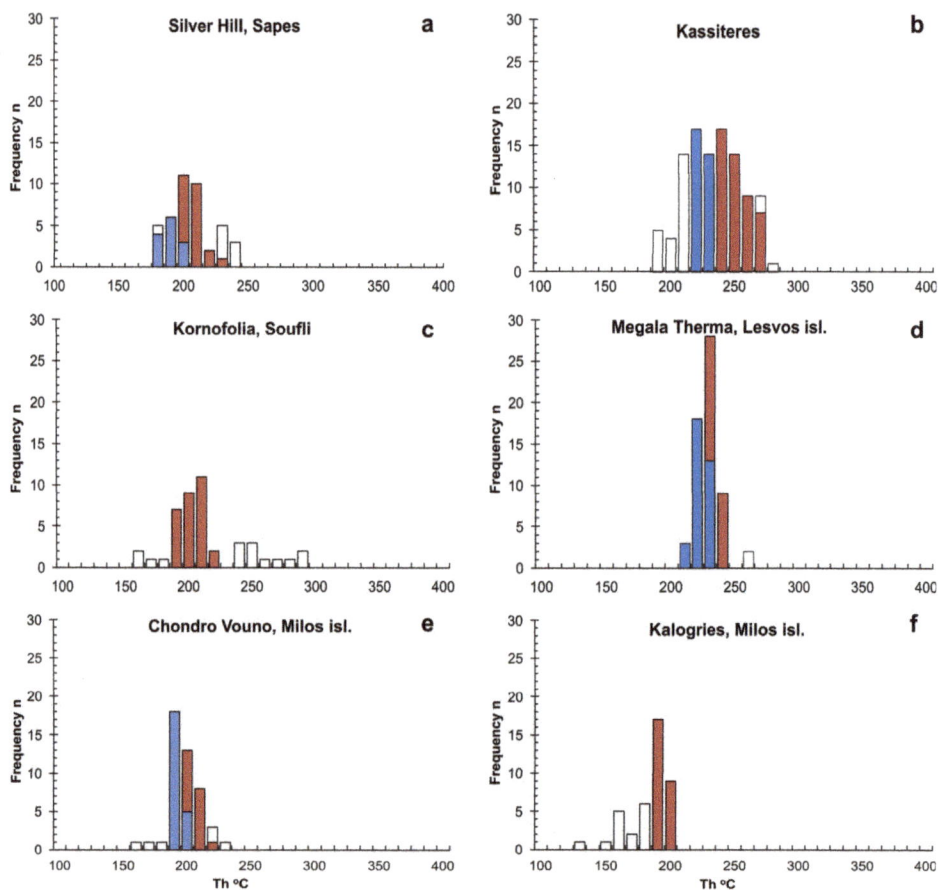

Figure 7. Distribution of the homogenization temperatures of fluid inclusions from various amethyst occurrences in volcanic rocks from Greece (**a**) Silver Hill from Sapes; (**b**) Kassiteres; (**c**) Kornofolia from Soufli; (**d**) Megala Therma from Lesvos island; (**e**) Chondro Vouno and (**f**) Kalogries from Milos island. Red bars represent older FIAs with higher T_h at the external parts of the amethysts and blue bars correspond to younger FIAs with the lower T_h at the inner crystal parts. White bars show T_h of fluid inclusions affected by post-entrapment processes and were not considered for any further interpretation.

The final ice melting temperatures of fluid inclusions ($T_{m(ice)}$ = −5.1 to −3.2 °C) in amethysts from Chondro Vouno on Milos Island reveal low to moderate salinities of 5.3 to 8.0 wt % NaCl equiv. The inclusions homogenized at temperatures between 164 and 232 °C (Figure 7e). A total of 8 FIAs of two-phase inclusions were selected for microthermometric analyses (Table 3). The higher temperature hydrothermal fluids from which the amethyst crystals grew is represented by two FIAs that contain inclusions with Th of 204 to 221 °C. A second and slightly cooler fluid is reported by four FIAs within a homogenization temperature range of 193 to 203 °C.

Low salinities (3.4 to 5.6 wt % NaCl equiv) obtained from the final ice melting temperatures ($T_{m(ice)}$ = −3.4 to −2.0 °C) were recorded in the fluid inclusions from the amethysts of Kalogries on Milos Island. The inclusions homogenized at temperatures between 139 and 209 °C, but the majority of homogenization temperatures are scattered from 190 to 200 °C (Figure 7f). This is also confirmed by the T_h observed in six FIAs which ranges from 190 to 207 °C. Lower homogenization temperatures (<193 °C) of the fluid inclusions are attributed to post-entrapment phenomena and were rejected from the dataset.

6. Amethyst Oxygen Isotopes

The oxygen isotopic composition has been analysed in hand-picked amethyst crystals from the thick sections used for fluid inclusion analyses (Table 4). The composition of the fluid in equilibrium with the quartz crystals has been calculated with the fractionation equilibrium equation of Sharp et al. [64], using the temperature from the fluid inclusion analyses. The estimated minimum trapping pressures calculated from the measured T_h for the studied amethysts range between 0.01 and 0.03 kbar, which corresponds to paleo-depths up to 150 m under lithostatic conditions. Due to the shallow (epithermal) environment of amethyst formation, the pressure correction is likely to be <10 °C (e.g., Bodnar et al. [65]), thus not significantly affecting the calculation of $\delta^{18}O$ values (see also discussion).

Table 4. Oxygen isotope compositions of amethyst quartz from various volcanic rocks in Greece (δ values in‰ relative to Standard Mean Ocean Water (SMOW)). The oxygen isotope values in the fluid ($\delta^{18}O_{Fl}$) in equilibrium with quartz have been calculated according to Sharp et al. [64], considering the range and average homogenization temperatures given by the fluid inclusions without any pressure correction.

Sample	Locality	$\delta^{18}O_{Qtz}$	T_h (°C) Range	$\delta^{18}O_{Fl}$ min	$\delta^{18}O_{Fl}$ max
KS1	Kassiteres	10.5	211–275	−0.5	2.6
SH2	Sapes	8.2	189–230	−4.3	−1.7
KIR1	Kirki	19.1	-	-	-
SF2	Kornofolia	20.5	194–223	8.4	10.2
LS2	Lesvos	3.3	219–246	−7.3	−5.8
M1	Milos	14.1	193–219	1.9	3.5
M2	Milos	13.4	190–207	1.0	2.1

The studied amethysts yield isotopic $\delta^{18}O$ values between 3.3 and 20.5‰ (Figure 8). Those from Kassiteres show a $\delta^{18}O$ value of 10.5‰ and a range of values between −0.5 and 2.6‰ for the associated fluid at equilibrium at temperatures from 211 to 275 °C, respectively. The Silver Hill amethyst at Sapes (north of Kassiteres), has a $\delta^{18}O$ value of 8.2‰, while the values for the associated fluid at equilibrium range between −4.3 and −1.7 for temperatures between 189 and 230 °C. The amethyst at Kornofolia, Evros district, yielded the highest $\delta^{18}O$ values (20.5‰), corresponding to values from 8.4 to 10.2‰ for the associated fluid at temperatures between 189 and 230 °C. The Lesvos amethyst has the lowest $\delta^{18}O$ value of all amethysts analysed (3.3‰), which corresponds to values between −7.3 to −5.8‰ for the associated fluid at temperatures from 219 to 246 °C. Finally, the two Milos amethyst samples have comparable $\delta^{18}O$ values of 13.4 and 14.1‰ and display a range from 1.9 to 3.5‰ for the associated fluid at temperatures between 193–219 °C (M1), and from 1.0 to 2.1‰ for a fluid at 190–207 °C (M2).

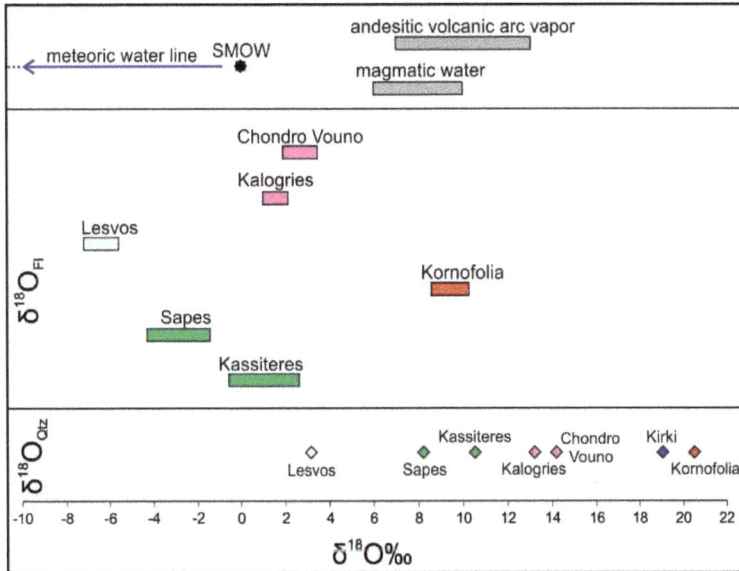

Figure 8. $\delta^{18}O$ values of amethyst from volcanic rocks in Greece (diamonds). The minimum and maximum and average values of the associated fluid at equilibrium at the temperature ranges as well as average values estimated from fluid inclusion analyses are also shown. The values for magmatic water and andesitic volcanic arc vapor are from Taylor [66] and Giggenbach [67], respectively.

7. Discussion

Volcanic rocks host the majority of amethyst deposits worldwide. In Europe, famous amethyst localities are among others those of Idar–Oberstein, Baden–Baden and Chemnitz in Germany [1,68], of Osilo in Sardinia (Italy) [1,69], of Příbram (Czech Republic) [1], Schemnitz and Kremnitz (Slovakia) [1], Roşia Montană, Săcărâmb, Baia Sprie and Cavnik (Romania) [1], and Madjarovo (Bulgaria) [1,8,70], all hosted in volcanic rocks of various ages from Paleozoic to Neogene. Quartz geodes in the basalts of Deccan, India [1], host amethyst associated with zeolites. Several amethyst localities in Japan are associated with epithermal Cu-Pb-Zn veins on Honshu Island [1]. Host-rocks for the amethyst in Brandberg (Namibia) [1,71], SW Nova Scotia, and Thunder Bay (Lake Superior, Canada) are silicified basalts [1,7,8]. Epithermal vein systems in Colorado, USA (Creede mining district and Cripple Creek) [72] and numerous similar occurrences in Mexico (e.g., at Guanajuato, Guerraro and Las Vigas districts) [1,73] are characterized by gem-quality amethyst crystals occurring mostly as the gangue of the veins and less common as typical geodes. Finally, huge amethyst geodes within basaltic lavas in Rio Grande do Sul, Brazil [74–80] and in Artigas, Uruguay [79,81–83] represent the most important resources of amethyst today. Amethyst crystallization conditions are still a matter of scientific debate. For the amethyst-bearing geodes in the basalts of Uruguay and Brazil, low temperatures of formation in the range from 50 to 120 °C were estimated from fluid inclusion data [74,75,82]. However, earlier works in Brazilian amethyst indicated homogenization temperatures of 152 to 238 °C and salinity of 0.9 to 2.6 wt % NaCl equiv and suggested a magmatic origin for the amethyst crystals [77].

Microthermometric data of the selected primary and pseudosecondary FIAs record the evolution of the hydrothermal fluids which were involved during amethyst formation. The plots of T_h versus salinity of the fluid inclusion assemblages (Figure 9) are discussed for this reason and may be used for genetic interpretations. In the absence of evident boiling, the homogenization temperatures only yield a minimum estimate of the temperature during fluid entrapment. Although stratigraphic reconstruction in the various volcanic environments does not allow a precise determination of the depth of amethyst formation, it is assumed, based from previous information as presented in the regional geology chapter, that pressure correction is insignificant by comparison to other shallow environments (e.g., Bodnar et al. [65]). In this case, the measured T_h corresponds to the temperature at which various parts of the amethysts grew. Accordingly, the homogenization temperatures of the studied amethysts are interpreted to be close to the formation temperatures.

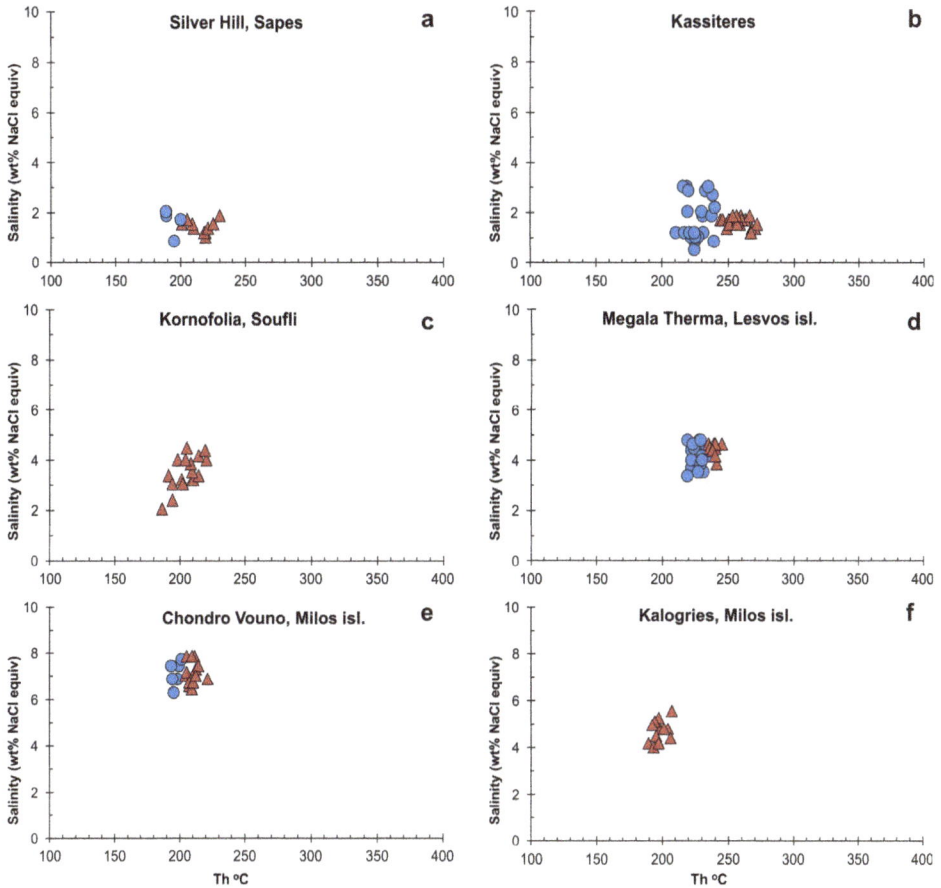

Figure 9. Homogenization temperatures versus salinity diagrams of fluid inclusions from various amethyst occurrences in volcanic rocks from Greece (**a**) Silver Hill from Sapes; (**b**) Kassiteres; (**c**) Kornofolia from Soufli; (**d**) Megala Therma from Lesvos island; (**e**) Chondro Vouno and (**f**) Kalogries from Milos island. Red triangles represent FIAs with higher T_h at the external parts of the amethysts and blue circles correspond to FIAs with the lower T_h at the inner crystal parts.

In the cases of Silver Hill in Sapes and Kassiteres, the evolution of the fluids show two different groups, the high T_h and the low T_h (Figure 9a,b) which demonstrates obvious cooling of the hydrothermal fluids during the formation of the amethyst. In Silver Hill, the low salinity fluid (0.9 to 2.1 wt % NaCl equiv) evolved from higher temperatures (230–201 °C) of the older amethyst close to the vein walls to lower temperatures (203–189 °C) for the younger amethyst at the centre of the banded veins, confirming previous results of Melfos [12]. The same situation is observed for the amethysts from Kassiteres where the majority of the FIAs homogenized at temperatures between 211 and 270 °C, reflecting gradual cooling of the fluid from the external with higher T_h (275–238 °C) to the inner parts of the veins with lower T_h (239–211 °C) and variable salinities.

The observed minor increase of salinity and decrease of temperature, especially in Kassiteres, is possibly associated with boiling in open system with steam loss [38,84]. Boiling produces large quantities of vapour and, in open systems, causes loss of H_2O and other volatile species. This mechanism results to the partitioning of salts into the liquid phase and the residual liquid becomes more saline with a gradually decreasing temperature [84,85]. However, boiling was not confirmed by the presence of vapour-rich fluid inclusions at FIAs in any of these two amethyst occurrences, although boiling processes could be responsible for the crystallization of amethyst together with adularia at Sapes and Kassiteres [38,86,87]. Calcite accompanying amethyst at Kirki has a platy habit, indicating that its deposition probably took place from boiling hydrothermal fluids (according to Simmons and Browne [86] and Simmons et al. [87]).

The trends of the fluid inclusions in amethysts from Kornofolia, Megala Therma, Chondro Vouno and Kalogries from higher to lower temperatures and salinities (Figure 9c–f) may indicate a dilution process due to mixing of moderately saline hydrothermal fluids with low temperature-low salinity fluids having roughly similar temperatures, as it is described by Hedenquist [88]. Since no evidence of phase separation, such as boiling assemblages or vapour-rich inclusions were observed, the "mixing" hypothesis is preferred, although "gentle boiling", such as described in Moncada et al. [89], cannot be fully excluded.

Fluid inclusions have been studied previously in the epithermal systems of Milos Island. The Profitis Ilias epithermal gold mineralization was formed at temperatures from 200 to 250 °C by fluids with diverse salinity (3–15 wt % NaCl equiv) under boiling conditions [90]. Fluid inclusion data in quartz from the Vani Mn deposit (about 1 km NE of Kalogries) and the Chondro Vouno Au-Ag deposit in Kilias et al. [90,91] and Naden et al. [92], show temperatures in the range of about 100 to 230 °C, boiling conditions, and involvement of seawater in addition to meteoric water for quartz deposition. Salinity in both areas shows a wide range from 0.1 to 17 wt % NaCl equiv and is best explained by boiling phenomena. Similarly, Smith et al. [93] suggested a fluid with increasing salinity (3–8 wt % NaCl equiv) and decreasing temperature 180–220 °C for the Triades deposit (lying between Chrondro Vouno and Kalogries), indicating an extensive boiling system. Homogenization temperatures of fluid inclusions in barite from Triades showed homogenization temperatures between 280 and 340 °C and low salinity from 2.14 to 5.62 wt % NaCl equiv [94], although the high temperatures could be attributed to leakage or stretching of the fluids during post-entrapment reequilibration.

Comparing the distribution of T_h (193–221 °C) and salinity (6.3 to 7.9 wt % NaCl equiv) of the studied amethysts in Chondro Vouno with the adjacent epithermal system of Profitis Ilias we can also assume by analogy a boiling system for the formation of amethyst. Similarly, at the Kalogries amethysts, the distribution of T_h (190–207 °C) and salinity (4.0–5.6 wt % NaCl equiv) are comparable with the nearby Vani Mn deposit where Kilias et al. [91], consider extreme boiling of seawater and mixing either with condensed boiled-off vapor or heated meteoric water to be the major control on mineralization. Alfieris et al. [31] also confirm the role of the vapor phase in the intermediate- to high-sulfidation state fluids under boiling conditions for the epithermal systems on Milos Island. Salinity variations can therefore be produced by continuous boiling mainly in restricted fractures [84] or by fluid mixing.

Homogenization temperatures and salinity measured here for amethyst from the intermediate-sulfidation epithermal deposits at Silver Hill and Kassiteres (Sapes area), Megala Therma (Lesvos Island) and Chondro Vouno (Milos Island) are in the range of the values obtained from other similar deposits like, for example, the Madjarovo ore field in Eastern Rhodope, Bulgaria, and the Amethyst vein system at Creede mining district, Colorado. At Madjarovo, amethyst crystallized from low salinity fluids (2.0–6.9 wt % NaCl equiv) in the range 160–240 °C [70]. At Creede, average homogenization temperatures and salinity in amethyst decrease from lower to upper levels of the mine, from an average $T_h = 238\,^{\circ}C$ and an average salinity of 9.8 wt % NaCl equiv to $T_h = 170\,^{\circ}C$ and an average salinity of 6.5 wt % NaCl equiv. The decrease in temperature and dilution of hydrothermal solutions were interpreted as a result of mixing with near-surface waters [72]. Stable isotope data revealed significant $\delta^{18}O$ variations in the studied amethyst. In general, most $\delta^{18}O$ values correspond to a mixing of magmatic and oceanic (and/or meteoric) water, with the highest magmatic component in Komofolia and the lowest in Lesvos.

The studied amethysts are genetically related to the development of epithermal systems, during the waning stages of Oligocene to Pleistocene volcanic activity. All the studied amethyst occurrences are related to intermediate- to low-sulfidation epithermal veins with well-developed hydrothermal alteration zoning (e.g., silicic alteration grading outward to adularia and/or sericitic alteration, then to argillic-, propylitic-, zeolitic alteration and finally to fresh volcanic rock).

Amethyst occurs in quartz veins crosscutting all the above alteration zones. The mineralogical data presented in this study are in accordance with those obtained through fluid inclusion measurements and may further be used to estimate the crystallization conditions of the studied amethyst. In Kirki, the coexistence of laumontite, analcime, heulandite and smectite indicate temperatures of about 175 °C. Similar temperatures of formation can be assumed through the coexistence of heulandite and clinoptilolite accompanying amethyst in Kornofolia. Application of the chlorite geothermometer after Cathelineau [95], suggests temperatures of amethyst formation for Megala Therma, Lesvos Island, from 223 to 234 °C. Mineralogical and geological information indicate that amethyst formation took place mainly from near neutral to alkaline fluids and in the stability field of adularia, calcite, chlorite and zeolites.

It is generally accepted that amethyst is formed through a process involving irradiation in which Fe^{3+} loses and electron and gives rise to a new color center, Fe^{4+}, which is responsible for its violet color [2]. Amethyst requires oxidizing conditions to incorporate Fe^{3+} and these may result from mixing of oxidized meteoric and/or seawater with upwelling hydrothermal fluids [96]. Natural radiation can probably be explained by the moderate U concentrations of the surrounding volcanic host rocks. Shallow submarine conditions are most likely to have prevailed in all areas, contributing sea- and meteoric water to the hydrothermal fluids, Sapes being probably under subaerial conditions.

8. Conclusions

Epithermally altered volcanic rocks in Greece host amethyst veins in association with various amounts of adularia, calcite, smectite, chlorite, sericite, pyrite, zeolites (laumontite, heulandite, clinoptilolite), analcime, barite, fluorite and goethite/lepidocrocite in the veins. Host rocks are Oligocene to Pleistocene lavas and pyroclastics of intermediate- to acidic composition and with a calc-alkaline to shoshonitic affinity. Precipitation of amethyst took place during the final stages of the magmatic-hydrothermal activity in the areas from near-neutral to alkaline fluids. The present study shows that the amethysts from six volcanogenic environments in Greece with diverse ages but similar geotectonic regimes with post-subduction extensional magmatism were formed by hydrothermal fluids with relatively low temperatures (~200–250 °C) and low to moderate salinity (1–8 wt % NaCl equiv). A genetic mechanism with a gradually cooling fluid from the external to the inner parts of the veins, possibly with subsequent boiling in an open system, is considered for the amethysts of Silver Hill in Sapes and of Kassiteres. Amethysts from Kornofolia, Megala Therma, Chondro Vouno and Kalogries were formed by mixing of moderately saline hydrothermal fluids with low-salinity fluids at relatively lower temperatures (~200 °C for Kornofolia, Chondro Vouno and Kalogries, and ~250 °C for Megala Therma), indicating the presence of dilution processes, although boiling in an open system is not to be excluded. Stable isotope studies revealed a mixing between magmatic and marine (and/or meteoric) waters for most of the samples. A major contribution of magmatic water is only evidenced for the Kornofolia amethyst. The oxidizing conditions required for the amethyst formation were probably the result of mixing between meteoric or seawater with upwelling hydrothermal fluids.

Ongoing geological and gemological work aims to investigate if the volcanic-hosted amethyst in Greece is of gemstone quality and to evaluate their potential for future exploitation.

Author Contributions: P.V. collected the studied samples. P.V. assisted by C.M., V. M., D.A. and I.P. evaluated the mineralogical and geochemical data. V.M. and P.V. conducted the fluid inclusion measurements, and evaluated them with A.T. A.T., J.G. and P.V. evaluated the isotopic data. P.V. and V.M. wrote the manuscript.

Acknowledgments: Authors would like to thank Evangelos Michailidis for their kind help on microanalyses in the University of Athens. Two anonymous reviewers are especially thanked for their constructive comments that greatly improved the manuscript.

Conflicts of Interest: The authors declare no conflict of interest.

References

1. Lieber, W. *Amethyst: Geschichte, Eigenschaften, Fundorte*; Christian Weise Verlag: München, Germany, 1994; 188p.
2. Rossman, G.R. Colored varieties of the silica minerals. *Rev. Mineral.* **1994**, *29*, 433–467.
3. Cox, R.T. Optical absorption of the d^4 ion Fe^{4+} in pleochroic amethyst quartz. *J. Phys. C Solid State Phys.* **1977**, *10*, 4631–4643. [CrossRef]
4. Fritsch, E.; Rossman, G.R. An update of colors in gems part 2: Colors involving multiple atoms and color centers. *Gems Gemol.* **1988**, *24*, 3–15. [CrossRef]
5. Cohen, A.J. Amethyst color in quartz, the result of radiation protection involving iron. *Am. Mineral.* **1985**, *70*, 1180–1185.
6. Scholz, R.; Chaves, M.L.S.C.; Krambrock, K.; Pinheiro, V.B.; Barreto, S.B.; de Menezes, M.G. Brazilian quartz deposits with special emphasis on gemstone quartz and its color treatment. In *Quartz: Deposits, Mineralogy and Analytics*; Götze, J., Möckel, R., Eds.; Springer: Berlin/Heidelberg, Germany, 2012; pp. 139–159.
7. Wilson, B.S. Colored gemstones from Canada. In *Geology of Gem Deposits*; Groat, L.A., Ed.; Mineralogical Association of Canada Short Course; Mineralogical Association of Canada: Québec, QC, Canada, 2007; Volume 37, pp. 255–270.
8. Kievlenko, E.Y. *Geology of Gems*; Ocean Pictures Ltd.: Romsey, UK, 2003; 468p.
9. Götze, J. Chemistry, textures and physical properties of quartz—Geological interpretation and technical application. *Mineral. Mag.* **2009**, *73*, 645–671. [CrossRef]
10. Götze, J.; Möckel, R. *Quartz: Deposits, Mineralogy and Analytics*; Springer: Berlin/Heidelberg, Germany, 2012.
11. Voudouris, P.; Katerinopoulos, A. New occurrences of mineral megacrysts in Tertiary magmatic-hydrothermal and epithermal environments in Greece. *Doc. Nat.* **2004**, *151*, 1–21.
12. Melfos, V. Study of fluid inclusions in amethysts from areas of Macedonia and Thrace: Sapes, Soufli, Nevrokopi. In Proceedings of the 2nd Congress of the Economic Geology Committee, Mineralogy & Petrology (GSG), Thessaloniki, Greece, 7–9 October 2005; pp. 219–228. (In Greek)
13. Maneta, V.; Voudouris, P. Quartz megacrysts in Greece: Mineralogy and environment of formation. *Bull. Geol. Soc. Greece* **2010**, *43*, 685–696. [CrossRef]
14. Voudouris, P.; Katerinopoulos, A.; Melfos, V. Alpine-type fissure minerals in Greece. *Doc. Nat.* **2004**, *151*, 23–45.
15. Voudouris, P.; Psimis, I.; Mavrogonatos, C.; Kanellopoulos, C.; Kati, M.; Chlekou, E. Amethyst occurrences in Tertiary volcanic rocks of Greece: Mineralogical and genetic implications. *Bull. Geol. Soc. Greece* **2013**, *47*, 477–486. [CrossRef]
16. Driesner, T.; Heinrich, C.A. The system H_2O-NaCl. I. Correlations for molar volume, enthalpy, and isobaric heat capacity from 0 to 1000 degrees C, 1 to 5000 bar, and 0 to 1 X-NaCl. *Geochim. Cosmochim. Acta* **2007**, *71*, 4880–4901. [CrossRef]
17. Bodnar, R.J. Revised equation and table for determining the freezing point depression of H_2O-NaCl solutions. *Geochim. Cosmochim. Acta* **1993**, *57*, 683–684. [CrossRef]
18. Mattey, D.P. LaserPrep: An Automatic Laser-Fluorination System for Micromass 'Optima' or 'Prism' Mass Spectrometers. *Micromass Appl. Note* **1997**, *207*, 8.
19. Jolivet, L.; Brun, J.P. Cenozoic geodynamic evolution of the Aegean region. *Int. J. Earth Sci.* **2010**, *99*, 109–138. [CrossRef]

20. Kydonakis, K.; Brun, J.-P.; Sokoutis, D.; Gueydan, F. Kinematics of Cretaceous subduction and exhumation in the western Rhodope (Chalkidiki block). *Tectonophysics* **2015**, *665*, 218–235. [CrossRef]

21. Menant, A.; Jolivet, L.; Tuduri, J.; Loiselet, C.; Bertrand, G.; Guillou-Frottier, L. 3D subduction dynamics: A first-order parameter of the transition from copper- to gold-rich deposits in the eastern Mediterranean region. *Ore Geol. Rev.* **2018**, *94*, 118–135. [CrossRef]

22. Fytikas, M.; Innocenti, F.; Manneti, P.; Mazzuoli, R.; Peccerillo, A.; Villari, L. Tertiary to Quaternary evolution of volcanism in the Aegean region. In *The Geological Evolution of the Eastern Mediterranean*; Dixon, J.E., Robertson, A.H.F., Eds.; Geological Society Special Publication: Oxford, UK, 1984; Volume 17, pp. 687–699.

23. Innocenti, F.; Kolios, N.; Manetti, O.; Mazzuoli, R.; Peccerillo, G.; Rita, F.; Villari, L. Evolution and geodynamic significance of the Tertiary orogenic volcanism in northeastern Greece. *Bull. Volcanol.* **1984**, *47*, 25–37. [CrossRef]

24. Christofides, G.; Pecskay, Z.; Soldatos, T.; Eleftheriadis, G.; Koroneos, A. The Tertiary Evros volcanic rocks (Greece): Petrology, K/Ar geochronology and volcanism evolution. *Geol. Carpath.* **2004**, *55*, 397–409.

25. Marchev, P.; Kaiser-Rohrmeier, M.; Heinrich, C.; Ovtcharova, M.; von Quadt, A.; Raicheva, R. Hydrothermal ore deposits related to post-orogenic extensional magmatism and core complex formation: The Rhodope Massif of Bulgaria and Greece. *Ore Geol. Rev.* **2005**, *27*, 53–89. [CrossRef]

26. Melfos, V.; Voudouris, P. Cenozoic metallogeny of Greece and potential for precious, critical and rare metals exploration. *Ore Geol. Rev.* **2017**, *59*, 1030–1057. [CrossRef]

27. Pe-Piper, G.; Piper, D.J.W. *The Igneous Rocks of Greece: The Anatomy of an Orogen*; Gebruder Borntraeger: Berlin, Germany, 2002; Volume 30.

28. Ottens, B.; Voudouris, P. *Griechenland: Mineralien-Fundorte-Lagerstätten*; Christian Weise Verlag: München, Germany, 2018; in press.

29. Fytikas, M.; Innocenti, F.; Kolios, N.; Manneti, P.; Mazzuoli, R.; Poli, G.; Rita, F.; Villari, L. Volcanology and petrology of volcanic products from the island of Milos and neighboring islets. *J. Volcanol. Geotherm. Res.* **1986**, *28*, 297–317. [CrossRef]

30. Stewart, A.L.; McPhie, J. Facies architecture and Late Pliocene–Pleistocene evolution of a felsic volcanic island, Milos, Greece. *Bull. Volcanol.* **2006**, *68*, 703–726. [CrossRef]

31. Alfieris, D.; Voudouris, P.; Spry, P.G. Shallow submarine epithermal Pb-Zn-Cu-Au-Ag-Te mineralization on western Milos Island, Aegean Volcanic Arc, Greece: Mineralogical, Geological and Geochemical constraints. *Ore Geol. Rev.* **2013**, *53*, 159–180. [CrossRef]

32. Peccerillo, A.; Taylor, S.R. Geochemistry of Eocene calc-alkaline volcanic rocks from the Kastamonu area, northern Turkey. *Contr. Mineral. Petrol.* **1976**, *58*, 63–81. [CrossRef]

33. Le Bas, M.J.; Le Maitre, R.W.; Streckeisen, A.; Zanettin, B. A chemical classification of volcanic rocks based on total alkali-silica diagram. *J. Petrol.* **1986**, *27*, 745–750. [CrossRef]

34. Irvine, T.N.; Baragar, W.R.W. A guide to chemical classification of the common volcanic rocks. *Can. J. Earth Sci.* **1971**, *8*, 523–548.

35. Miyashiro, A. Volcanic Rock Series in Island Arcs and Active Continental Margins. *Am. J. Sci.* **1974**, *274*, 321–355. [CrossRef]

36. Shand, S.J. *The Eruptive Rocks*, 2nd ed.; John Wiley: New York, NY, USA, 1943.

37. Liakopoulos, A. Hydrothermalisme et Minéralisations Métallifères de l' île de Milos (Cyclades, Grèce). Ph.D. Thesis, Université Pierre et Marie Curie, Paris, France, 1987.

38. Voudouris, P. Mineralogical, Geochemical and Fluid Inclusion Studies on Epithermal Vein Type Gold/Silver Mineralizations at Kassiteres/Sapes, (NE-Greece). Ph.D. Thesis, University of Hamburg, Hamburg, Germany, 1993.

39. Voudouris, P. The minerals of Eastern Macedonia and Western Thrace: Geological framework and environment of formation. *Bull. Geol. Soc. Greece* **2005**, *37*, 62–77.

40. Shawh, A.J.; Constantinides, D.C. The Sappes gold project. *Bull. Geol. Soc. Greece* **2001**, *34*, 1073–1080. [CrossRef]

41. Voudouris, P.; Velitzelos, D.; Velitzelos, E.; Thewald, U. Petrified wood occurrences in western Thrace and Limnos island: Mineralogy, geochemistry and depositional environment. *Bull. Geol. Soc. Greece* **2007**, *40*, 238–250. [CrossRef]

42. Voudouris, P.; Alfieris, D. New porphyry-Cu ± Mo occurrences in northeastern Aegean/Greece: Ore mineralogy and transition to epithermal environment. In *Mineral Deposit Research: Meeting the Global Challenge*; Mao, J., Bierlein, F.P., Eds.; Springer: Berlin, Germany, 2005; pp. 473–476.

43. Alfieris, D. Geological, Geochemical and Mineralogical Studies of Shallow Submarine Epithermal Mineralization in an Emergent Volcanic Edifice, at Western Milos Island, Greece. Ph.D. Thesis, University of Hamburg, Hamburg, Germany, 2006.

44. Coffey, J. *The Stratigraphy and Palaeontology of Cape Vani, Milos, Greece*; Honours Research Report; School of Earth Sciences, University of Melbourne: Parkville, VIC, Australia, 2005.

45. Hey, M.H. A new review of the chlorites. *Mineral. Mag.* **1954**, *30*, 277–292. [CrossRef]

46. Coombs, D.S.; Alberti, A.; Armbruster, T.; Artioli, G.; Colella, C.; Galli, E.; Grice, J.D.; Liebau, F.; Mandarino, J.A.; Minato, H.; et al. Recommended nomenclature for zeolite minerals: Report of the Subcommittee on Zeolites of the Mineralogical Association, Commission on New Minerals and Mineral Names. *Eur. J. Mineral.* **1998**, *10*, 1037–1081. [CrossRef]

47. Larsen, A.O.; Nordrum, F.S.; Döbelin, N.; Armbruster, T.; Petersen, O.V.; Erambert, M. Heulandite-Ba, a new zeolite species from Norway. *Eur. J. Mineral.* **2005**, *17*, 143–153. [CrossRef]

48. Voudouris, P.; Maneta, V. *Quartz in Greece*; CreateSpace Publ. and Amazon.com, Inc.: Seattle, WA, USA, 2017. (In Greek)

49. Roedder, E. Fluid inclusions. In *Reviews in Mineralogy*; Mineralogical Society of America: Chantilly, VA, USA, 1984; Volume 12, p. 644.

50. Van den Kerkhof, A.M.; Hein, U.F. Fluid inclusion petrography. *Lithos* **2001**, *55*, 27–47. [CrossRef]

51. Goldstein, R.H.; Reynolds, T.J. *Systematics of Fluid Inclusions in Diagenetic Minerals*; SEPM Short Course; Society for Sedimentary Geology: Broken Arrow, OK, USA, 1994; Volume 31.

52. Audétat, A.; Günther, D. Mobility and H_2O loss from fluid inclusions in natural quartz crystals. *Contrib. Mineral. Petrol.* **1999**, *137*, 1–14. [CrossRef]

53. Tarantola, A.; Diamond, L.W.; Stünitz, H. Modification of fluid inclusions in quartz by deviatoric stress I: Experimentally induced changes in inclusion shapes and microstructures. *Contrib. Mineral. Petrol.* **2010**, *160*, 825–843. [CrossRef]

54. Tarantola, A.; Diamond, L.W.; Stünitz, H.; Thust, A.; Pec, M. Modification of fluid inclusions in quartz by deviatoric stress III: Influence of principal stresses on inclusion density and orientation. *Contrib. Mineral. Petrol.* **2012**, *164*, 537–550. [CrossRef]

55. Diamond, L.W.; Tarantola, A. Interpretation of fluid inclusions in quartz deformed by weak ductile shearing: Reconstruction of differential stress magnitudes and pre-deformation fluid properties. *Earth Planet. Sci. Lett.* **2015**, *417*, 107–119. [CrossRef]

56. Stünitz, H.; Thust, A.; Heilbronner, R.; Behrens, H.; Kilian, R.; Tarantola, A.; Fitz Gerald, J.D. Water redistribution in experimentally deformed natural milky quartz single crystals-Implications for H_2O-weakening processes. *J. Geophys. Res. Solid Earth* **2017**, *122*, 866–894. [CrossRef]

57. Bakker, R.J.; Jansen, J.B.H. Preferential water leakage from fluid inclusions by means of mobile dislocations. *Nature* **1990**, *345*, 58–60. [CrossRef]

58. Bakker, R.J.; Jansen, J.B.H. A mechanism for preferential H_2O leakage from fluid inclusions in quartz, based on TEM observations. *Contrib. Mineral. Petrol.* **1994**, *116*, 7–20. [CrossRef]

59. Goldstein, R.H. Fluid inclusions: Analysis and interpretation. In *Fluid Inclusions: Analysis and Interpretation*; Samson, I.M., Anderson, A.J., Marshall, D.D., Eds.; SEPM Short Course; Society for Sedimentary Geology: Broken Arrow, OK, USA, 2003; Volume 32, pp. 9–53.

60. Shepherd, T.; Rankin, A.; Alderton, D. *A Practical Guide to Fluid Inclusion Studies*; Blackie and Son: Glasgow, UK, 1985.

61. Bodnar, R.J. Introduction to fluid inclusions. In *Fluid Inclusions: Analysis and Interpretation*; Samson, I.M., Anderson, A.J., Marshall, D.D., Eds.; SEPM Short Course; Society for Sedimentary Geology: Broken Arrow, OK, USA, 2003; Volume 32, pp. 1–8.

62. Sterner, S.M.; Bodnar, R.J. Synthetic fluid inclusions-VII. Re-equilibration of fluid inclusions in quartz during laboratory-simulated metamorphic burial and uplift. *J. Metamorph. Geol.* **1989**, *7*, 243–260. [CrossRef]

63. Sterner, S.M.; Hall, D.L.; Keppler, H. Compositional re-equilibration of fluid inclusions in quartz. *Contrib. Mineral. Petrol.* **1995**, *119*, 1–15. [CrossRef]

64. Sharp, Z.D.; Gibbons, J.A.; Maltsev, O.; Atudorei, V.; Pack, A.; Sengupta, S.; Shock, E.L.; Knauth, L.P. A calibration of the triple oxygen isotope fractionation in the SiO_2–H_2O system and applications to natural samples. *Geochim. Cosmochim. Acta* **2016**, *186*, 105–119. [CrossRef]

65. Bodnar, R.J.; Reynolds, T.J.; Kuehn, C.A. Fluid-Inclusion systematics in epithermal systems. *Rev. Econ. Geol.* **1985**, *2*, 73–97.

66. Taylor, H.P., Jr. Oxygen and hydrogen isotope relationships in hydrothermal mineral deposits. In *Geochemistry of Hydrothermal Ore Deposits*, 2nd ed.; Barnes, H.L., Ed.; Wiley: New York, NY, USA, 1979; pp. 236–318.

67. Giggenbach, W.F. The origin and evolution of fluids in magmatic-hydrothermal systems. In *Geochemistry of Hydrothermal Ore Deposits*, 3nd ed.; Barnes, H.L., Ed.; Wiley: New York, NY, USA, 1997; pp. 737–796.

68. Ettig, F. Sachsischer Amethyst. *Aufschluss* **1952**, *3*, 52–53.

69. Bringe, H.H.; Grubessi, O. Amethyst und Yugawaralith von Osilo. *Aufschluss* **1982**, *33*, 41–44.

70. Breskovska, V.; Tarkian, M. Mineralogy and Fluid Inclusion Study of Polymetallic Veins in the Madjarovo Ore Field, Eastern Rhodope, Bulgaria. *Mineral. Petrol.* **1993**, *49*, 103–118. [CrossRef]

71. Henn, J.; Lieber, W. Amethyst von Brandberg, Namibia. *Lapis* **1993**, *18*, 44–48.

72. Robinson, R.W.; Norman, D.I. Mineralogy and Fluid Inclusion Study of the Southern Amethyst Vein System, Creede Mining District, Colorado. *Econ. Geol.* **1984**, *79*, 439–447. [CrossRef]

73. Wallace, T. Die violetten Schätze von Guerrero und Veracruz, Mexico. *Extra Lapis* **2015**, *49*, 80–99.

74. Gilg, H.A.; Morteani, G.; Kostitsyn, Y.; Preinfalk, C.; Gatter, I.; Strieder, A.J. Genesis of amethyst geodes in basaltic rocks of the Serra Geral Formation (Ametista do Sul, Rio Grande do Sul, Brazil): A fluid inclusion, REE, oxygen, carbon, and Sr isotope study on basalt, quartz, and calcite. *Miner. Depos.* **2003**, *38*, 1009–1025. [CrossRef]

75. Gilg, H.A.; Krüger, Y.; Taubald, H.; van den Kerkhof, A.M.; Frenz, M.; Morteani, M. Mineralisation of amethyst-bearing geodes in Ametista do Sul (Brazil) from low-temperature sedimentary brines: Evidence from monophase liquid inclusions and stable isotopes. *Miner. Depos.* **2014**, *49*, 861–877. [CrossRef]

76. Proust, D.; Fontaine, C. Amethyst-bearing lava flows in the Paraná Basin (Rio Grande do Sul, Brazil): Cooling, vesiculation and formation of the geodic cavities. *Geol. Mag.* **2007**, *144*, 53–65. [CrossRef]

77. Proust, D.; Fontaine, C. Amethyst geodes in the basaltic flow from Triz quarry at Ametista do Sul (Rio Grande do Sul, Brazil): Magmatic source of silica for the amethyst crystallizations. *Geol. Mag.* **2007**, *144*, 731–740. [CrossRef]

78. Commin-Fischer, A.; Berger, G.; Polvé, M.; Dubois, M.; Sardini, P.; Beaufort, D.; Formoso, M. Petrography and chemistry of SiO_2 filling phases in the amethyst geodes from the Serra Geral Formation deposit, Rio Grande do Sul, Brazil. *J. South Am. Earth Sci.* **2010**, *29*, 751–760. [CrossRef]

79. Hartmann, L.A.; da Cunha Duarte, L.; Massonne, H.J.; Michelin, C.; Rosenstengel, L.M.; Bergmann, M.; Theye, T.; Pertille, J.; Arena, K.R.; Duarte, S.K. Sequential opening and filling of cavities forming vesicles, amygdales and giant amethyst geodes in lavas from the southern Paraná volcanic province, Brazil and Uruguay. *Int. Geol. Rev.* **2012**, *54*, 1–14. [CrossRef]

80. Hartmann, L.A.; Antunes, L.M.; Rosenstengel, L.M. Stratigraphy of amethyst geode-bearing lavas and fault-block structures of the Entre Rios mining district, Paraná volcanic province, southern Brazil. *Ann. Braz. Acad. Sci.* **2014**, *86*, 187–198. [CrossRef]

81. Duarte, L.C.; Hartmann, L.A.; Vasconcellos, M.A.Z.; Medeiros, J.T.N.; Theye, T. Epigenetic formation of amethyst-bearing geodes from Los Catalanes gemological district, Artigas, Uruguay, southern Paraná Magmatic Province. *J. Volcanol. Geotherm. Res.* **2009**, *184*, 427–436. [CrossRef]

82. Morteani, G.; Kostitsyn, Y.; Preinfalk, C.; Gilg, H.A. The genesis of the amethyst geodes at Artigas (Uruguay) and the paleohydrology of the Guarani aquifer: Structural, geochemical, oxygen, carbon, strontium isotope and fluid inclusion study. *Int. J. Earth Sci.* **2010**, *99*, 927–947. [CrossRef]

83. Duarte, L.C.; Hartmann, L.A.; Ronchi, L.H.; Berner, Z.; Theye, T.; Massonne, H.J. Stable isotope and mineralogical investigation of the genesis of amethyst geodes in the Los Catalanes gemological district, Uruguay, southernmost Parana volcanic province. *Miner. Depos.* **2011**, *46*, 239–255. [CrossRef]

84. Wilkinson, J.J. Fluid inclusions in hydrothermal ore deposits. *Lithos* **2001**, *55*, 229–272. [CrossRef]

85. Hedenquist, J.W.; Henley, R.W. The importance of CO_2 on freezing point measurements of fluid inclusions; evidence from active geothermal systems and implications for epithermal ore deposition. *Econ. Geol.* **1985**, *80*, 1379–1406. [CrossRef]

86. Simmons, S.F.; Browne, P.R.L. Hydrothermal minerals and precious metals in the Broadlands-Ohaaki geothermal system: Implications for understanding low-sulfidation epithermal environments. *Econ. Geol.* **2001**, *95*, 971–999. [CrossRef]

87. Simmons, S.F.; White, N.W.; John, D.A. Geological characteristics of epithermal precious and base metal deposits. *Econ. Geol.* **2005**, *100*, 485–522.

88. Hedenquist, J.W. Boiling and dilution in the shallow portion of the Waiotapu geothermal system, New Zealand. *Geochim. Cosmochim. Acta* **1991**, *55*, 2753–2765. [CrossRef]

89. Moncada, D.; Mutchler, S.; Nieto, A.; Reynolds, T.J.; Rimstidt, J.D.; Bodnar, R.J. Mineral textures and fluid inclusion petrography of the epithermal Ag-Au deposits at Guanajuato, Mexico: Application to exploration. *J. Geochem. Explor.* **2012**, *114*, 20–35. [CrossRef]

90. Kilias, S.P.; Naden, J.; Cheliotis, I.; Shepherd, T.J.; Constandinidou, H.; Crossing, J.; Simos, I. Epithermal gold mineralisation in the active Aegean volcanic arc: The Profitis Ilias deposit, Milos island, Greece. *Miner. Depos.* **2001**, *36*, 32–44. [CrossRef]

91. Kilias, S.P.; Detsi, K.; Godelitsas, A.; Typas, M.; Naden, J.; Marantos, Y. Evidence of Mn-oxide biomineralization, Vani Mn deposit, Milos, Greece. In *Digging Deeper, Proceedings of the Ninth Biennial Meeting of the Society for Geology Applied to Mineral Deposits, Dublin, Ireland 20–23 August 2007*; Irish Association of Economic Geologists: Dublin, Ireland, 2007; pp. 1069–1072.

92. Naden, J.; Kilias, S.P.; Darbyshire, D.B.F. Active geothermal systems with entrained seawater as analogues for transitional continental magmato-hydrothermal and volcanic-hosted massive sulfide mineralization-the example of Milos Island, Greece. *Geology* **2005**, *33*, 541–544. [CrossRef]

93. Smith, D.J.; Naden, J.; Miles, A.J.; Bennett, H.; Bicknell, S.H. Mass wasting events and their impact on the formation and preservation of submarine ore deposits. *Ore Geol. Rev.* **2018**, *97*, 143–151. [CrossRef]

94. Vavelidis, M.; Melfos, V. Fluid inclusion evidence for the origin of the barite silver-gold-bearing Pb-Zn mineralization of the Triades area, Milos island, Greece. *Bull. Geol. Soc. Greece* **1998**, *32*, 137–144.

95. Cathelineau, M. Cation size occupancy in chlorites and illites as a function of temperature. *Clay Miner.* **1988**, *23*, 471–485. [CrossRef]

96. Fournier, R.O. The behaviour of silica in hydrothermal solutions. Geology and geochemistry of epithermal systems. *Rev. Econ. Geol.* **1985**, *2*, 45–61.

minerals   **MDPI**

Article

Deciphering Silicification Pathways of Fossil Forests: Case Studies from the Late Paleozoic of Central Europe

Steffen Trümper [1,2,*], Ronny Rößler [1,2] and Jens Götze [3]

[1]   Museum of Natural History Chemnitz, Moritzstraße 20, D-09111 Chemnitz, Germany; roessler@naturkunde-chemnitz.de

[2]   Institute of Geology, TU Bergakademie Freiberg, Bernhard-von-Cotta-Straße 2, D-09599 Freiberg, Germany

[3]   Institute of Mineralogy, TU Bergakademie Freiberg, Brennhausgasse 14, D-09599 Freiberg, Germany; jens.goetze@mineral.tu-freiberg.de

*   Correspondence: truemper@naturkunde-chemnitz.de

Received: 5 July 2018; Accepted: 26 September 2018; Published: 1 October 2018

Abstract: The occurrence and formation of silicified wood from five late Paleozoic basins in Central Europe was investigated. Fossil wood from diverse geological settings was studied using field observations, taphonomic determinations as well as mineralogical analyses (polarizing microscopy, cathodoluminescence (CL) microscopy and spectroscopy). The results indicate that silicification is either a monophase or multiphase process under varying physico-chemical conditions. In particular, CL studies revealed complex processes of silica accumulation and crystallization. The CL characteristics of quartz phases in silicified wood can mostly be related to blue (390 and 440 nm), yellow (580 nm), and red (650 nm) emission bands, which may appear in different combinations and varying intensity ratios. Yellow CL is typical for initial silicification, reflecting quick precipitation under oxygen-deficient conditions caused by initial decay of the organic material. Blue CL is predominantly of secondary origin, resulting from replacement of precursor phases by a secondary hydrothermal quartz generation or subsequent silicification of wood. The red CL can be related to a lattice defect (non-bridging oxygen hole center—NBOHC).

Keywords: petrified wood; petrifaction; cathodoluminescence; quartz; environment

1. Introduction

Both permineralized and petrified plants with preserved cellular details are among the most fascinating remains of the flora and their evolution during Earth history [1–4]. Corresponding fossil occurrences are distributed worldwide in sedimentary and volcanic rocks encompassing million year timespans, beginning with the colonization of the terrestrial realm by higher plants in the Devonian [5] until today's environments, in which ongoing mineralization processes can be observed [6]. There is a large variety of geological settings that yield anatomically preserved plant fossils, their dispersed organs or whole fossil forests preserved in situ. Volcanically-affected landscapes [7], which might be expected to be more prone to rapidly preserve plant remains, are commonly represented in the fossil record. However, in epiclastic sedimentary environments petrifactions frequently occur, even in quite unexpected cases, such as low latitude tropical rainforests, which usually show the largest recycling rates of any organic matter [8]. Most permineralized or petrified wood is silicified, viz. preserved by SiO_2 modifications. This may be due to the overall availability of silica as a result of predominating rocks rich in SiO_2 and associated weathering processes.

The desire to decipher the secrets of petrifaction developed over centuries and has involved manifold attempts and curious experiments. Several questions and phenomena remained so far unexplained, although there is some progress towards understand single aspects of petrifaction, e.g., referring to the conspicuous colors found in fossil wood [9] or the steps and mechanisms of fossilization [10,11].

Fundamental studies and experiments [12–15], both under natural conditions and in the laboratory, have contributed towards the understanding of quartz precipitation, plant tissue impregnation and the related time frame [2,16–18]. However, many fossil woods and their occurrences cannot be explained though uniformitarian comparisons, such as postvolcanic hotsprings [19,20]. The application of analytical methods, like scanning electron microscopy (SEM) with element contrast imaging by backscattered electrons (BSE) or orientation contrast imaging by electron backscatter diffraction (EBSD), additionally permits a study at the nanoscale of quartz crystallites and their interplay with cell wall structures [21].

Long used successfully in diverse raw material studies and sedimentary petrology [22–25] cathodoluminescence (CL) microscopy also seems a promising approach to analyze silicified wood, because it helps to visualize trace elements or internal textures, such as growth zoning, secondary alteration, or recrystallization [26–28]. Accordingly, the formation history of different quartz generations in fossil wood could be reflected [29–32]. More recently, spatially and time-resolved analysis of CL emission colors and spectra provided new insights into CL variation of silica and its modifications [23,33,34].

The goal of this study is to recognize CL patterns associated with silicification pathways by examining fossil woods preserved in different geological settings. Paleobotanical examination combined with analytical techniques such as polarizing microscopy, cathodoluminescence (CL) microscopy, and spectroscopy, as well as scanning electron microscopy (SEM), are applied.

2. Material and Methods

Specimens investigated during this study are stored in the Paleontological Collection of the Museum für Naturkunde Chemnitz (K2097, K6556, K6562, K6563), except for one, which belongs to the Museum für Naturkunde Berlin (1977/7). The Chemnitz collection represents a remarkable set of anatomically preserved plant fossils from different sites worldwide and encompasses material from various geological settings and stratigraphic levels [35]. In this approach, the focus is on silicified wood-bearing localities, which have been investigated recently by the authors in the field and under the microscope. Additionally, any comparisons and discussions of the selected localities benefit from their well-known geological settings and stratigraphy both in a regional and global context [36].

Samples were documented macroscopically by using a Nikon D5300 digital camera, and microscopically by using NIS-Elements D (version 3.2) software under a Nikon SMZ 1500 microscope (all Nikon Cooperation, Tokyo, Japan). Thin sections were investigated under plane-polarized (PPL), cross-polarized transmitted light (XPL) and cathodoluminescence (CL). CL analyses were carried out on polished standard thin sections which were coated with carbon in order to prevent any build-up of electrical charge during CL operation. The thin (ca. 5 nm) carbon layer ensures transparency for both transmitted light and CL observations, and has no influence on the microscopic investigations. The CL microscope was a "hot cathode" CL microscope HC1-LM (designed by U. Zinkernagel and modified by R. Neuser at the Ruhr-University, Bochum, Germany), which used an acceleration voltage of 14 kV and a beam current of 0.2 mA [37]. CL images were taken with an Olympus DP72 digital camera (Olympus Cooperation, Tokyo, Japan). CL spectra in the wavelength range of 370–900 nm were recorded with an Acton Research SP-2356 spectrograph (Princeton Instruments, Acton, MA, USA), linked to the microscope via an optical glass-fiber guide. CL spectra were recorded under standardized conditions with a spot size of about 30 μm in diameter by using a pinhole aperture. An Hg-halogen lamp was used for wavelength calibration. The measured CL spectra are presented as curves in wavelength-CL intensity diagrams. Due to the light diffraction, a CL spectrum results from interference of CL signals derived from an area larger than the spot. Accordingly, a spectrum contains several CL signals caused by different defect structures in the corresponding silica phases.

3. Geological Setting

3.1. Late Paleozoic Environments Capable of Preserving Silicified Woods

Petrified wood dating back to the late Paleozoic is known from numerous localities in Central Europe, covering a wide range of lithologies and depositional settings (Figure 1). This abundance and diversity is connected to climatic and geotectonic conditions in tropical central Pangaea, which were conducive for wood petrifaction. Initiated by climatic change from tropical ever-wet to subtropical seasonally wet conditions from the late Moscovian into the Permian (Figure 2), wood-forming gymnosperms (i.e., cordaitaleans, conifers) flourished and expanded from extra-basinal areas prone to erosion, to the basin centers [38,39]. Contemporaneously, late- to post-collisional collapse of the Variscan Mountains in Central Europe facilitated the formation of closely arranged intramontane basins, where fossils were preserved [40,41]. Rapid denudation filled the newly formed accommodation space with thick, predominantly siliciclastic successions [42]. Syndepositional explosive volcanism using intersecting faults as conduits favored rapid burial of plants, including the entombment of entire ecosystems known as "T^0 assemblages" [7,43,44].

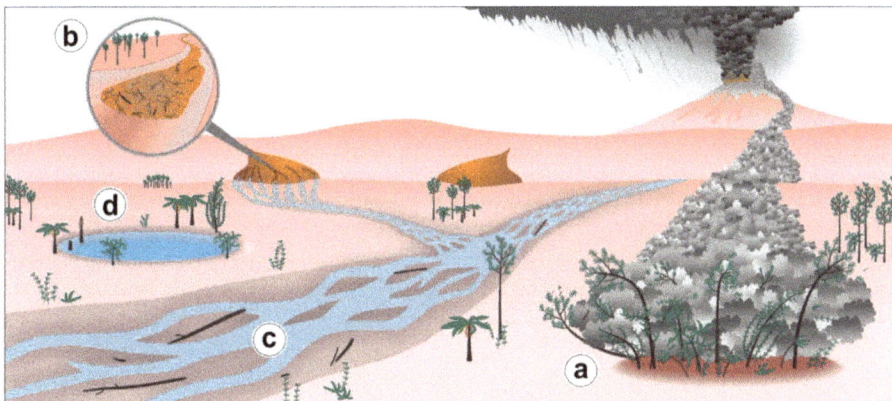

Figure 1. Petrified-wood forming environments of the late Paleozoic in Central Europe. (**a**) In situ burial in pyroclastic deposits (Flöha, Wendishain, Chemnitz). (**b**) Alluvial transport and burial by mass flows with/without volcanic material (Winnweiler). (**c**) Reworking and burial of stems in fluvial deposits (Kyffhäuser). (**d**) Silicification in lacustrine and palustrine environments (Manebach).

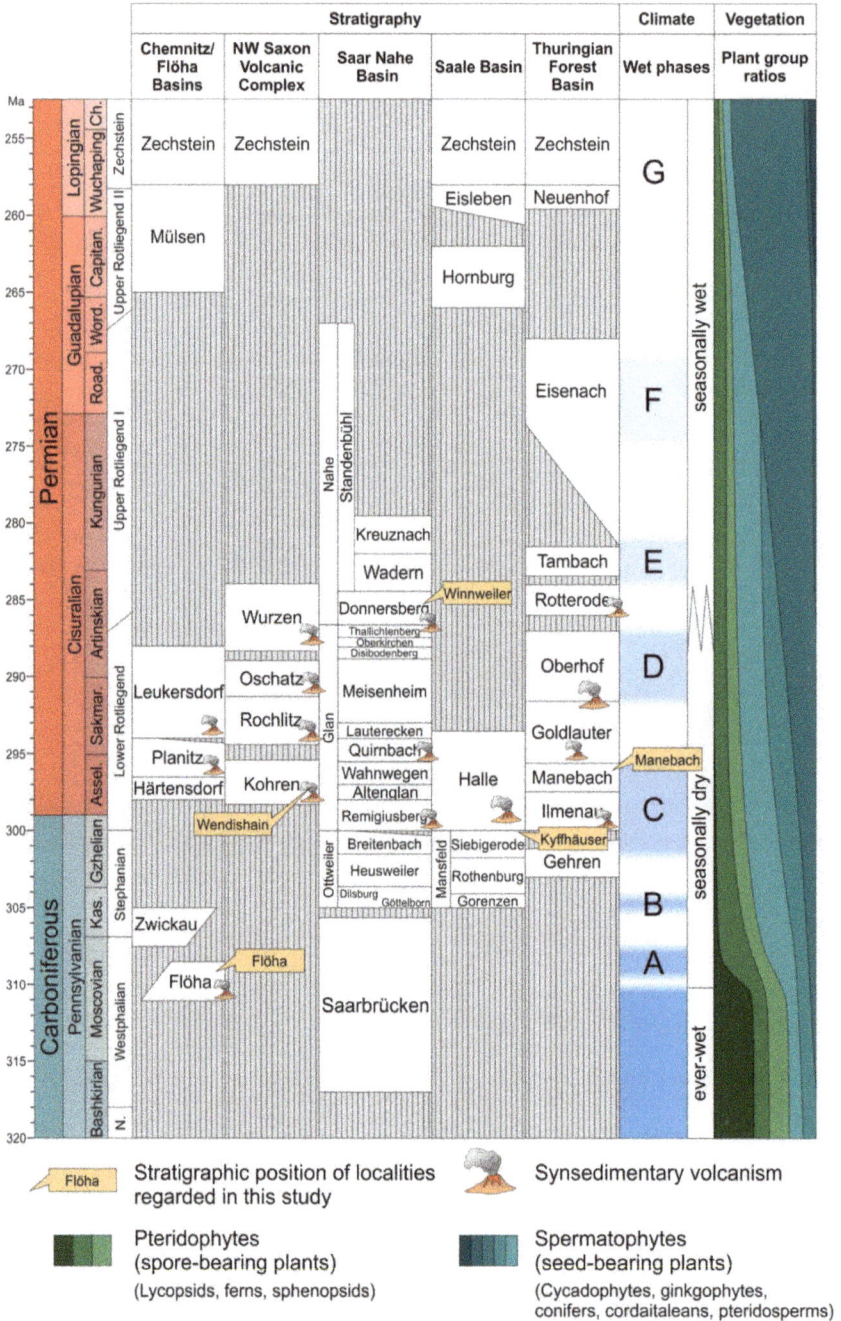

Figure 2. Lithostratigraphy of selected late Paleozoic intramontane basins. Tectonic basin formation around the Carboniferous-Permian boundary is accompanied by volcano-sedimentary deposits and abundant hiati in the stratigraphic record. From the middle Pennsylvanian onwards, gradual aridization favored the expansion and diversification of seed-bearing plants including wood-forming gymnosperms (conifers, cordaitaleans, ginkgophytes). Compiled and modified after [36,44–46]. Vegetation data are based on [38,39,47] and personal observations.

3.2. Sampling Locations of Investigated Silicified Woods

The positions of the investigated occurrences of silicified wood in Central Europe are presented in Figure 3.

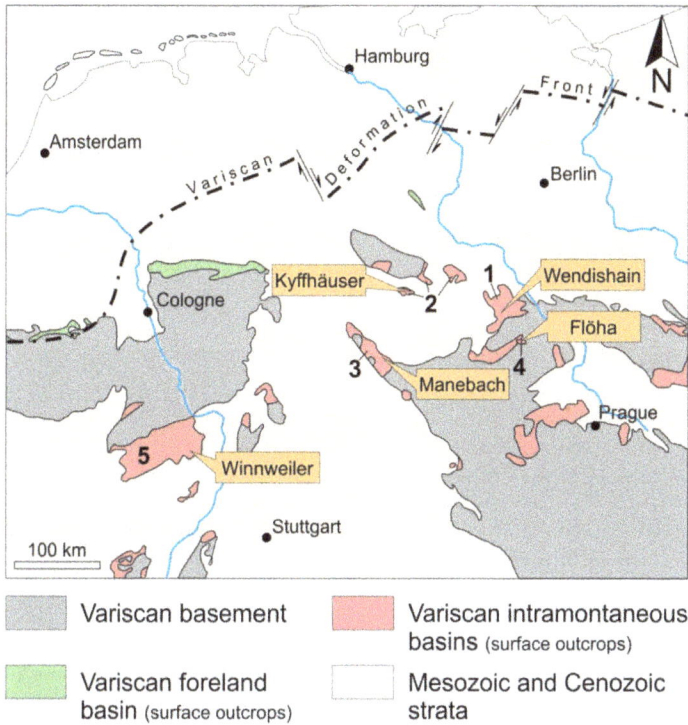

Figure 3. Generalized geological map of Central Europe showing the position of the selected localities. Modified after [48]. 1—NW Saxon Volcanic Complex; 2—Saale Basin; 3—Thuringian Forest Basin; 4—Flöha Basin; and 5—Saar-Nahe Basin.

Sample K6563 was found in Holocene colluvial sediments of the Struth Forest near Falkenau, located in the foreland of the Erzgebirge in East Central Germany. Based on the adhering rock, the stem was originally embedded in sediments accompanying the Schweddey Ignimbrite of the late Carboniferous Flöha Formation (Moscovian, Flöha Basin, Figures 2 and 3). This feldspar-rich pyroclastic rock is attributed to a pyroclastic density current entombing and fragmenting late Carboniferous coal-forming vegetation including woody trunks [49–51]. The age of the Schweddey Ignimbrite was recently constrained to 310 ± 2 Ma based on radiometric U/Pb measurements obtained from different outcrops, and a Bolsovian macroflora [49,50,52].

Sample K6562 from Wendishain is derived from a 48 × 40 cm sized block containing an incomplete silicified stem section measuring 30 × 40 cm in diameter, and its adhering pyroclastic host rock. This block was found parautochthonously on a field in Wendishain. The adhering rock consists of a greenish wood-bearing quartz-rich ignimbrite. Based on reworked lithics derived from the Leisnig Porphyry, the stem was originally embedded in the topmost up to 10 m thick beds of the Kohren Formation (Asselian, NW Saxon Volcanic Complex, Figures 2 and 3) overlying the Leisnig Porphyry. The Kohren Formation recorded deposition by alluvial fans, meandering rivers and lacustrine to palustrine sediments in abandoned channels and swamps, accompanied by synsedimentary volcanism [53–55].

Sample K6556 from Winnweiler was removed directly from alluvial volcano-clastic deposits of the Donnersberg Formation (middle to late Artinskian, Saar-Nahe Basin, Figures 2 and 3). Its sediments reflect deposition in alluvial fans, braided and meandering rivers, including associated floodplains, within the half-graben like Saar-Nahe Basin [56]. Synsedimentary volcanism resulted in the formation of voluminous subvolcanic, effusive, pyroclastic, and epiclastic rocks [57]. The silicified wood-hosting strata at Winnweiler are of low maturity with respect to sediment composition and texture. The poorly sorted and non-stratified wood-bearing debrites range from coarse-grained pebbly sandstones to sandy conglomerates accompanied by high contents of clay and silt in the matrix. The matrix may be red or light green in color, and contains reworked tuffaceous material.

Sample K2097 was found allochthonously in alluvial fan deposits formed at the northern margin of the Kyffhäuser SW of Tilleda during the Pleistocene. Silicified woods at Tilleda are derived from pebbly, fine- to coarse-grained quartz arenites of the Siebigerode Formation (latest Gzhelian, Saale Basin, Figures 2 and 3) [58–61] exposed in the Kyffhäuser. These log-bearing sandstones are highly mature with respect to composition, expressed by quartz contents ranging from 70 to 90% volume. The Siebigerode Formation reflects deposition in medium- to large-scaled braided rivers [59,62]. Wood logs derived from extrabasinal areas were transported in the fluvial system during floods, and subsequently settled on barforms [59].

Sample 1977/7 has been found in colluvial deposits derived from uphill outcrops of the Manebach Formation (Asselian, Thuringian-Forest Basin, Figures 2 and 3) in the Schulzental valley close to Manebach. The up to 200 m thick, greyish Manebach Formation shows a diverse lithology, comprising fluvial channel sandstones, *Scoyenia*-bearing siltstones, lacustrine claystones, hydromorphic paleosols, palustrine coals, and silicified peats [63,64]. Its host rock represents silicified peat intercalated in sandstones and conglomerates. The Manebach Formation is correlated climatostratigraphically with the late Carboniferous-early Permian wet phase C (Figure 2). Silicified plants from Manebach recently attracted scientific attention due to the exceptional preservation of fungal endophytes, epiphytic ferns and pteridosperms, as well as coprolites of parasitic arthropods within their tissues [65,66].

4. Results

4.1. Flöha

The specimen is a 7 cm long fragment, and possesses an oval shape in cross-section measuring 9.5 × 5.5 cm in diameter (Figure 4b,c). Anatomically, the stem only consists of wood showing a green color (Figure 4c–g). The latter is caused by a green mineral phase occupying the former cell walls or partially filling the cell lumina (Figure 4e).

Silicification is promoted by clear quartz filling the cell lumina (Figure 5a). Quartz crystals are restricted to 1–4 tracheids (Figure 5b). The wood is variably preserved, i.e., disintegrated into single cells (Figure 5d), plastically deformed (Figures 4g and 5d), or fractured (Figure 4f,g and Figure 5d,e). In areas of appropriate anatomical preservation, the 28 to 50 μm wide cells are oval to polygonal in cross section d,e, and Figure 5a,b, and point to a cordaitalean affinity. Under CL, the silicified wood predominantly displays a short-lived blue CL (Figure 5c,f). Spectral measurements (Figure 5g) show that the blue band is very broad and consists of at least two overlapping component bands centered at 390 and 450 nm. Due to the spectral range of the analytical arrangement, the 390 nm band is only visible as a shoulder in the CL spectra. After three minutes of irradiation, CL intensity decreases by 60% (Figure 5g). Next to blue luminescent areas, yellow CL (main emission band at ca. 570 nm) and red CL (main emission band at 650 nm) occur randomly within the cell lumina (Figure 5c,e,f). Areas of blue CL on the one hand, and yellow and red CL, on the other hand, do not differ with regard to anatomical preservation.

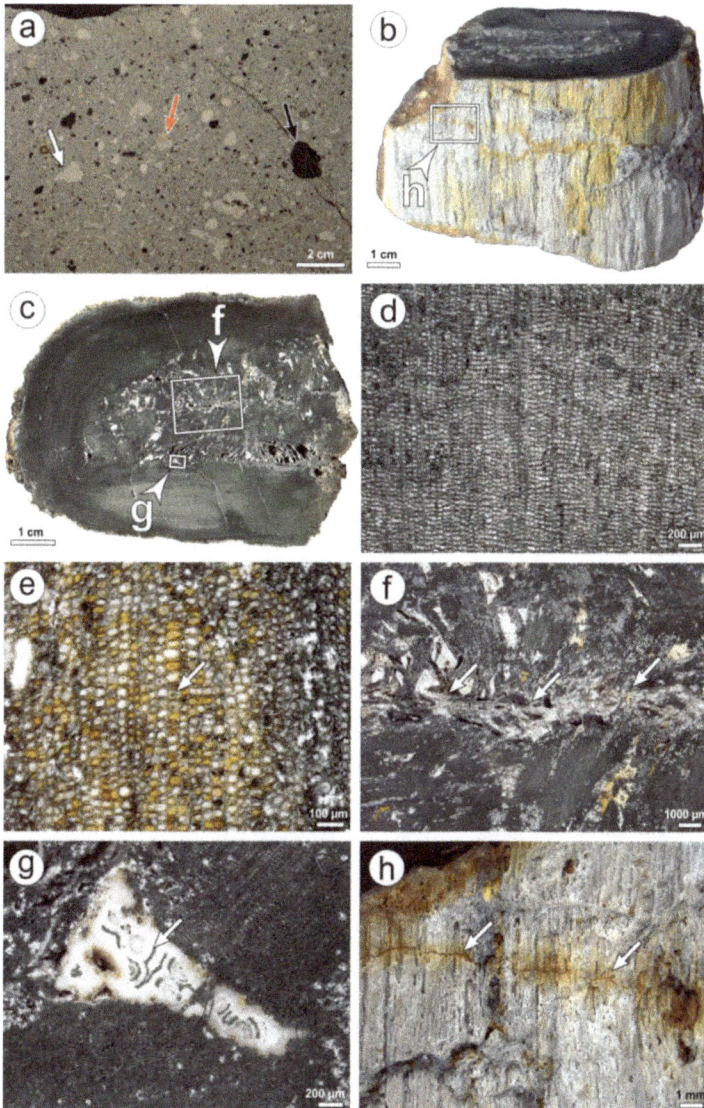

Figure 4. Features of K6563 and the Schweddey Ignimbrite from Flöha. (**a**) Schweddey Ignimbrite, containing carbonized plant remains (black arrow), weakly welded pumice (white arrow) and cm-sized feldspars (here plagioclase incorporating an orthoclase, red arrow). (**b**) Stem, from which K6563 is derived from. (**c**) Stem in cross section. Note the inner part of the stem showing abundant cavities filled by white agate. (**d**) Green-colored wood in cross section. (**e**) Close-up of the wood showing polygonal shapes of the tracheids in cross-section. Cell walls and parts of the cell lumina (arrow) are filled by a green mineral phase. (**f**) Close-up of the stem's center. The former pith cavity is compressed to a white line (arrows). (**g**) Cavity filled by white agate. Note intercalations of a green mineral phase (arrow). (**h**) Limonite-bearing fractures (arrows) as pathways for oxidation caused by weathering.

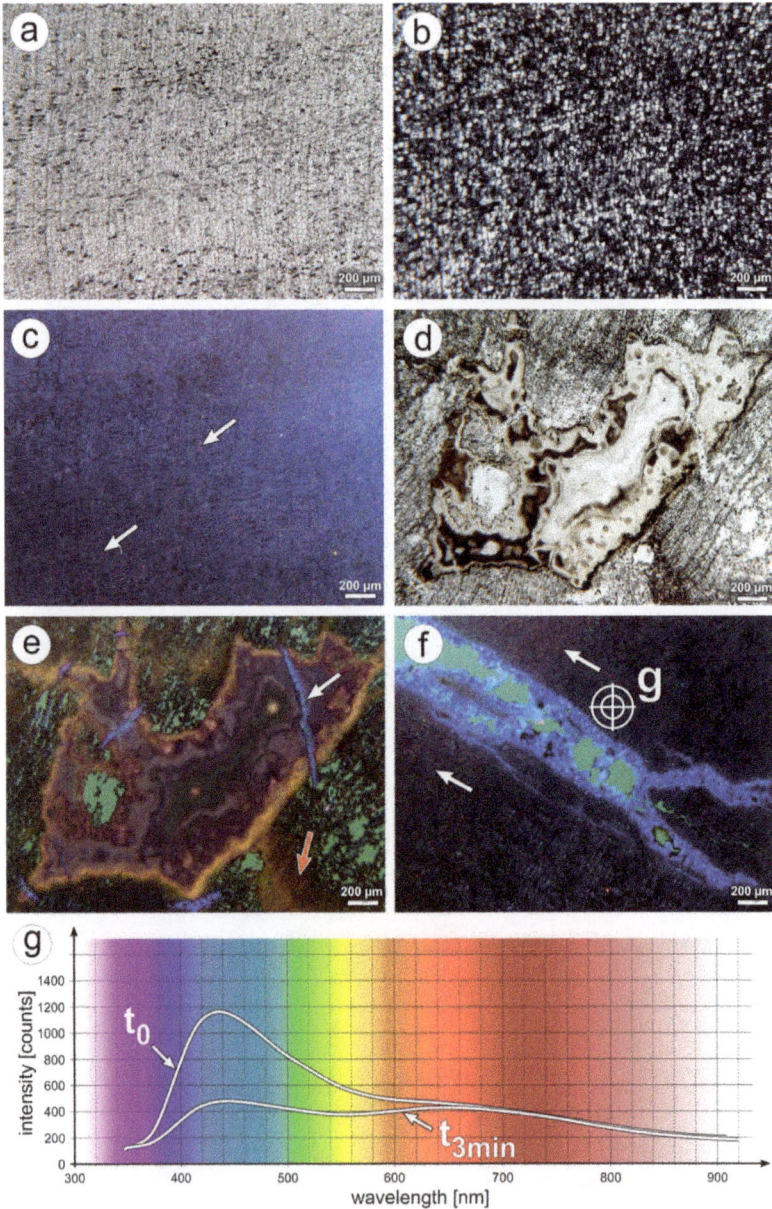

Figure 5. CL of K6563. (**a**) Silicified wood; PPL image. (**b**) Same view as in (a); XPL image. Note that quartz crystals comprise single tracheids or tracheid groups. (**c**) Same view as in (a), CL image, showing a short-lived blue CL interspersed with centers of red CL (arrows). (**d**) Agate-filled cavity; PPL image. (**e**) Same view as in (d); CL image. The agate possesses a brown CL, whereas surrounding intercellular space is filled by green-luminescing epoxy. Fillings of tight fractures dissecting the agate have a blue CL (white arrow). Neighboring areas show a yellow CL (red arrow). (**f**) Fracture filled by euhedral quartz displaying a short-lived blue CL; CL image. Note diffusively occurring red CL in cell lumina (arrow). (**g**) CL spectra taken at (f). Note decreasing intensity of blue CL after three minutes of irradiation, whereas red CL remains constant.

From inwards outwardly, the center is characterized by a 30 mm long line orientated parallel to the stem's largest diameter (Figure 4f), and resulted from pith compression. It is surrounded by a 1.5 cm thick zone of inner wood showing abundant cavities, each a couple of millimeters in length (Figure 4c,f). The cavities are empty or filled by white agate, and are aligned sub-parallel to the stem's smallest diameter (Figure 4c,f,g). Internally, the agate shows a concentric layering, often accompanied by intercalations of the green mineral phase (Figure 4g). In places, botryoidal structures can be observed. Fragmented wood files occurring in isolation within the cavities are encrusted by agate layers. Under CL, the agate shows a reddish-violet CL (Figure 5e). Along cavity margins, orange CL can be observed (Figure 5e). As a special feature, the agate-filled cavities are dissected by 100 μm wide fractures filled by quartz. These fractures end at the edge of the agate-filled cavities, and show a short-lived blue CL (Figure 5e). Wood around the cavities is disintegrated. The intercellular space is exclusively filled by epoxy, derived from thin section preparation and reveal intense, homogeneous green CL (Figure 5e).

The outer wood, a 2 cm thick zone, lacks any agate-filled cavities (Figure 4c). Radially aligned, up to 0.5 mm wide fractures occurring here are incompletely filled by euhedral quartz blades rooting at the fracture-wood interface (Figure 5f). This euhedral quartz shows a short-lived blue CL identical to fillings in fractures occurring within agate-filled cavities (Figure 5f). Along fractures and across the surface of K6563, hematite formation reflects oxidation probably connected to weathering (Figure 4h).

It should be mentioned here in addition that silicified wood from other outcrop areas of the Schweddey Ignimbrite (e.g., the Oderan Forest site) show a short-lived blue CL throughout the wood.

4.2. Wendishain

The rock adhering to the stem consists of crystals, wood fragments, pumice, and volcanic lithics in order of decreasing abundance, embedded in a green-colored cryptocrystalline siliceous matrix (Figure 6a). The rock is non-stratified and poorly sorted and, thus, is classified as ignimbrite. The crystal content comprises angular to splintery, 1 or 2 mm large quartz crystals displaying a short-lived blue CL at 440 nm (Figure 7a), and black, columnar amphibols. The angular, pale-green to violet wood fragments range in diameter from 3 mm to 2.5 cm. The white pumice fragments are poorly welded, and reach sizes of a couple of millimeters up to 3 cm (Figure 6a). Lithics are restricted to less than 2.6 cm large red volcanics showing abundant altered feldspars, amphibols, and rare quartz (Figure 6a). The matrix is cryptocrystalline and revealed a short-lived blue CL (Figure 7a).

The stem possesses an intensive orange to red color and comprises a sector of the originally complete diameter, which is supposed to have represented ca. 70 to 90 cm (Figure 6b). Anatomically, this fragment only consists of wood, and adjoins the embedding pyroclastic rock with an uneven, but macroscopically sharp interface (Figure 6c). Internally, the stem displays a color zonation, comprising three zones. From outside inwards, there is a greyish-white to rose-colored, 0.5 to 2 cm thick zone 1, followed by a predominant 14 to 20 cm thick red to orange zone 2 showing grey dots, and a dark-grey to brown zone 3 (Figure 6c,d). A thin section of K6562 has been cut from the margin, comprising the embedding tuff and zones 1 and 2. Color zonation is barely reflected in the anatomical preservation both macroscopically and microscopically. The wood's preservation shows cell walls being diffusely outlined under transmitted light (Figure 6e). However, cells are more clearly recognizable under CL. Silicification is promoted by coarse crystalline to cryptocrystalline quartz (Figure 6f,g).

On a microscopic scale, there is a fluent transition from ignimbrite to wood, based on gradual disintegration of the tissue (Figure 6e). Silicified wood from zone 1 displays a time-dependent CL, being initially blue (decreasing 450 and 390 nm bands), and turning into red (increasing emission band at 650 nm) (Figure 7b). Fracture fillings consisting of coarse-crystalline quartz are almost non-luminescent (Figures 6g and 7c,d). Towards zone 2, initially blue CL of the cell walls is gradually replaced by yellow CL (emission band at ca. 570 nm) in areas distant to fractures (Figure 7c,d). In zone 2, finally, cell walls have a yellow CL, and cell lumina are characterized by red to brown CL (main CL emission band around 650 nm) (Figure 7e,f). Fractures in areas transitional to zone 1 and within zone 2 are filled by cryptocrystalline silica showing a reddish-brown CL similar to the surrounding wood (Figure 7f).

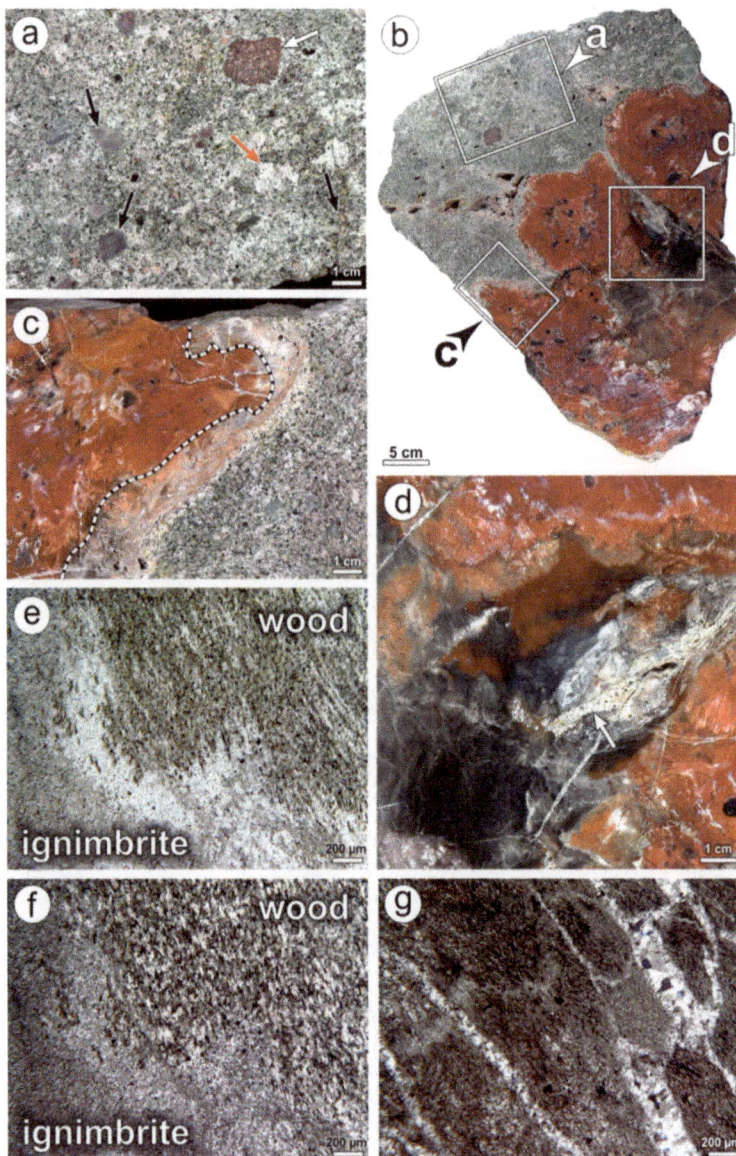

Figure 6. Features of K6562 and its host rock from Wendishain. (**a**) Ignimbrite consisting of wood fragments (black arrows), pumice (red arrow) and lithics (white arrow). (**b**) Cross-section showing silicified wood (red) with adhering ignimbrite (green). (**c**) Color zonation of the wood with an outer greyish-white to rose-colored zone 1, and an inner orange zone 2 (dotted line as border). (**d**) Color zonation of the wood with orange zone 2, and dark grey to brown zone 3. Note fracture fillings with ignimbrite (arrow). (**e**) Gradual transition from wood into ignimbrite; PPL image. (**f**) Same view as (**e**); XPL image. (**g**) Zone 2; XPL image. Note coarse-crystalline quartz in the fractures.

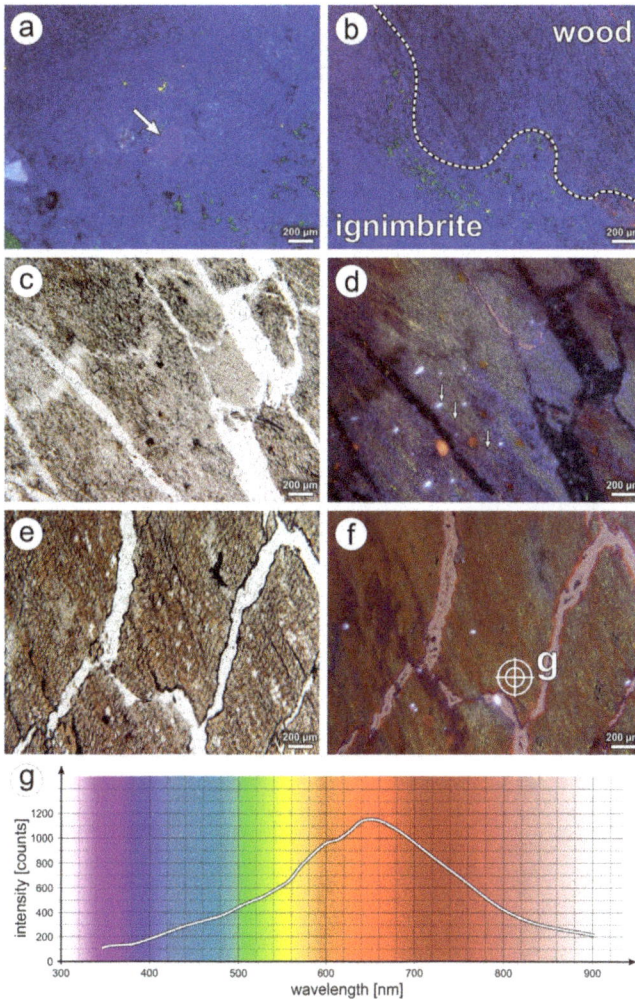

Figure 7. CL of K6562. (**a**) CL image of the ignimbrite showing short-lived blue CL of the matrix and a quartz grain (arrow). (**b**) Same view as in 8e; CL image. Note identical CL of both wood in zone 1 and ignimbrite. (**c**) Transition of zone 1 and 2 PPL image. (**d**) Same view as in (c); CL image. Yellow CL is restricted to areas distant to fractures (arrows). (**e**) Zone 2; PPL image. (**f**) Same view as in (e); CL image. Note that blue CL has been replaced by red CL in the cell lumina. Fractures are filled by red-luminescent cryptocrystalline silica. (**g**) CL spectrum taken at (f).

4.3. Winnweiler

K6556 contains a 2.9 cm thick silicified wood. It consists of two xylem strands but lacks any pith. Based on the aforementioned anatomical characteristics, the wood is identified as a gymnosperm root. The root is embedded in a matrix made up of 0.2 to 2 cm large wood fragments, red chalcedony, pale agate and calcite in order of decreasing abundance (Figure 8b,d). In general, silicified wood occurs in two preservation types, which can also be distinguished by their CL: (1) anatomically well preserved showing yellow and red CL (emission bands at 580 and 650 nm, respectively; Figures 8d,e and 9a–d), and (2) anatomically poor-preserved displaying a short-lived blue CL (Figures 8f–h and 9e,f). The first type is limited to the innermost 20 mm of the gymnosperm root. Here, cells are preserved without visible deformation (Figures 8e and 9a). The tracheids possess a circular cross cut, and decrease in diameter from 64 µm in the root's center to 39 µm in the outer wood (Figure 8d). According to [67], a coniferophyte origin can be assumed. The general decline of tracheid diameter is superimposed by rhythmic variations of tracheid diameters and wall thicknesses, reflecting some kind of concentric growth zonation (Figure 8d). In general, the inner part of the root exposes different CL, each of them showing a more or less distinct spatial distribution (Figure 9c,d). Red CL is predominantly restricted to silica either filling cell lumina or incompletely covering the inner surface of fractures running through the wood (Figure 9c). Rarely, both cell lumina and walls show a red CL. By contrast, yellow CL is strictly limited to the former cell walls (Figure 9c). Areas characterized by yellow and red CL merge laterally into such being almost non-luminescent (Figure 9d).

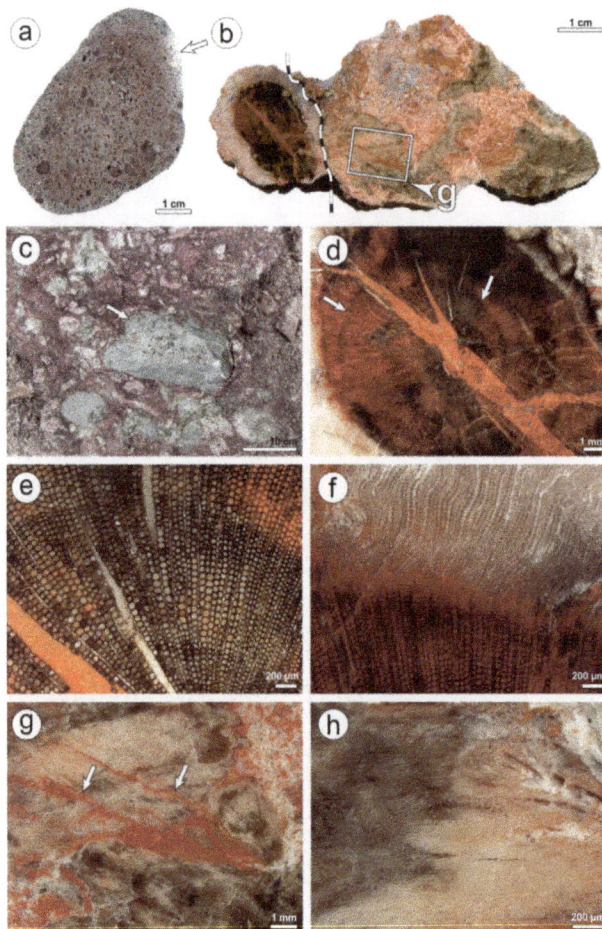

Figure 8. Features of K6556 and its host rock from Winnweiler. (**a**) Rhyolite clast from alluvial debrites of the Donnersberg Fm. Note roundness and bleaching (arrow). (**b**) K6556. The dotted white line marks

the interface of the root cross-cut (left) and the surrounding matrix containing wood fragments (right). (**c**) Typical debrite of the middle Donnersberg Fm. Note crucial features of alluvial mass flows: missing sorting and bedding, and polymict composition with lithics (here green tuffs; arrow). (**d**) Close-up of the inner root in cross-section showing growth zonation. (**e**) Round-shaped tracheid cross-cuts. (**f**) Border of well-preserved inner root (below) and poor-preserved outer root rim (above). (**g**) Close-up of wood fragments surrounding the root. Note the jigsaw-like match of neighboring fragments (white arrows). (**h**) Frayed wood in the matrix surrounding the root.

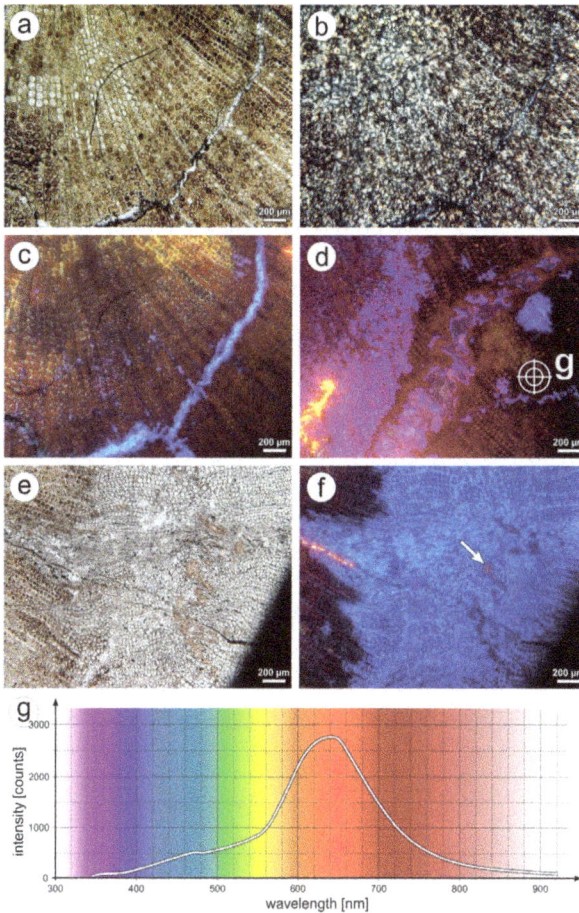

Figure 9. CL of K6556. (**a**) Anatomically well-preserved wood of the inner part of the root (type 1; PPL image). (**b**) Same view as in (a); XPL image. (**c**) CL image of (a). Note limitation of yellow CL to the former cell walls, whereas red CL occurs within the cells. Fracture filling shows a short-lived blue CL. (**d**) All CL spectra detected in K6556; image taken from the inner root. Note that anatomical preservation within irregularly shaped occurrences of short-lived blue CL is lower than around. (**e**) Anatomically poor-preserved wood of the outer rim of the root (type 2; PPL). (**f**) CL image of (e). Brown-luminescing agate fills holes in the wood (arrow). The occurrence of short-lived blue CL is strictly connected to cellular preservation. (**g**) CL spectrum taken at (d).

Type 2 showing a short-lived blue CL is demonstrated in the following parts of K6556: (1) fractures running radially and concentrically through the wood in the inner part of the root (Figure 9c), (2) irregularly to circular-shaped patches of the wood within the inner part of the root (Figure 9d), (3) the approximately 10 mm thick outer rim of the gymnosperm root (Figure 9f), and (4) wood fragments surrounding the root. Anatomical preservation is generally weak, as indicated by highly deformed cells (Figure 8f–h). In places, the wood is disintegrated into tracheid rows and even single tracheids (Figure 8h). Irregularly, but sharply shaped holes within type-2-silicifed wood are filled by brown-luminescing agate (Figure 9f).

A couple of tracheids along the outer margin of the root, as well as fractures within, are completely or partially filled by calcite indicated by intense orange CL (emission band at 620 nm; Figure 9d,f).

4.4. Kyffhäuser

K2097 (Figure 10d) is a fragment of silicified *Agathoxylon*-type wood showing the typical "pointstone preservation" of Kyffhäuser specimens. In cross section, the eponymous "points" are circular to oval or rhombic, darkish brown areas of silicified wood, ranging from 1.5 to 3 mm in diameter (Figure 10e). If non-circular in shape, the "points" are orientated with their largest diameter in a radial or tangential direction. Within the "points", preservation of cell structures follows a gradient. Non-deformed cells and well recognizable cell walls only occur in the innermost area, which encompasses approximately one-third to one-half of the "point's" complete diameter (Figures 10g and 11a). Here, each former cell is occupied by one quartz crystal, which may be either transparent or milky and macroscopically opaque (Figure 11b). In some "points", a concentric alternation of transparent and milky cell lumina fillings resulted in the formation of rings (Figures 10e and 11a,c). In contrast to the "point's" center, cells in the outermost part are sheared, resulting in oval shapes (Figures 10f,h and 11a). Cell walls are blurred, but still act as boundaries for quartz crystals, which—in contrast to the innermost area—cover two or three cells instead of one. Concerning CL, the "points" exhibit a weak yellow CL (main emission band at 580 nm, Figure 11c). As a feature special to Kyffhäuser woods, such areas are strewn with 50 μm large dots, showing a much more intense yellow to orange CL. These dots occur separately, but also rarely form clusters of two or three dots (Figure 11c,e,f).

The "points" either occur as tight clusters (Figures 10d and 11d), or are separated from each other by transparent, fine- to macro-crystalline quartz (e.g., having grain sizes of 0.5 to 3 mm; Figure 10d–f). In the latter case, the "points" are surrounded by radially aligned euhedral quartz blades. In these quartz blades, increasing deformation of cell structures, which starts in the outermost area of the "points", culminates in herringbone-like arrangements of poorly preserved cell walls (Figure 10f,h and Figure 11a,b). With increasing distance to the "points", quartz crystals become anhedral, clear and free from any cell structures.

Areas of crystalline quartz in between the "points", where anatomical preservation is low or cells are not present, are characterized by intense, short-lived blue CL resulting in non-luminescence after a few minutes irradiation (Figure 11e,f).

In a longitudinal section, the previously described "pointstone preservation" appears as a banded structure (Figure 10f). Hence, the "points" just represent two-dimensional cross-cuts of actually spheroidal to tube-like petrifaction bodies.

Conspicuously, the previously described anatomical and CL features can be observed in silicified woods from other localities situated in the Siebigerode Formation across the whole Saale Basin, e.g., Siebigerode and Wettin, which are located 28 km or 52 km, respectively, NE of the Kyffhäuser.

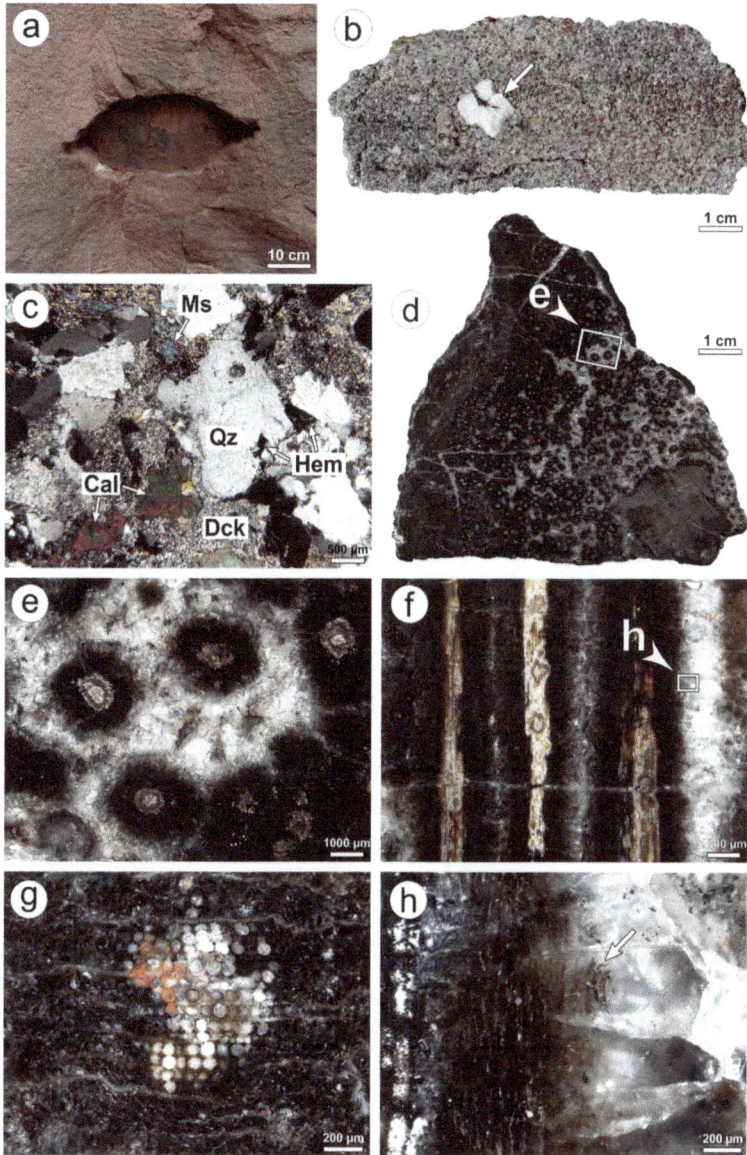

Figure 10. Features of K2097 and its host rock from the Kyffhäuser. (**a**) Appearance of logs in the field: cross-cut of a log horizontally embedded in medium-grained sandstones. Note vertical compaction by oval geometry of the cross-cut. (**b**) Typical reddish white kaolin sandstones of the Siebigerode Fm. The log-bearing fine- to medium-grained sandstones are mostly well sorted and only contain rarely pebbles (arrow). (**c**) XPL image of B showing the mineralogical composition of the sandstones: quartz (Qz) and muscovite (Ms) as detrital grains, whereas intergranular space is filled by dickite (Dck), calcite (Cal) and hematite (Hem) in order of decreasing abundance. (**d**) Cross-cut of silicified *Agathoxylon*-type wood preserved as "pointstone". (**e**) Close-up of the "points". (**f**) "Pointstone" preservation in longitudinal section revealing a tube-like or spheroidal geometry. (**g**) Center of a "point" in cross-section. Note the increasing deformation from the center outwardly. (**h**) Close-up of F showing large quartz blades, which incorporate plastically deformed wood cells.

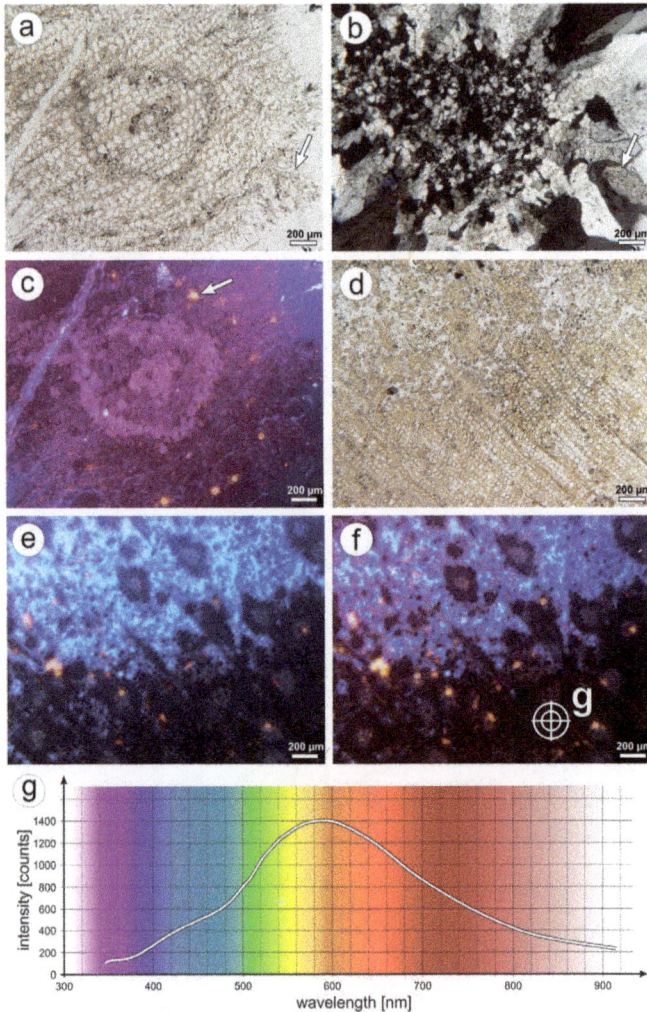

Figure 11. CL of K2097. (**a**) "Point" in PPL. Note highly deformed cells surrounding the "point" (arrow). (**b**) Same view as in (a); XPL image. Each cell is occupied by one quartz crystal in the inner part of the "points". (**c**) Same view as in a; CL image. The macroscopically visible rings yield different CL intensities under CL. The "point" is rich in yellow halos (arrow). (**d**) "Pointstone" fabric; PPL image. (**e** + **f**) Same views as in (d). CL images taken initially (**e**) and after three minutes of irradiation (**f**). Note decreasing intensity of blue CL in between the "points". (**g**) CL spectrum taken at (f).

4.5. Manebach

1977/7 consists of an *Agathoxylon*-type stem, 2.5 × 3.5 cm in diameter, which is encrusted by a 1.2 cm thick stromatolite (Figure 12b). In cross section, the stem is almost complete, but lacking bark and pith (Figure 12c). The wood is permineralized, and its black color resulted from preserved carbon in the former cell walls (Figures 12d and 13a). Radial tracheid walls are covered by biseriate pits (Figure 12d). Cell lumina are filled with white silica, which appears to be homogeneous under polarized transmitted light. CL images of the cell fillings, however, reveal two different emission types: a first type with predominant, low-intense, time-dependent blue-red CL, and a second type with more or less stable yellow CL (emission band around 580 nm; Figure 13c–e). The first type is visually almost non-luminescent because of the low CL intensity, but extended exposure times and spectroscopic measurements reveal initial dark blue CL, gradually shifting to red (increasing 650 nm band) after three minutes of irradiation. Areas of yellow CL occur randomly, without apparent connections to wood anatomy or cellular preservation.

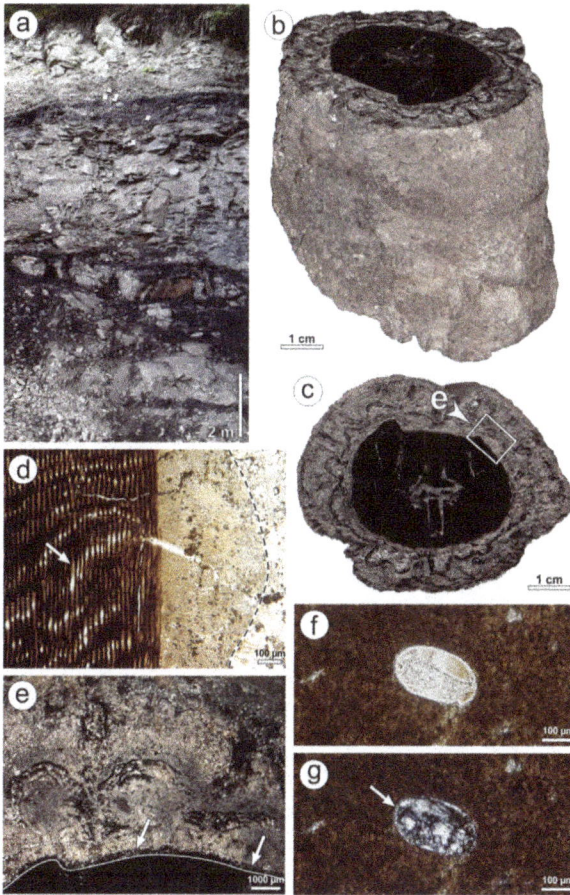

Figure 12. Features of 1977/7 and its host rock from Manebach. (**a**) Typical section of the Manebach Fm. showing a vertical succession of floodplain siltstones (grey), fluvial sandstones (pale yellow) and intercalated coal seams (black). (**b**) Permineralized *Agathoxylon*-type stem, encrusted completely by a stromatolite; lateral view (1977/7). (**c**) Cross-section of B. Note the uniform thickness of the stromatolitic mantle. (**d**) Interface of stem (left) and silicified stromatolite (right); radial view; PPL image. Exceptional preservation in Manebach is reflected by delicate anatomical features, such as pits in the tracheids (arrow), and microbial filaments in silicified parts of the stromatolite (left of the dotted line). (**e**) Interface of permineralized stem (black) and stromatolite. Whereas calcite prevails in the stromatolite, silicification is limited to single black laminae, and a fringe around the stem (arrows and continuous line). (**f**) Ostracod within the stromatolite; PPL image. (**g**) Same as (f); XPL image. The ostracod is completely filled by quartz, which also replaced calcite in the shell. Residual calcite is indicated by interference colors of higher orders (arrow).

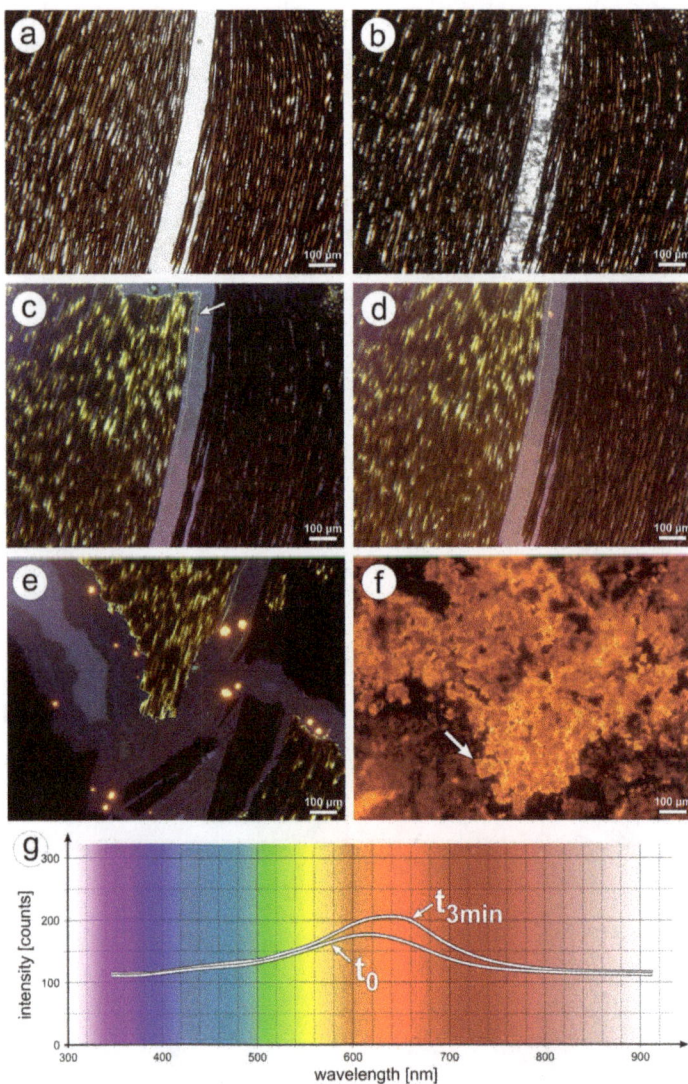

Figure 13. CL of 1977/7. (**a**) Radial section of the wood; PPL image. A longitudinally aligned fracture separates the wood along the tracheids. (**b**) Same photo as in (**a**); XPL image. Due to the strong inherent color of the organically preserved cell walls, the wood appears brown. The fracture is filled by two generations of quartz. (**c**) Initial CL image of a. The wood left of the fracture shows a yellow CL, whereas the fracture filling and the wood on the right side display a low intensity blue CL. Different generations of quartz within the fracture are indicated by yellow CL along their interface (arrow). (**d**) CL image after three minutes of irradiation. Note the shift to a red CL, whereas yellow CL is comparatively stable. (**e**) Initial CL image of a fracture dissecting the wood. The fracture filling is rich in less than 50 µm large calcite crystals showing an intense orange CL at 620 nm. (**f**) Initial CL image of the calcareous stromatolite encrustation, which is made up of 30 µm large euhedral calcite crystals (arrow). (**g**) CL spectra of silica showing a lowly intense, time-dependent color taken initially (t_0) and after three minutes of irradiation (t_{3min}). Each curve resulted from interference of three different luminescences: blue (440 nm), yellow (580 nm; diffracted radiation from neighboring areas) and red (650 nm). Note increasing intensity of red CL with time.

Up to 1 mm wide fractures, crossing the wood in longitudinal direction, are filled with one or two silica generations (Figure 13a–e). Surprisingly, all generations show the same CL, and belong to the time-dependent first type. In CL images, different generations of fracture fillings can only be distinguished by their borders, which possess a yellow CL similar to type two (Figure 13c). As a striking feature, the silica within the fractures encloses isolated, 10 to 70 μm large, subhedral or anhedral calcite crystals (Figure 13e).

The wood borders on the stromatolite with sharp contact (Figure 12d). The stromatolite is predominantly calcified, but also partially silicified. Silicifications are either limited to single stromatolithic laminae, or occur as a 300 μm thick fringe around the stem (Figure 12e). Figure 12d shows the fringe in radial section, in which microbial filaments still seem to be obvious. In contrast to silicified parts of the stromatolite, calcified parts do not exhibit any microbial structures. Calcite forms aggregates of up to 50 μm large rhombohedral crystals (Figure 13f). In places, calcified parts of the stromatolite contain up to 0.5 mm large ostracods, either occurring in cavities within the stromatolite, or being completely encrusted (Figure 12f). They are preserved with both valves still in occlusion and, therefore, point to life positions. Even if completely encrusted by calcified stromatolite, the ostracods are completely silicified, including their shell. Rarely, relic calcite can be recognized (Figure 12g).

5. Discussion

5.1. Specific Characteristics of the Silicified Wood Localities

5.1.1. Flöha

Petrified wood from Flöha shows a three-phase silicification. After sedimentation, a quick first-phase silicification promoted by hydrothermal fluids, most likely connected to the Schweddey Ignimbrite, affected the wood throughout the whole stem. This process is reflected by randomly distributed blue, yellow, and red CL of the corresponding quartz. Silicification was interrupted or accompanied, respectively, by deposition of the green mineral phase within the wood. Given the abundance of the same green mineral phase in the originally vitreous, nowadays clayish matrix of the Schweddey Ignimbrite [49,50], a clay mineral affiliation can be suggested. The green color points to reduced iron, which is typical for clay minerals of the mica, smectite, and/or chlorite groups.

Clay mineral deposition was accompanied by agate formation, which is reflected by intercalations of green clay minerals and white agate in cavities (Figure 4g). Both represent the second phase of silicification. The third and last phase of silicification affected fractures formed both in the wood and in agate-filled cavities. This last phase was initiated by hydrothermal fluids.

Silicification was preceded by sedimentation and mechanical compaction of the stem due to pith degradation. Compression of the horizontally embedded stem resulted in the formation of cavities, which represent fractures aligned sub-parallel to the direction of compaction. These processes were followed by a multiphase silicification, given variously luminescing quartz in the wood, agate-filled cavities and fractures dissecting the wood (Figures 4 and 5).

The **first phase of silicification** was accomplished by quartz precipitation in the wood. A clear chronological order of the differently luminescencing parts cannot be deduced based on the random distribution of yellow and red CL (Figure 5c,e,f). Additionally, a synchronous formation of blue-, yellow- and red-luminescing quartz should not be excluded. In particular, the short-lived blue CL and the yellow CL are typical for quartz formed from aqueous solution under low-temperature hydrothermal conditions [25,27]. Accordingly, the observed luminescence characteristics reflect quick silicification accompanied with hydrothermal fluids during the first phase of silicification.

Based on the observation that green clay minerals filled the remaining space in the cell lumina left from quartz precipitation (Figure 4e), indicates clay mineral impregnation following on the first phase of silicification. Additionally, based on the occurrence of clay minerals in the former cell walls (Figure 4d,e), the organic framework may still have existed during first-phase silicification. Afterwards, organic components were apparently decomposed and replaced by clay minerals. This process was accompanied and followed by agate formation in cavities, as intercalations of clay minerals within botryoidal agate suggest (Figure 4g). Hence, agate formation represents the **second phase of silicification** and reflects quick precipitation under high silica concentrations [33].

Silicification ceased with the formation of fractures both in the wood and in the agate-filled cavities (**third phase of silicification**; Figure 5e,f). By providing pathways for hydrothermal silica-bearing fluids, these fractures were filled by euhedral quartz (Figure 5f).

5.1.2. Wendishain

Petrified wood from Wendishain displays a two-phase-silicification. After burial by a pyroclastic flow, the embedding ignimbrite provided silica at high concentrations resulting in rapid first silicification of the complete stem. Whereas cell lumina were filled by red-luminescing silica, local oxygen deficiency in the cell walls resulted in the formation of yellow CL. Contemporaneously, or shortly afterwards, agate precipitation in fractures occurred. A second phase promoted by hydrothermal silica-bearing fluids entering the stem via fractures followed. Fractures were filled by coarse-crystalline quartz being almost non-luminescent.

A chronological order of formation of the differently luminescent silica phases cannot be deduced from preservation quality only. The fact that short-lived blue CL in the wood is limited to its outer margin bordering the ignimbrite (zone 1; Figure 7b), or to halos around fractures (Figure 9d), suggests a secondary origin of blue-luminescing silica. Consequently, red and yellow luminescing areas of zone 2 have to be regarded as representing the **first phase of silicification**, whereas short-lived blue CL displays a second phase of silicification.

Yellow and red CL of the silica phases reflect rapid precipitation from supersaturated solutions during the **first phase of silicification** [23,34,68]. This process was accompanied by local oxygen deficiency, as yellow CL of the former cell walls suggests (Figure 7d,f) [34]. Fractures were synchronously, or shortly afterwards filled, by red-luminescing agate (Figure 7f), displaying rapid precipitation from supersaturated solutions [33]. Conditions for first-phase silicification may be provided by alteration of the embedding ignimbrite.

Relict agate-filled fractures in blue-luminescing wood of zone 1, as well as yellow-luminescing cell walls in wood transitional from zone 1 to zone 2 (Figure 7d) indicates that, originally, the stem was completely affected by first-phase silicification. This is supported by an almost equal level of anatomical preservation throughout the stem (Figure 6e,g). **Second-phase silicification** has started at the outer margin of the stem, and penetrated the already silicified tissue afterwards via fractures. This process was accompanied by restructuring of the silica lattice and incorporation of alkali-compensated $[AlO_4/M^+]$ centers responsible for blue CL [25,27]. The observed time-dependent CL shifting from blue to red CL emission is typical for silica formed from hydrothermal solutions [23].

5.1.3. Winnweiler

CL analysis of K6556 from alluvial deposits of the Donnersberg Formation revealed a two-phase silicification, followed by calcification. During the first phase of silicification, intact wood characterized by wider tracheids was silicified rapidly under oxygen deficiency resulting in yellow and red CL. The second (hydrothermal) phase affected still non-silicified parts of the wood, which prior to that have been subjected to advanced decomposition. Agate was precipitated finally in holes left in the wood. Petrifaction finished with calcite formation in some tracheids, and within fractures crossing the wood. Given the composition and low permeability of the embedding epiclastic rocks, alteration of volcanogenic material is regarded as the probable source of the silica at the Winnweiler locality.

According to both preservation types observed in K6556, which also show a different CL, a two-phase silicification can be assumed. A first phase, which was restricted to the inner anatomically well-preserved part of the root (type 1), and a second phase, which affected the anatomically poorly preserved outer rim of the root and surrounding wood fragments (type 2).

The well-preserved tissues in the central part of the root indicate a negligible degree of decomposition during the **first phase of silicification**. Yellow and red CL of the corresponding silica phases (Figure 9c,d) refer to rapid precipitation at high concentrations of silica and oxygen deficiency [34,68]. The spatial separation of both luminescence types—red CL of silica filling cell lumina and fractures, and yellow CL of the former cell walls (Figure 9c)—may indicate an asynchronous formation of both, and thus sub-phases within first-phase silicification. However, which CL—the red or yellow one—was formed first, cannot be answered based on the data observed in K6556.

Widespread plastic deformation of wood cells showing a short-lived blue CL (Figure 9e,f) suggests an advanced state of decomposition during the **second phase of silicification**. Such features would not be preserved in case of an origin of the blue-luminescing silica by replacement or recrystallization of pre-existing silica phases. Consequently, an initially incomplete impregnation or silicification of the wood must be assumed. As silica impregnation of wood depends, amongst other things, on the tracheid diameter [23,69], first-phase silicification of the root's interior could have been favored by its wider tracheids. Short-lived blue CL indicates the origin of the silica from hydrothermal solutions [23,28]. Finally, holes in the wood left from hydrothermal silicification after the second phase were filled by agate showing a red to brown CL (Figure 9f). Red to brown CL is abundant in cryptocrystalline silica [23,33], and points to precipitation from supersaturated solution [33]. Calcification is regarded as final stage of wood petrifaction.

5.1.4. Kyffhäuser

Results point to a two-phase silicification for the Kyffhäuser logs. Silica precipitation was initiated by reducing, acidic conditions within the decomposing wood after burial. However, as initial impregnation of the wood with silica was incomplete, first-phase silicification was limited to single tracheids, where anatomical preservation therefore is best. Wood decomposition was faster than silicification, resulting in gradients of anatomical preservation nowadays known as "pointstone preservation". The model explains abundant vertical compaction of logs in the Kyffhäuser section, which is a consequence of sediment burden of decomposing, slowly silicifying wood in a highly aggrading fluvial system [59]. Silicification was completed by hydrothermal precipitation of quartz showing a short-lived blue CL in empty spaces left from decomposition and fractures in the wood.

CL proves a two-phase silicification of woods from the Kyffhäuser: one phase characterized by yellow CL, and another characterized by short-lived blue CL. However, which phase occurred first is reflected by anatomical preservation. As cell structures are best-preserved within the "points", silicification certainly started there, suggesting that the yellow luminescing areas represent the **first phase**. After initial silicification of single tracheids in the center of the later "points", silicification must have propagated into the surrounding wood tissue. This is supported by gradually decreasing anatomical preservation from the center of the "points" outwardly. Ongoing silicification was accompanied by simultaneous degradation of the still non-petrified wood tissue, mirrored by plastic deformation of cell walls due to probable lignin decomposition. Wood decomposition is in accordance with yellow CL of the corresponding quartz, which—according to [34,68]—is connected to rapid crystallization under conditions of oxygen deficiency and high concentrations of silica. Anyway, both anatomical preservation and yellow CL indicate reducing conditions within the buried wood, which stands in contrast to oxidizing conditions in the embedding hematite-bearing, originally permeable sediments. Reducing conditions were probably accompanied by low pH values, given the fact that oxygen-deficiency favors the formation of organic acids in decomposing wood. A reducing, acidic environment facilitates the precipitation of silica in the wood. Besides SiO_2, uranium was probably mobilized and accumulated within the wood material. The characteristic, up to 50 μm large radiation halos developed around radioactive inclusions within the silica due to radiation damage [70,71]. However, the first phase of silicification ceased from formation of radially aligned quartz blades around the "points". At this time, wood decomposition had progressed enough, that quartz could grow euhedrally and permeate and shear residual wood tissue.

The **second phase** of silicification affected the space in between the "points", as well as fractures crossing the wood. The observed short-lived blue CL is typical for a hydrothermal origin [28,34]. Contrary to the first phase, any wood at places of second-phase silicification has already been decomposed completely prior to quartz crystallization.

5.1.5. Manebach

Sample 1977/7 revealed two types of the occurring silica phases with yellow, and a time-dependent low-intensity CL, respectively. The CL signatures point to silicification from hydrothermal solutions [23,28]. The occurrence of yellow CL in between the other quartz generations indicates changing physico-chemical conditions during silicification (Figure 13c).

However, both luminescence types indicate a silicification by hydrothermal fluids, which must have occurred quite early and rapidly, given the yellow CL and the following reasons: 1. the generally exceptional preservation (Figures 12d and 13a); 2. the organic preservation of tissues [11], including the potential for preserving delicate endophytic fungi [66]. Moreover, silica precipitation is supposed to have been accomplished under stable conditions, which is reflected by several quartz generations in fractures with identical CL (Figure 13c,e), and the overall homogeneous CL found in other samples.

Silicification of the stem followed on from stromatolite formation. This is supported by silica fringes around the silicified stem (Figure 12e), isolated calcite crystals completely enveloped by silica in fractures of the wood (Figure 13e), and the replacement of calcite by silica in ostracod shells (Figure 12g). Additionally, observations by [72] on recent caldera lakes prove the preservation of microbial filaments in siliceously preserved stromatolites as being typical for a secondary silicification. Stromatolite formation must have affected the stem in growth position, as the uniform thickness of the encrustation suggests (Figure 12c). Additionally, lacking bark and pith refer to the stem as being already dead during stromatolite formation. Whether the death of the stem was connected to flooding prior to stromatolite formation cannot be answered with certainty. It should, however, be noted that early conifers of the late Paleozoic preferred dry sites. Accordingly, conifers are comparably rare elements in floral assemblages of the fluvial Manebach Fm. [73].

The taphonomic pathway for 1977/7 can be reconstructed as follows: (1) Death of the tree, accompanied by loss of both the bark and the pith tissues; (2) flooding of the still upright tree and in situ encrustation with stromatolites; (3) rapid silicification of the stem and partial replacement of calcite by silica in the stromatolite; and (4) entombment of the stem. The results point to in situ calcification with subsequent silicification of upright stems in hydrothermal ponds or lakes.

5.2. General CL Characteristics of Silicified Wood

The localities regarded in this study generally suggest that wood silicification can be accomplished as a monophase (Manebach), or multiphase (Flöha, Wendishain, Winnweiler, Kyffhäuser) process. A biphasic silicification appears to be the most common case. These observations are reflected in other studies [29–31] and are already indicated by different silica generations and modifications visible under transmitted light. Silica in petrified wood displays only a small number of CL colors, comprising the composite blue (390 and 440 nm), yellow (580 nm), and red (650 nm) emission bands. Other studies [29–31,74] found a similar range, except for very rare occurrences of green CL [31]. Hence, petrified wood CL is apparently less diverse than luminescences identified in quartz-bearing sedimentary, magmatic, and metamorphic rocks, and hydrothermal veins, in general [23,29]. However, blue CL in the considered localities is restricted to micro- and macrocrystalline quartz, whereas both microcrystalline (quartz) and cryptocrystalline silica varieties (agate, chalcedony) can additionally display yellow and red CL. Similar CL was observed in previous studies [29–31,74].

With respect to petrified woods from the late Paleozoic in Central Europe (this study; [29–31,74]), parts of the wood affected by initial silicification exclusively show yellow and/or red luminescences. Especially the formation of yellow luminescing silica appears to be inevitable in early silicification, given the occurrence of yellow CL in all selected localities and depositional settings. This observation could be connected to the fact that oxygen-deficient organic decay is always present in recently buried wood. Both yellow and red CL point to initial silicification as being rapid and derived from supersaturated silica-bearing fluids under low-thermal (<250 °C) conditions [23,29,34].

Blue luminescing silica in petrified wood is predominantly of secondary origin, i.e., formed by later crystallization in fractures or non-silicified wood (Winnweiler, Flöha, Kyffhäuser, Hawk Mountains/Intrasudetic Basin according to [74]), or replacement of precursing phases (Chemnitz/[29], Wendishain, Flöha). Amongst the investigated localities, Manebach (and nearby localities occurring in the Manebach Formation, e.g., Crock) is exceptional as blue CL forms part of the first phase of silicification.

5.3. Occurrence-Specific CL Characteristics of Silicified Wood

First of all, CL characteristics of petrified wood can be specific either to a locality, or to certain taphonomic circumstances (i.e., host rock/depositional environment). The discrimination of both factors requires a good taphonomic knowledge of all preservation states occurring at a site and in different depositional settings. However, this study, as well as previous studies [29–31,74], indicate characteristic CL spectra according to their host rocks. These differences can be characterized as follows:

1. Petrified wood preserved in pyroclastic rocks (Figure 14):

    (a) multiphase (Figure 14a). First phase with yellow and/or red CL; second phase with blue CL. Both phases do not differ in anatomical preservation. Additionally, the first phase is replaced by, or even interspersed with the second phase. Anatomical preservation ranges from poorly preserved to well-preserved, with moderate to high preservation quality on average. Examples: Wendishain, Flöha (both this study), Chemnitz [29], petrified woods from tuffs intercalated in the žaltman Arkoses/Czech Republic [31].

    (b) monophase (Figure 14d), promoted by short-lived blue CL.

2. Petrified wood preserved in epiclastic rocks (Figure 14): two subtypes:

    (c) multiphase (Figure 14b). First phase with yellow and red CL; second phase with blue CL. Both phases differ remarkably in anatomical preservation, i.e., the second phase is accompanied by poor preservation of the corresponding wood. Anatomical preservation in the same range as (1). Example: Winnweiler (this study).

    (d) monophase (Figure 14e), promoted by time-dependent blue-red CL. CL intensity conspicuously low. Example: petrified woods buried in-situ in epiclastic fluvial deposits of the Triassic Fremouw Formation/Antarctica [31] and Winnweiler.

3. Petrified wood preserved in siliciclastic rocks (Figure 14): two subtypes:

    (e) multiphase (Figure 14c). First phase with yellow CL; second phase with blue CL. Both phases differ remarkably in anatomical preservation, i.e., the second phase is accompanied by poor preservation of the corresponding wood. Anatomical preservation ranges from poorly to moderately preserved, but is generally lower than in (1) and (2). Examples: Kyffhäuser, Wettin, Siebigerode (all this study).

    (f) monophase (Figure 14f), promoted by time-dependent blue-red CL; in places accompanied with yellow CL. CL intensity conspicuously low. Broad range of anatomical preservation, ranging from moderate to high (including permineralizations with preservation of different intracellular fungi, [65,66]). Examples: Manebach, Crock (both this study), Balka/Czech Republic, petrified plants from the štikov Arkoses/Czech Republic and Tocantins/Brazil (the latter three in [31]).

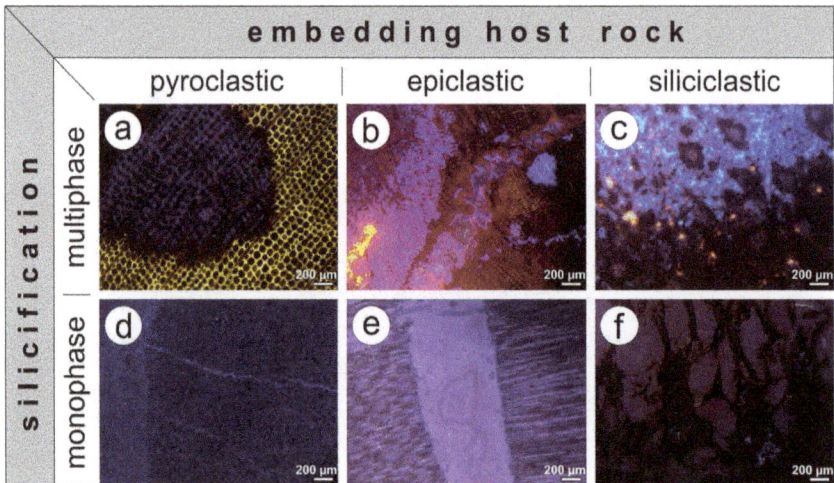

Figure 14. CL types of anatomically preserved (silicified) wood from the late Paleozoic of Central Europe based on [29–31,74], and this study. (**a**) Multiphase-pyroclastic type (K4815 from Chemnitz). Note replacement of primary yellow/red CL by blue CL. (**b**) Multiphase-epiclastic type (K6556 from Winnweiler). Note the reduced anatomical preservation in areas of blue CL. (**c**) Multiphase-siliciclastic type (K2097 from the Kyffhäuser). Note the generally low preservation quality and yellow CL of the initially silicified wood. (**d**) Monophase-pyroclastic type (K6573 from Flöha). (**e**) Monophase-epiclastic type (K6404 from Winnweiler). Note the homogeneous, initially blue CL. (**f**) Monophase-siliciclastic type (K6561 from Crock).

Based on the previous classification, multiphase and monophase silicifications occur in each type of embedding host rock (Figure 14). Accordingly, silicification can be accomplished stepwise (multiphase) or rapidly and completely (monophase) from the beginning.

Focusing on multiphase silicifications, differences with respect to the host rock comprise of (1) the CL colors occurring in the initial phase of silicification, and (2) the timing and process of second-phase silicification. Regarding the first point, red CL is only detected in petrified woods embedded in volcaniclastic deposits (Figures 9c,d and 14a,b), [30,74] described red CL in initially-silicified parts of the wood from the latest Pennsylvanian fluvial žaltman Arkoses of the Intrasudetic Basin. However, [31] specified this occurrence later as being embedded in tuffs (Table 1). Additionally, all finds showing a red CL were made in recent streams and colluvial deposits lacking their exact stratigraphic occurrence. Hence, there is no example to date for initially formed red CL in anatomically preserved plants embedded in siliciclastic rocks.

Concerning the timing of second-phase silicification: In case of woods embedded in epiclastic and siliciclastic rocks, an interruption of silicification is indicated by poor preservation of wood showing blue CL (Figure 7f, 9f and Figure 11f). This break was apparently accompanied by ongoing degradation. Woods embedded in pyroclastic rocks, by contrast, show almost no differences in anatomical preservation between primarily and secondarily silicified parts (Figure 7d,f). Here, a hydrothermal overprint probably occurred in close succession to the initial silicification. Moreover, replacement of yellow and red CL by blue CL proves an originally complete silicification of the wood during the initial phase of silicification. However, with decreasing content of volcanic material in the embedding rocks, anatomical preservation decreases and the time interval between initial and final silicification increases. This relation could be caused by the fact that volcanogenic material is capable of producing high amounts and concentrations of silica within a short time span. Accordingly, silicification rate is enhanced favoring a rapid and complete silicification.

Monophase silicifications can consist of short-lived blue CL or homogeneous, low-intense, time-dependent CL. Both are due to early silicification promoted by hydrothermal fluids. Given the results, this CL type can be found in almost all host rocks and, thus, might be connected to different processes. In the case of Manebach and Crock, silica precipitation accompanied by time-dependent CL was obviously caused by volcanic activity, i.e., silica-bearing springs. Temperatures were low regarding well-preserved tissues of the permineralized stems including their endophytic biota. Litho- and biofacies of the Manebach Formation in general point to environments unfavorable for stromatolite formation, e.g., moderate to high rates of siliciclastic deposition and abundant metazoans in a differentiated ecosystem [64,75–77]. Hence, silica-bearing springs could represent niches for stromatolites. However, the results reflect an early rapid silicification for both Manebach and Crock, which is regarded here as being responsible for the preservation of organic tissues. This conclusion is supported by the latest studies on permineralizations [11].

6. Conclusions

Since 2000, several studies have demonstrated the vital contribution of CL microscopy in elucidating the diverse processes of wood silicification. In particular, CL has become a powerful tool in revealing silicification steps, as well as the physicochemical conditions of silica precipitation based on the visualization of lattice defects in the corresponding silica phases.

Results of the present study emphasize the use of multidisciplinary approaches, but offer the following suggestions for further petrified wood research:

- Interpreting the silicification of anatomically preserved wood requires a well-established knowledge of the geological background and stratigraphy, the mineralogy of the wood-bearing host rocks, and possible diagenetic influences.
- CL features and spectra of corresponding silica phases are indispensable to decipher the pathways of silicification and to reveal primary and secondary processes. Features like corroded detrital quartz in siliciclastic host rocks reflect diagenetic overprints capable of providing and/or dissolving silica. Hence, diagenetic processes might be reflected in the CL properties of petrified wood and can indicate possible secondary hydrothermal silicification.
- By now, CL replacement patterns were only observed in petrified woods embedded in ignimbrites. However, such patterns are missing in epiclastic deposits rich in volcanogenic components. Further research is necessary to verify this observation. A possible factor responsible for replacements during silicification of wood could be the temperature, which can be much higher in ignimbrites than in epiclastic/siliciclastic rocks.

The complexity and variability of CL features in silicified wood detected in this and previous studies emphasize the necessity of identifying CL patterns useful for reconstructing silicification processes. This concern not only refers to comparative studies of several localities from different depositional environments, but also demands for more detailed investigations of each single occurrence. In situ preserved T^0 assemblages, for instance, like the early Permian Chemnitz Fossil Forest, are known to deliver different preservational forms occurring in various stacked host rocks [78]. However, petrified wood CL offers the possibility of distinguishing in which host rock the silicification took place. Steps of silicification can be revealed and described based on specific CL patterns observed in the corresponding silica phases. Additionally, this study has demonstrated that only a combination of paleobotanical and taphonomical features with CL patterns and spectra in petrified wood is crucial for a single occurrence. A site-specific "taphonomic fingerprint" based on CL patterns could break new grounds in petrified wood analysis, e.g., provenance studies of anatomically preserved plants found as reworked clasts in fluvial or glacial deposits.

Author Contributions: Investigation: S.T.; methodology: J.G.; project administration: S.T.; supervision: R.R.; visualization: S.T.; writing—original draft: S.T.; writing—review and editing: R.R. and J.G.

Funding: This research was funded by the Deutsche Forschungsgemeinschaft (DFG grant RO 1273/4-1 to RR).

Acknowledgments: We are grateful to Manfred Barthel, Stephan Schultka, both Berlin, and Robert Noll, Tiefenthal, Frank Löcse, St. Egidien, Ralf Puschmann, Frankenberg, and Jörg W. Schneider, Freiberg, for useful discussion and providing crucial specimens and photos. Special thanks are dedicated to Michael Magnus, Gudrun Geyer, Ronny Ziesemann, and Anja Obst, Freiberg, for thin section preparation. Reinhard Kleeberg, Freiberg, benefited the study with clay mineral analytics. Mathias Merbitz, Chemnitz, Anne Förster, and Marcel Hübner, both Freiberg, supported the preparation of the investigation material. We thank Jason A. Dunlop, Berlin, for linguistic corrections and the anonymous reviewers for their comprehensive help to improve the manuscript.

Conflicts of Interest: The authors declare no conflict of interest.

References

1. Dernbach, U.; Tidwell, W.D.I. *Secrets of Petrified Plants—Fascination from Millions of Years*; D'ORO Publishers: Heppenheim, Germany, 2002; p. 232. ISBN 978-3932181047.

2. Daniels, F.J.; Dayvault, R.D. *Ancient Forests. A Closer Look at Fossil Wood*; Western Colorado Publishing Company: Grand Junction, CO, USA, 2006; p. 450. ISBN 978-0966293814.

3. Schüssler, H.; Simon, T. *Aus Holz Wird Stein. Kieselhölzer aus Dem Keuper Frankens*; Verlag und Offsetdruck Eppe GmbH: Aulendorf, Germany, 2007; p. 192. ISBN 978-3890890913.

4. Niemirowska, A. *Skamieniałe Drewno Drzew Gatunków Lisciastych, Iglastych Oraz Paproci*; Wydawnictwo Poligraf: Brzezia Łaka, Poland, 2013; p. 384. ISBN 978-8378560838.

5. Kerp, H. Organs and tissues of Rhynie chert plants. *Philos. Trans. R. Soc. B* **2017**, *373*. [CrossRef] [PubMed]

6. Akahane, H.; Furuno, T.; Miyama, H.; Yoshikawa, T.; Yamamoto, S. Rapid wood silicification in hot spring water: An explanation of silicification of wood during the Earth's history. *Sediment. Geol.* **2004**, *169*, 219–228. [CrossRef]

7. Rößler, R.; Zierold, T.; Feng, Z.; Kretzschmar, R.; Merbitz, M.; Annacker, V.; Schneider, J.W. A snapshot of an early Permian ecosystem preserved by explosive volcanism: New results from the Chemnitz Petrified Forest, Germany. *Palaios* **2012**, *27*, 814–834. [CrossRef]

8. Philippe, M.; Boonchai, N.; Ferguson, D.K.; Hui, J.; Songtham, W. Giant trees from the Middle Pleistocene of Northern Thailand. *Quat. Sci. Rev.* **2013**, *65*, 1–4. [CrossRef]

9. Mustoe, G.E.; Acosta, M. Origin of petrified wood color. *Geosciences* **2016**, *6*, 25. [CrossRef]

10. Mustoe, G.E. Late Tertiary Petrified Wood from Nevada, USA: Evidence of multiple silicification pathways. *Geosciences* **2015**, *5*, 286–309. [CrossRef]

11. Mustoe, G.E. Wood Petrifaction: A New View of Permineralization and Replacement. *Geosciences* **2017**, *7*, 119. [CrossRef]

12. St. John, R.N. Replacement vs. impregnation in petrified wood. *Econ. Geol.* **1927**, *22*, 729–739. [CrossRef]

13. Buurman, P. Mineralization of fossil wood. *Scr. Geol.* **1972**, *12*, 1–35.

14. Leo, R.F.; Barghoorn, E.S. Silicification of wood. *Harvard Univ. Bot. Mus. Leafl.* **1976**, *25*, 1–47.

15. Sigleo, A.C. Geochemistry of silicified wood and associated sediments, Petrified Forest National Monument, Arizona. *Chem. Geol.* **1979**, *26*, 151–163. [CrossRef]

16. Ballhaus, C.T.; Gee, K.; Greef, K.; Bockrath, C.; Mansfield, T.; Rhede, D. The silicification of trees in volcanic ash: An experimental study. *Geochim. Cosmochim. Acta* **2012**, *84*, 62–74. [CrossRef]

17. Hellawell, J.; Balhaus, C.; Gee, C.T.; Mustoe, G.E.; Nagel, T.J.; Wirth, R.; Rethemeyer, J.; Tomaschek, F.; Geisler, T.; Greef, K.; Mansfeldt, T. Incipient silicification of recent conifer wood at a Yellowstone hot spring. *Geochim. Cosmochim. Acta* **2015**, *149*, 79–87. [CrossRef]

18. Dietrich, D.; Viney, M.; Lampke, T. Petrifactions and wood-templated ceramics: Comparisons between natural and artificial silicification. *IAWA J.* **2015**, *36*, 167–185. [CrossRef]

19. Channing, A.; Edwards, D. Experimental silicification of plants in Yellowstone hot-spring environments. *Trans. R. Soc. Edinb. Earth Sci.* **2004**, *94*, 503–521. [CrossRef]

20. Channing, A.; Edwards, D. Wetland megabias: Ecological and ecophysiological filtering dominates the fossil record of hot spring floras. *Palaeontology* **2013**, *56*, 523–556. [CrossRef]

21. Dietrich, D.; Lampke, T.; Rößler, R. Microstructure study on silicified wood from the Permian Petrified Forest of Chemnitz. *Paläontol. Z.* **2013**, *87*, 397–407. [CrossRef]

22. Sippel, R.F. Sandstone petrology, evidence from luminescence petrography. *J. Sediment. Petrol.* **1968**, *38*, 530–554. [CrossRef]

23. Götze, J.; Plötze, M.; Habermann, D. Origin, spectral characteristics and practical applications of the cathodoluminescence (CL) of quartz: A review. *Mineral. Petrol.* **2001**, *71*, 225–250. [CrossRef]

24. Richter, D.K.; Götte, T.; Götze, J.; Neuser, R.D. Progress in application of cathodoluminescence (CL) in sedimentary petrology. *Mineral. Petrol.* **2003**, *79*, 127–166. [CrossRef]

25. Götze, J. Chemistry, textures and physical properties of quartz—Geological interpretation and technical application. *Mineral. Mag.* **2009**, *73*, 645–671. [CrossRef]

26. Zinkernagel, U. Cathodoluminescence of quartz and its application to sandstone petrology. *Contrib. Sedimentol.* **1978**, *8*, 1–69.

27. Ramseyer, K.; Baumann, J.; Matter, A.; Mullis, J. Cathodoluminescence colours of alpha-quartz. *Mineral. Mag.* **1988**, *52*, 669–677. [CrossRef]

28. Ramseyer, K.; Mullis, J. Factors influencing short-lived blue cathodoluminescence of alpha quartz. *Am. Mineral.* **1990**, *75*, 791–800.

29. Götze, J.; Rößler, R. Kathodolumineszenz-Untersuchungen an Kieselhölzern—I. Silifizierungen aus dem Versteinerten Wald von Chemnitz (Perm, Deutschland). *Veröff. Mus. Naturkunde Chemnitz* **2000**, *23*, 35–50.

30. Matysova, P.; Leichmann, J.; Grygar, T.; Rößler, R. Cathodoluminescence of silicified trunks from the Permo-Carboniferous basins in eastern Bohemia, Czech Republic. *Eur. J. Mineral.* **2008**, *20*, 217–231. [CrossRef]

31. Matysová, P.; Rössler, R.; Götze, J.; Leichmann, J.; Forbes, G.; Taylor, E.L.; Sakala, J.; Grygar, T. Alluvial and volcanic pathways to silicified plant stems (Upper Carboniferous-Triassic) and their taphonomic and palaeoenvironmental meaning. *Palaeogeogr. Palaeoclimatol. Palaeoecol.* **2010**, *292*, 127–143. [CrossRef]

32. Matysová, P.; Booi, M.; Crow, M.C.; Hasibuan, F.; Perdono, A.P.; Van Waveren, I.M.; Donovan, S.K. Burial and preservation of a fossil forest on an early Permian (Asselian) volcano (Merangin River, Sumatra, Indonesia). *Geol. J.* **2017**, *154*, 1–19. [CrossRef]

33. Götze, J.; Plötze, M.; Fuchs, H.; Habermann, D. Defect structure and luminescence behaviour of agate—Results of electron paramagnetic resonance (EPR) and cathodoluminescence (CL) studies. *Mineral. Mag.* **1999**, *63*, 149–163. [CrossRef]

34. Götze, J.; Pan, Y.; Stevens-Kalceff, M.; Kempe, U.; Müller, A. Origin and significance of the yellow cathodoluminescence (CL) of quartz. *Am. Mineral.* **2015**, *100*, 1469–1482. [CrossRef]

35. Rößler, R.; Zierold, T. Back to the roots of palaeobotany—Chemnitz and its palaeontological collection. In *Paleontological Collections of Germany, Austria and Switzerland*; Beck, L.A., Joger, U., Eds.; Springer International Publishing: Cham, Switzerland, 2018; ISBN 978-3-319-77400-8.

36. Schneider, J.W.; Werneburg, R.; Rößler, R.; Voigt, S.; Scholze, F. Example for the description of basins in the CPT Nonmarine-Marine Correlation Chart Thuringian Forest Basin, East Germany. *Permophiles* **2015**, *61*, 29–35.

37. Neuser, R.D.; Bruhn, F.; Götze, J.; Habermann, D.; Richter, D.K. Kathodolumineszenz: Methodik und Anwendung. *Zentral. Geol. Paläontol.* **1995**, *1*, 287–306.

38. DiMichele, W.A.; Aronson, R.B. The Pennsylvanian-Permian vegetational transition: A terrestrial analogue to the onshore-offshore hypothesis. *Evolution* **1992**, *46*, 807–824. [CrossRef] [PubMed]

39. DiMichele, W.A.; Pfefferkorn, H.W.; Gastaldo, R.A. Response of late Carboniferous and early Permian plant communities to climate change. *Annu. Rev. Earth Planet. Sci.* **2001**, *29*, 461–487. [CrossRef]

40. McCann, T.; Pascal, C.; Timmermann, M.J.; Krzywiec, P.; López-Gómez, J.; Wetzel, A.; Krawczyk, C.M.; Rieke, H.; Lamarche, J. Post-Variscan (end Carboniferous–Early Permian) basin evolution in Western and Central Europe. In *European Lithosphere Dynamics*; Gee, G., Stephenson, R.A., Eds.; Geological Society Memoir No. 32; Geological Society: London, UK, 2006; pp. 355–388. ISBN 9781 86239 212 0.

41. Kroner, U.; Romer, R.L. Two plates—Many subduction zones: The Variscan orogeny reconsidered. *Gondwana Res.* **2013**, *24*, 298–329. [CrossRef]

42. Schulmann, K.; Schaltegger, U.; Jezek, J.; Thompson, A.B.; Edel, J.-B. Rapid burial and exhumation during orogeny: Thickening and synconvergent exhumation of thermally weakened and thinned crust (Variscan Orogen in Western Europe). *Am. J. Sci.* **2002**, *302*, 856–879. [CrossRef]

43. Rößler, R.; Barthel, M. Rotliegend taphocoenoses preservation favoured by rhyolithic explosive volcanism. *Freib. Forsch.* **1998**, *C474*, 59–101.

44. Luthardt, L.; Rößler, R.; Schneider, J.W. Palaeoclimatic and site-specific conditions in the early Permian fossil forest of Chemnitz—Sedimentological, geochemical and palaeobotanical evidence. *Palaeogeogr. Palaeoclimatol. Palaeoecol.* **2016**, *441*, 627–652. [CrossRef]

45. Roscher, M.; Schneider, J.W. Permo-Carboniferous climate: Early Pennsylvanian to Late Permian climate development of central Europe in a regional and global context. In *Non-Marine Permian Biostratigraphy and Biochronology*; Lucas, S.G., Cassinis, G., Schneider, J.W., Eds.; Geological Society London Special Publications: London, UK, 2006; Volume 265, pp. 95–136.

46. Schneider, J.W.; Körner, F.; Roscher, M.; Kroner, U. Permian climate development in the northern peri-Tethys area—The Lodève basin, French Massif Central, compared in a European and global context. *Palaeogeogr. Palaeoclimatol. Palaeoecol.* **2006**, *240*, 161–183. [CrossRef]

47. Barthel, M. *Die Rotliegendflora des Thüringer Waldes*; Naturhist. Mus. Schleusingen: Schleusingen, Germany, 2009; p. 170.

48. Schneider, J.W.; Romer, R.L. The late Variscan molasses (late Carboniferous to late Permian) of the Saxo-Thuringian Zone. In *Pre-Mesozoic Geology of Saxo-Thuringia—From the Cadomian Active Margin to the Variscan Orogen*; Linnemann, U., Romer, R.L., Eds.; E. Schweizerbart'sche Verlagsbuchhandlung: Stuttgart, Germany, 2010; pp. 323–346. ISBN 3510652592.

49. Löcse, F.; Meyer, J.; Klein, R.; Linnemann, U.; Weber, J.; Rößler, R. Neue Florenfunde in einem Vulkanit des Oberkarbons von Flöha—Querschnitt durch eine ignimbritische Abkühlungseinheit. *Veröff. Mus. Naturkunde Chemnitz* **2013**, *36*, 85–142.

50. Löcse, F.; Linnemann, U.; Schneider, G.; Annacker, V.; Zierold, T.; Rößler, R. 200 Jahre *Tubicaulis solenites* (Sprengel) Cotta—Sammlungsgeschichte, Paläobotanik und Geologie eines oberkarbonischen Baumfarn-Unikats aus dem Schweddey-Ignimbrit vom Gückelsberg bei Flöha. *Veröff. Mus. Naturkunde Chemnitz* **2015**, *38*, 5–46.

51. Löcse, F.; Zierold, T.; Rößler, R. Provenance and collection history of *Tubicaulis solenites* (Sprengel) Cotta—A unique fossil tree fern and its 200-year journey through the international museum landscape. *J. Hist. Collect.* **2018**, *30*, 241–251. [CrossRef]

52. Gothan, W. Die Altersstellung des Karbons von Flöha i. Sa. im Karbonprofil auf Grund der Flora. *Abh. Sächs. Geol. Landesamt* **1932**, *12*, 5–16.

53. Walter, H.; Rößler, R. Ein großer Kieselholz-Stamm aus dem Rotliegend Sachsens (Kohren-Formation, Nordwestsächsische Senke). *Veröff. Mus. Naturkunde Chemnitz* **2006**, *29*, 177–188.

54. Röllig, G. Zur Petrogenese und Vulkanotektonik der Pyroxenquarzporphyre (Ignimbrite) des Nordsächsischen Vulkanitkomplexes. *Jahrb. Geol.* **1976**, *5/6*, 175–268.

55. Walter, H. Rotliegend im Nordwestsächsischen Becken. In *Stratigraphie von Deutschland X. Rotliegend. Teil I: Innervariscische Becken*; Lützner, H., Kowalczyk, G., Eds.; E. Schweizerbart'sche Verlagsbuchhandlung: Stuttgart, Germany, 2012; pp. 517–529. ISBN 3510492250.

56. Haneke, J.; Lorenz, V.; Stollhofen, H. Donnersberg-Formation. In *Stratigraphie Von Deutschland X. Rotliegend. Teil I: Innervariscische Becken*; Lützner, H., Kowalczyk, G., Eds.; E. Schweizerbart'sche Verlagsbuchhandlung: Stuttgart, Germany, 2012; pp. 254–377. ISBN 3510492250.

57. Lorenz, V.; Haneke, J. Relationship between diatremes, dykes, sills, laccoliths, intrusive-extrusive domes, lava flows, and tephra deposits with unconsolidated water-saturated sediments in the late Variscan intermontane Saar-Nahe Basin, SW Germany. *Geol. Soc. Lond. Spec. Publ.* **2004**, *234*, 75–124. [CrossRef]

58. Charpentier, J.F.W. *Mineralogische Geographie der Chursächsischen Lande Mit Kupfern*; Verlag Siegfried Lebrecht Crusius: Leipzig, Germany, 1778; p. 432.

59. Trümper, S.; Rößler, R. Neues vom Kyffhäuser: Geologische Untersuchungen erhellen die Taphonomie vom Wald zum Holzlagerplatz (Oberkarbon, N-Thüringen). *Geowiss. Mitt.* **2017**, *67*, 20–21.

60. Mägdefrau, K. Die Kieselhölzer im obersten Oberkarbon des Kyffhäusergebirges. *Ber. Deutsch. Bot. Ges.* **1958**, *71*, 133–142.

61. Meister, J. Sedimentpetrographische und lithologische Untersuchungen im Siles des Kyffhäusers. *Hallesches Jahrb. Mitteldtsch. Erdgesch.* **1967**, *9*, 75–92.

62. Schneider, J.W.; Rößler, R.; Gaitzsch, B.G.; Gebhardt, U.; Kampe, A. Saale-Senke. In *Stratigraphie von Deutschland X. Rotliegend. Teil I: Innervariscische Becken*; Lützner, H., Kowalczyk, G., Eds.; E. Schweitzerbart'sche Verlagsbuchhandlung: Stuttgart, Germany, 2012; pp. 447–460. ISBN 3510492250.

63. Lützner, H. Sedimentologie der Manebach-Formation in den fossilführenden Aufschlüssen bei Manebach. *Beitr. Geol. Thüring.* **2001**, *8*, 67–91.

64. Lützner, H.; Andreas, D.; Schneider, J.W.; Voigt, S.; Werneburg, W. Stefan und Rotliegend im Thüringer Wald und seiner Umgebung. In *Stratigraphie von Deutschland X. Rotliegend. Teil I: Innervariscische Becken*; Lützner, H., Kowalczyk, G., Eds.; E. Schweizerbart'sche Verlagsbuchhandlung: Stuttgart, Germany, 2012; pp. 418–487. ISBN 3510492250.

65. Barthel, M.; Krings, M.; Rößler, R. Die schwarzen Psaronien von Manebach, ihre Epiphyten, Parasiten und Pilze. *Semana* **2010**, *25*, 41–60.

66. Krings, M.; Harper, C.J.; White, J.F.; Barthel, M.; Heinrichs, J.; Taylor, E.L.; Taylor, T.N. Fungi in a *Psaronius* root mantle from the Rotliegend (Asselian, Lower Permian) of Thuringia, Germany. *Rev. Palaeobot. Palynol.* **2017**, *239*, 14–30. [CrossRef]

67. Noll, R.; Rößler, R.; Wilde, V. 150 Jahre Dadoxylon. Zur Anatomie fossiler Koniferen- und Cordaitenhölzer aus dem Rotliegend des euramerischen Florengebietes. *Veröffentlichungen des Museums für Naturkunde* **2005**, *28*, 29–48.

68. Heaney, P.J. A proposed mechanism for the growth of chalcedony. *Contrib. Mineral. Petrol.* **1993**, *115*, 66–74. [CrossRef]

69. Götze, J.; Möckel, R.; Langhof, N.; Hengst, M.; Klinger, M. Silicification of wood in the laboratory. *Ceram. Silik.* **2008**, *52*, 268–277.

70. Owen, M.R. Radiation-damage halos in quartz. *Geology* **1988**, *16*, 529–532. [CrossRef]

71. Meunier, J.D.; Sellier, E.; Pagel, M. Radiation-damage rims in quartz from uranium-bearing sandstones. *J. Sediment. Petrol.* **1990**, *60*, 53–58. [CrossRef]

72. Kremer, B.; Kazmierczak, J.; Łukomska-Kowalczyk, M.; Kempe, S. Calcification and silicification: Fossilization potential of cyanobacteria from stromatolites of Niuafo'ou's Caldera Lakes (Tonga) and implications for the early fossil record. *Astrobiology* **2012**, *12*, 535–548. [CrossRef] [PubMed]

73. Barthel, M. Pflanzengruppen und Vegetationseinheiten der Manebach-Formation. *Beitr. Geol. Thüring.* **2001**, *8*, 93–123.

74. Mencl, V.; Matysová, P.; Sakala, J. Silicified wood from the Czech part of the Intra Sudetic Basin (Late Pennsylvanian, Bohemian Massif, Czech Republic): Systematics, silicification, and palaeoenvironment. *Neues Jahrb. Geol. Paläontol. Abh.* **2009**, *253*, 269–288. [CrossRef]

75. Pratt, B.R. Stromatolite decline—A reconsideration. *Geology* **1982**, *10*, 512–515. [CrossRef]

76. Werneburg, R. Ein See-Profil aus dem Unter-Rotliegend des (Unter-Perm) von Manebach (Thüringer Wald). *Veröff. Naturhist. Mus. Schleus.* **1997**, *12*, 63–67.

77. Werneburg, R. Ein Pelycosaurier aus dem Rotliegend des Thüringer Waldes. *Veröff. Naturhist. Mus. Schleus.* **1999**, *14*, 55–58.

78. Luthardt, L.; Hofmann, M.; Linnemann, U.; Gerdes, A.; Marko, L.; Rößler, R. A new U-Pb zircon age and a volcanogenic model for the early Permian Chemnitz Fossil Forest. *Int. J. Earth Sci.* **2018**, *107*, 2465–2489. [CrossRef]

minerals

MDPI

Article

Textural Characteristics of Noncrystalline Silica in Sinters and Quartz Veins: Implications for the Formation of Bonanza Veins in Low-Sulfidation Epithermal Deposits

Tadsuda Taksavasu [1],*, Thomas Monecke [1] and T. James Reynolds [2]

[1]   Center for Mineral Resources Science, Department of Geology and Geological Engineering,
     Colorado School of Mines, 1516 Illinois Street, Golden, CO 80401, USA; tmonecke@mines.edu
[2]   FLUID INC., 1401 Wewatta St. #PH3, Denver, CO 80202, USA; fluidinc@comcast.net
*   Correspondence: ttaksavasu@mines.edu; Tel.: +1-334-587-1917

Received: 25 June 2018; Accepted: 31 July 2018; Published: 2 August 2018

Abstract: Silica sinters forming at the Wairakei geothermal power plant in New Zealand are composed of noncrystalline opal-A that deposited rapidly from cooling geothermal liquids flashed to atmosphere. The sinter is laminated with alternating layers of variably compacted silicified filamentous microbes encased by chains of fused silica microspheres. Microscopic inspection of bonanza quartz vein samples from the Buckskin National low-sulfidation epithermal precious metal deposit in Nevada showed that colloform bands in these veins exhibit relic microsphere textures similar to those observed in the silica sinters from the Wairakei power plant. The textural similarity suggests that the colloform bands were originally composed of noncrystalline opal-A that subsequently recrystallized to quartz. The colloform bands contain dendrites of electrum and naumannite that must have grown in a yielding matrix of silica microspheres deposited at the same time as the ore minerals, implying that the noncrystalline silica exhibited a gel-like behavior. Quartz bands having other textural characteristics in the crustiform veins lack ore minerals. This suggests that ore deposition and the formation of the colloform bands originally composed of compacted microspheres of noncrystalline silica are genetically linked and that ore deposition within the bonanza veins was only episodic. Supersaturation of silica and precious metals leading to the formation of the colloform bands may have occurred in response to transient flashing of the hydrothermal liquids. Flashing of geothermal liquids may thus represent a key mechanism in the formation of bonanza precious metal grades in low-sulfidation epithermal deposits.

Keywords: silica sinter; opal-A; recrystallization; colloform quartz; epithermal gold–silver deposits

---

1. Introduction

Low-sulfidation epithermal deposits are an important source of gold and silver. The deposits form in the shallow (<2 km) subsurface in association with subaerial geothermal systems. Ore formation takes place from near-neutral chloride waters at temperatures below ~300 °C [1–6]. The high-grade ores of many low-sulfidation epithermal deposits are contained in quartz veins that developed along subvertical faults [6–8]. It is well established that there is a close association between fluid immiscibility and mineral deposition in the epithermal environment [2,3,9–14].

The quartz veins in low-sulfidation epithermal deposits exhibit a wide range of textural characteristics [14–17]. Previous works [16,18–20] have proposed that some of the quartz textures encountered in epithermal veins are of secondary origin and formed as a result of recrystallization of a noncrystalline silica precursor phase. In particular, colloform quartz, which is characterized by the presence of continuous bands that are rounded or botryoidal, is invoked to represent a recrystallization texture. Saunders [19,20] showed that gold in low-sulfidation epithermal quartz veins from the Sleeper deposit in Nevada forms dendrites intergrown with fine-grained colloform quartz. He proposed that the gold and silica were originally precipitated as colloidal particles in the deeper part of the system and then mechanically transported upward by the ore-forming hydrothermal fluids. The presence of sedimentary structures suggests that the silica originally deposited in the vein was soft and gel-like. Transformation to microcrystalline quartz may have occurred under hydrothermal conditions immediately after deposition, as suggested by experimental studies [21–24].

This study describes the textural characteristics of silica sinter formed at the Wairakei geothermal power plant in New Zealand using a combination of optical microscopy and scanning electron microscopy. It is shown that the noncrystalline silica sinter is composed of high-porosity opal-A laminae formed by filamentous microbes and alternating low-porosity laminae of densely packed and merged opal-A microspheres. The textural characteristics of the low-porosity laminae in the silica sinter are strikingly similar to those of colloform quartz bands in bonanza ore veins from the Buckskin National low-sulfidation epithermal gold–silver deposit in Nevada. Both silica deposits are composed of silica microspheres and show a similar distribution of micropores between densely packed spheres. The textural observations of the present study lend support to the hypothesis that colloform quartz in low-sulfidation epithermal veins can indeed form through recrystallization of microspheres of a noncrystalline silica precursor deposited during fluid immiscibility.

2. Silica Sinter from the Wairakei Geothermal Power Plant in New Zealand

2.1. Geological Background

The Wairakei geothermal field is located 8 km north of Taupo in the Taupo Volcanic Zone of North Island, New Zealand (Figure 1a). Exploration drilling for geothermal power development at Wairakei commenced in 1949, with development culminating in the commissioning of the Wairakei power plant in 1958, which at the time was only the second commercial geothermal operation in the world and the first to exploit a wet geothermal resource. Historically, Wairakei has been exploited in four production areas (Figure 1b), with today's installed capacity reaching in excess of ~350 MWe. A total of 54 production wells generate hydrothermal fluids at temperatures of up to ~260 °C. About half of the separated water is reinjected. To the southeast, Wairakei connects to the adjacent Tauhara geothermal field, which is currently exploited by a binary power station having an additional capacity of 28 MWe.

The Wairakei geothermal field is located over a broad, deep depression in Jurassic basement graywacke [25] that is filled by Quaternary volcanic and sedimentary rocks [26]. The regional basement structure [27] and basement drilling in other geothermal fields located to the northeast [28] suggest the presence of a northeast-striking, westward-deepening basement graben beneath Wairakei [29]. Active extension in the Taupo Volcanic Zone (2–8 mm/year) and related fault activation may be critical in maintaining fluid pathways in the basement and the overlying volcanic and sedimentary rocks [30–33].

The Jurassic greywacke basement at Wairakei is overlain by the Tahorakuri Formation, which is composed of thick (>650 m) pumiceous lithic tuff with intercalated partially welded ignimbrite (Figure 1c). These deposits are overlain by the ~0.32–0.34-Ma-old, crystal-rich, moderately welded Wairakei Ignimbrite [29,34,35]. The Wairakei Ignimbrite is of variable thickness, with one deep well encountering a thickness of ~1000 m (Figure 1c). The overlying Waiora Formation contains the main production aquifers of the Wairakei geothermal field. It has a minimum thickness of 400 m but reaches up to 2100 m in thickness. The volcanic and sedimentary deposits of the Waiora Formation are host to several large rhyolite units which have high fracture permeability in their brecciated margins (Figure 1c). The top of the Quaternary volcanic and sedimentary succession at Wairakei consists of the Huka Falls Formation and superficial deposits (Figure 1c). The Huka Falls formation comprises up to 300-m-thick lacustrine sediments and water-deposited tuffs that were accumulated in a long-lived shallow lake stretching northeastward for over 50 km from modern Lake Taupo. Superficial deposits include young pyroclastic fall and flows as well as their sedimentary and pedogenetic derivatives [29,34].

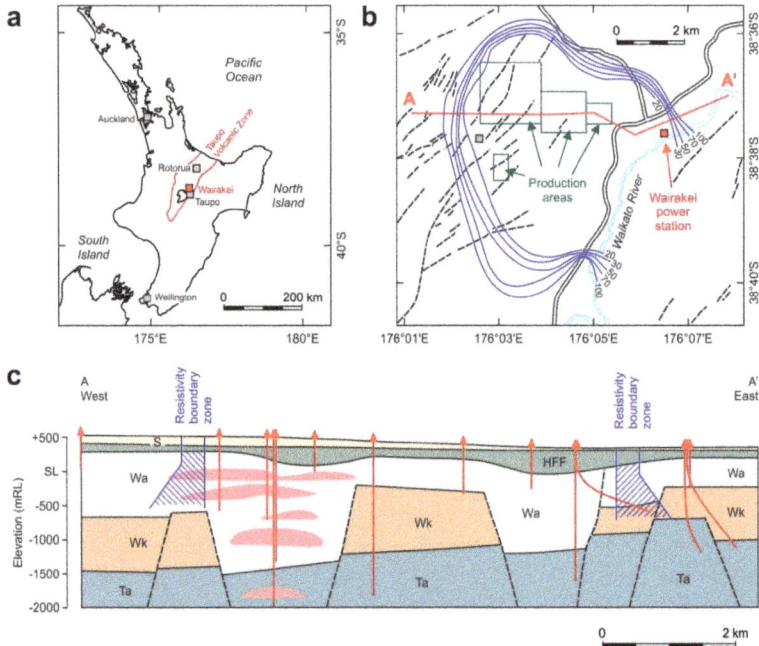

Figure 1. Wairakei geothermal field in New Zealand. (**a**) Map of North Island, New Zealand. The locations of the Taupo Volcanic Zone and the Wairakei geothermal field are highlighted. (**b**) Map of electrical resistivity (in Ωm) in the Wairakei geothermal field (modified from Hunt [36]). (**c**) Geological interpretation of structural elements at the Wairakei geothermal field based on well stratigraphy and seismic reflection data (modified from Rosenberg et al. [34]). The location of the section is given in B. No vertical exaggeration. Ta = Tahorakuri Formation; Wk = Wairakei Ignimbrite; Wa = Waiora Formation; HFF = Huka Falls Formation; S = Surficial deposits; mRL = meters above sea level.

Hydrothermal alteration of the volcanic and sedimentary succession at Wairakei in general increases in rank and intensity with increasing depth [34]. Argillic alteration characterized by the presence of smectite and minor illite/smectite is found in near-surface units. The presence of smectite indicates alteration temperatures <140 °C [37]. Propylitic alteration is the predominant alteration style below the argillic cap. Epidote occurs in small veins but is more common as a pervasive replacement of primary feldspars and other phenocrysts. The presence of wairakite in the propylitic-altered rocks indicates alteration temperatures of above 210 °C [37,38]. The highest rank alteration assemblage recognized at Wairakei includes wairakite, epidote, and prehnite, indicating that alteration occurred at temperatures above 240–280 °C [34].

2.2. Silica Deposits

At the Wairakei geothermal field, silica sinter is formed at a high-temperature (~99 °C) outlet sourced from flash plant 14 and feeds the bathing pool known as Honeymoon Pool. At the outlet, the geothermal liquid is flashed to atmosphere and ponds in a steel-lined concrete pool, referred to as a weirbox, that drains into a small channel. The silica sinter forms several centimeters thick crusts at the bottom and walls of the pool and around the inlet pipe (Figure 2a). Within the channel, tens of centimeters thick silica deposits have formed (Figure 2b). The channel is cleaned out on a regular basis to maintain downhill flow. Representative sampling of the silica sinter formed in the steel-lined concrete pool was conducted in 2017.

Macroscopically, the sampled silica sinter is layered and consists of alternating laminae of 3–12-mm-thick highly porous and friable white silica showing palisade textures and 1–2-mm laminae of nonporous, smooth silica that is slightly gray, vitreous, and translucent. The white silica laminae with palisade textures locally show patchy overprints where the silica is vitreous in hand specimen. The alternating laminae having different textures are wavy on the hand specimen scale and are laterally continuous over tens of centimeters (Figure 2c).

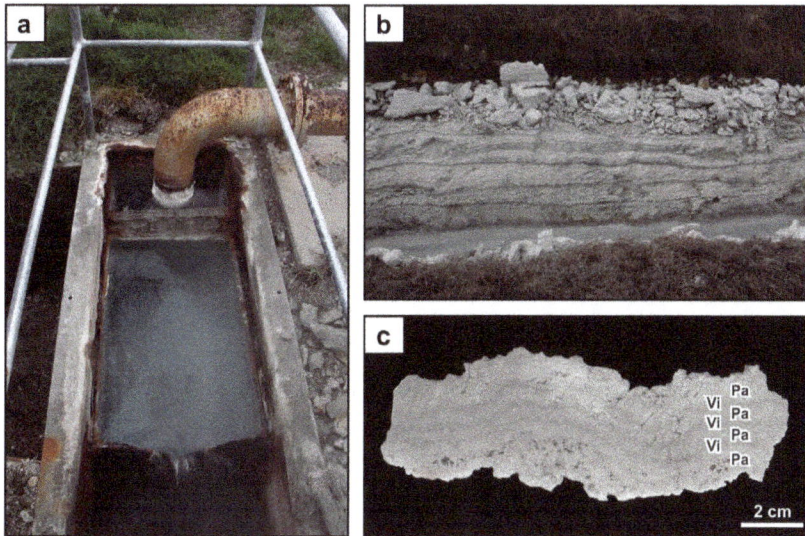

Figure 2. Photographs of silica sinters formed at the Wairakei power plant, New Zealand. (**a**) High-temperature (~99 °C) outlet at flash plant 14. The geothermal liquid is flashed to atmosphere and ponds in a steel-lined concrete pool. (**b**) Drainage channel that is partially filled with silica sinter formed from the cooling geothermal liquid. (**c**) Hand specimen of silica sinter showing alternating laminae of highly porous and friable white silica with palisade textures (Pa) and nonporous, smooth silica laminae that are gray, vitreous, and translucent (Vi).

2.3. X-ray Diffraction Analysis

The sampled silica sinter was cut to be able to better trace the alternating bands of white and gray silica. Using a dentist drill, both types of bands were sampled. The obtained material was powdered and then analyzed by X-ray diffraction analysis. Step-scan XRD data (15–50°2θ, 0.02°2θ step width, 1.0°2θ/min) of the powdered material were obtained using a Scintag XDS-2000 theta/theta diffractometer with a 2.2-kW sealed copper radiation source. An accelerating voltage of 40 kV was used, with a filament current of 40 mA and 0.5 and 0.3 mm of receiving slits.

The XRD patterns of both sinter bands are typified by broad diffraction bands. The highly porous and friable white silica showing palisade textures gave a diffraction band centered on 22.8°2θ (3.90 Å) with a full-width-at-half-maximum value of 4.7 Δ°2θ (0.80 Δd Å). The nonporous, smooth silica shows a diffraction band centered on 21.7°2θ (4.01 Å) with a full-width-at-half-maximum of 5.8 Δ°2θ (1.10 Δd Å). The diffraction experiments indicate that both types of laminae are composed of opal-A.

2.4. Water Content

The water content of the silica sinter from the Wairakei power plant was determined by gravimetry. A powdered aliquot of the silica sinter was dried at 105 °C overnight. The sample experienced a weight loss of 2.46 wt. %, which is interpreted to represent water absorbed by the noncrystalline material. Following further heating at 950 °C for 12 h, the sample experienced an additional weight loss of 5.64 wt. %.

2.5. Textural Characteristics

In thin section, the silica sinter is composed of variably compacted silica laminae. The highly porous and friable white silica laminae showing palisade textures are composed of erect, silicified filamentous microbes that range from 10 to over 300 μm in length and are 2–5 μm in diameter. Complex arrays of subparallel to slightly radiating, connected, and twisted filaments define the palisade texture visible in hand specimen (Figure 3a). Arrays of oriented filaments are connected by randomly oriented filaments forming a spider's web-like texture that exhibit incomplete framework patterns (Figure 3b). Cauliflower-like patterns are locally present. Individual filaments of the filamentous microbes are composed of heavily included, cloudy cores that are surrounded by a shell of fused microspheres of opal-A. The filaments are locally overgrown or cemented by globular aggregates composed of 20–50 individual silica spheres. These spheres are approximately 1–2.5 μm in diameter, with some being <1 μm in size.

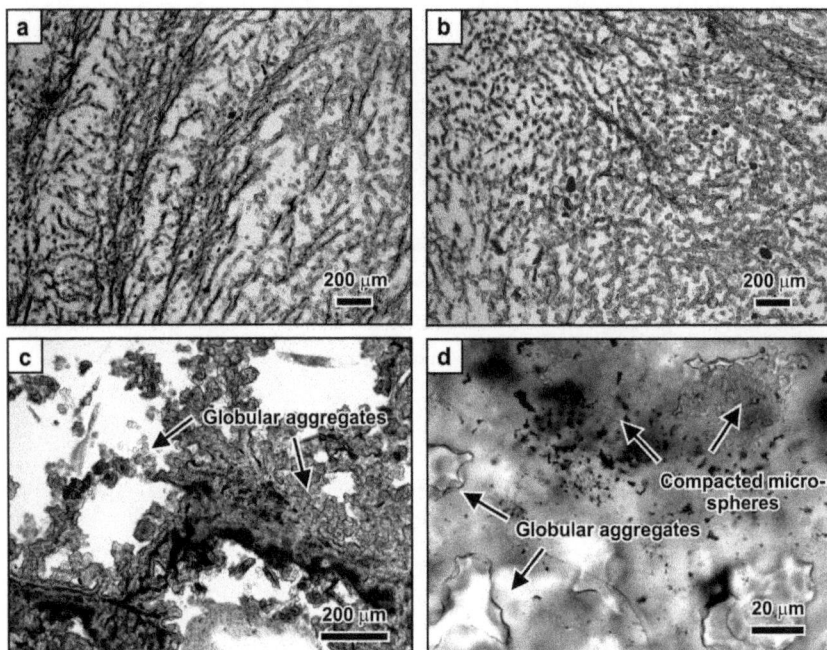

Figure 3. Transmitted light photomicrographs of silica sinter textures from the Wairakei geothermal power plant, New Zealand. (**a**) Palisade texture comprised of arrays of parallel silicified filamentous microbes. (**b**) Spider's web-like texture comprised of arrays of parallel silicified filamentous microbes. (**c**) Low-magnification image showing areas of compacted and noncompacted global aggregates composed of opal-A. (**d**) High-magnification image of compacted microspheres and globular aggregates. The outlines of the microspheres are highlighted in some cases by void space that appears dark.

The nonporous, smooth silica laminae that are gray, vitreous, and translucent in hand specimen are compositionally similar to the highly porous and friable silica laminae. However, the silicified filamentous microbes are densely packed and cemented together by globular aggregates of opal-A microspheres (Figure 3c). Dense vitreous silica layers are present that are composed of closely packed and fused silica spheres. In the vitreous silica layers, only faint outlines of the former opal-A microspheres can be recognized as the void space between the microspheres appears dark in plane-polarized light (Figure 3d). The compacted silica laminae contain small plant fragments and abundant Pinus pollen. The bisaccate pollen grains measure about 80–90 μm, with the pollen grain body measuring approximately 50–60 μm. The Pinus pollen most likely derived from the pine plantations (*Pinus radiata*) surrounding the sampling site at the Wairakei geothermal power plant. The pollen grains are overgrown and cemented by aggregates of microspherical opal-A.

Small pieces of the silica sinter were mounted on aluminum stubs and carbon coated for scanning electron microscopy. Imaging of small-scale textural relationships was conducted using a TESCAN MIRA3 LMH Schottky field-emission scanning electron microscope in secondary electron mode. A working distance of 12 mm and an acceleration voltage of 15.0 kV were used.

Scanning electron microscopy showed that the silicified filamentous microbes are composed of chains of fused silica microspheres. The chains can be subparallel or form spider's web structures as observed in thin section (Figure 4a,b). The filaments are overgrown by individual microspheres or globular groups of microspheres (Figure 4a). High-resolution imaging showed that the spheres in these globular aggregates are connected by small connection pads that consist of silica nanospheres (Figure 4c). These connection pads are visible where a microsphere has separated from the globular aggregate. In areas of low porosity, massive zones of silica occur that consist of fused microspheres. Locally present void space defines the outline of fused microspheres in these areas (Figure 4d).

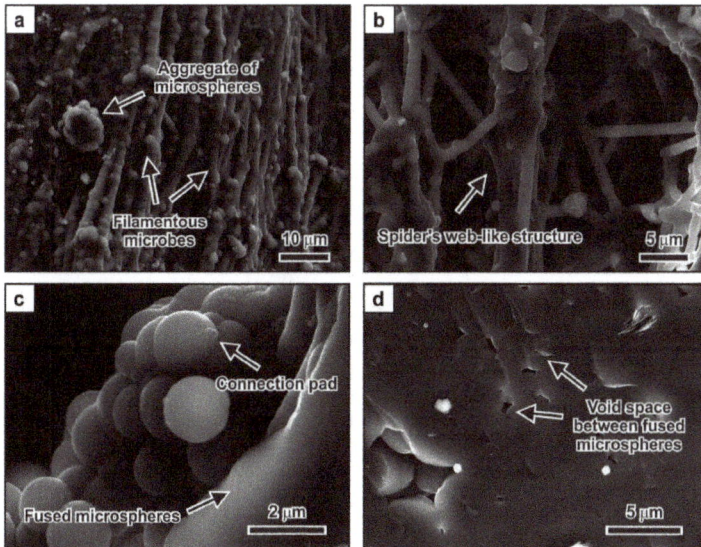

Figure 4. Scanning electron images of silica sinter from the Wairakei geothermal power plant, New Zealand. (**a**) Complex arrays of silicified filamentous microbes consisting of subparallel filaments. The silicified filamentous microbes are overgrown by globular aggregates consisting of a large number of individual silica spheres. (**b**) Spider's web-like structure consisting of silicified filamentous microbes. (**c**) Cluster of silica microspheres. Individual spheres are connected by pads consisting of small silica microspheres. (**d**) Fused silica microspheres. Note the shape of the voids between the fused silica microspheres.

3. Epithermal Veins from the Buckskin National Deposit in Nevada

3.1. Geological Background

The Buckskin National deposit is located at the National district of Humboldt County, Nevada [39–42]. The deposit was mined intermittently from 1906 to 1941, yielding 24,000 ounces of gold and 300,000 ounces of silver from 34,000 tons of ores. The mine was developed on a bonanza-type, low-sulfidation epithermal vein referred to as the Bell vein [40].

Located on the eastern slope of Buckskin Mountain, the Buckskin National deposit is hosted by ~700-m-thick succession of Early Miocene (16.57 ± 0.03 to 16.11 ± 0.03 Ma; [42]) massive rhyolite and associated volcaniclastic facies (Figure 5). The top of Buckskin Mountain is capped by a 30-m-thick carapace of finely laminated silica sinter and silicified epiclastic deposits. The reddish to gray-black silica sinter cropping out in an area that is 420 × 230 m in size contains high Hg concentrations [42,43]. The silicified epiclastic deposits are stratified and are moderately to well-sorted. Individual beds range from several millimeters to centimeters in thickness. Bedding in the epiclastic rocks dips 15–20° to the northeast [42].

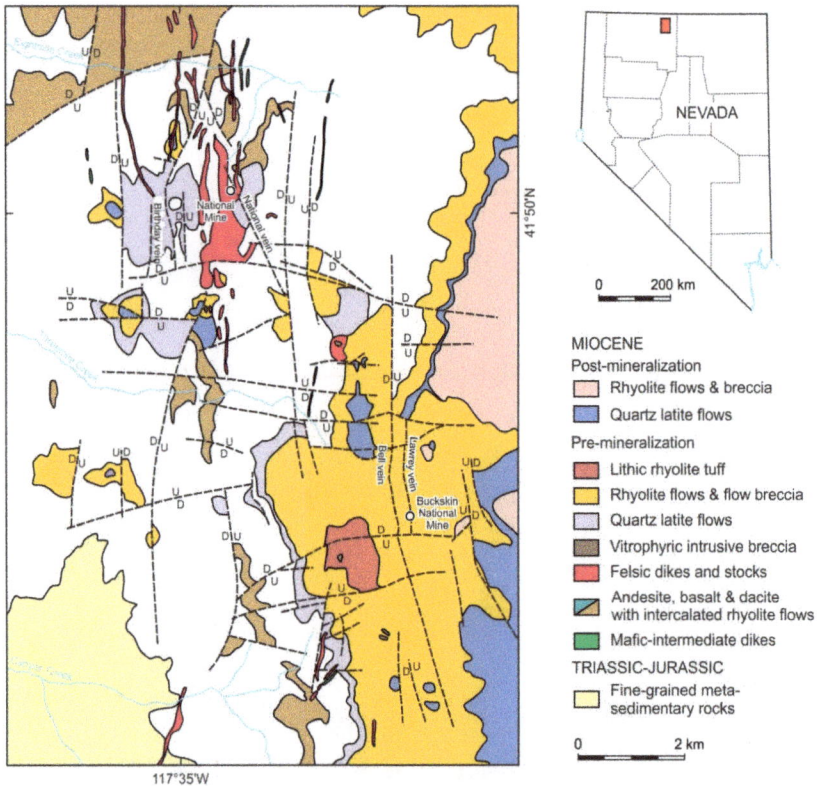

Figure 5. Geological map of the National district, Nevada (modified from Vikre [40]). The location of the Buckskin National deposit is highlighted.

The silica sinter cropping out on Buckskin Mountain represents the surface expression of the hydrothermal system that formed the low-sulfidation epithermal deposit. The Bell vein strikes N-S and dips 75° to the west. The vein has an average thickness of 1.8 m. Mining of the Bell vein has occurred over a strike length of 1.3 km [40]. Based on drilling, the vein is known to extend to a depth of at least ~790 m below the paleosurface [42]. The Bell vein consists primarily of quartz with adularia being the second most abundant gangue mineral. 40Ar/39Ar dating of adularia yielded an age of 16.06 ± 0.03 Ma [42]. Bladed calcite replaced by quartz is locally present. The main ore minerals include electrum, acanthite, miargyrite, pyrargyrite, aguilarite, clausthalite, naumannite, arsenopyrite, galena, pyrite, marcasite, sphalerite, and tetrahedrite. Stibnite is present in the vein within ~150 m below the paleosurface [40,42]. Fluid inclusion studies showed that most of the gold and silver in the Bell vein was precipitated at 200–250 °C [40,42]. Gangue mineral textures suggest that the hydrothermal liquids experienced phase separation during vein formation [40,42].

3.2. Bonanza-Type Quartz Veins

Vein textures were studied in 40 samples archived at the Colorado School of Mines as well as new samples collected from the waste dumps of the former Buckskin National mine in 2016. The vein samples are symmetrically or asymmetrically banded in hand specimen. Most bands are crustiform, consisting of alternating layers of quartz showing subtle differences in textures, colors, and grain sizes as described below. The layers are colorless, milky, or yellowish-cream to grayish-black in color and 1–5 mm in thickness (Figure 6). In addition to the crustiform bands, bands of massive gray quartz or euhedral quartz crystals occur in some samples. The centers of the veins are sutured or vuggy.

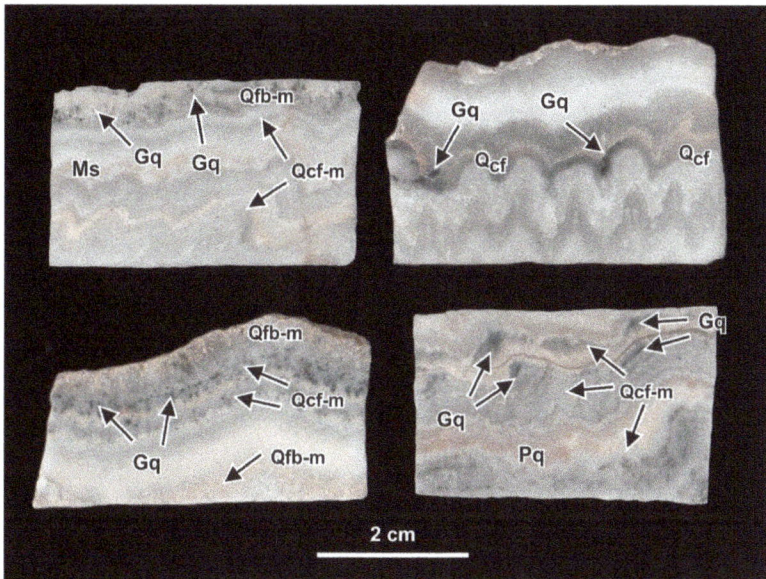

Figure 6. Photographs of crustiform quartz veins from the Buckskin National deposit in Nevada. The samples are characterized by the presence of colloform quartz layers that have spherical, botryoidal, reniform, or mammillary surfaces. Note that ore minerals only occur in some of the colloform bands while layers showing other quartz types are barren. Gq = dark gray quartz that hosts abundant ore minerals; Pq = pink quartz containing abundant adularia that is largely replaced by muscovite; Q_{cf} = colloform quartz; Q_{cf-m} = colloform-mosaic quartz; Q_{fb-m} = quartz showing fibrous or mosaic microtextures.

3.3. X-ray Diffraction Analysis

Using a dentist drill, one of the colloform bands in a bonanza-type quartz vein from the Buckskin National deposit was sampled. The obtained material was powdered using a mortar and pestle and used for X-ray diffraction analysis. Step-scan XRD data (5–60°2θ, 0.02°2θ step width, 1.0°2θ/min) were obtained using a Scintag XDS-2000 theta/theta diffractometer with a 2.2-kW sealed copper radiation source. An accelerating voltage of 40 kV was used, with a filament current of 40 mA and 0.5 and 0.3 mm of receiving slits. Peak-matching revealed that the colloform band is entirely composed of quartz. In addition to quartz, the sample contained small amounts of K-feldspar and muscovite.

3.4. Quartz Textures

Petrographic inspection on thin sections using an Olympus BX51 microscope showed that individual layers in the crustiform bands of the bonanza vein samples exhibit a wide range of textural characteristics. Following Dong et al. [16], the observed quartz textures can be classified as being primary growth textures, recrystallization textures, and replacement textures.

Colloform quartz represents the most common primary growth texture. Colloform bands are most commonly asymmetrical with their spherical, botryoidal, reniform, or mammillary surfaces pointing towards the center of the veins. The layers are laterally continuous and typically 1–5 mm in thickness. Within the colloform bands, the quartz typically appears murky or cloudy and sometimes is almost opaque in thin section (Figure 7a). Optical microscopy at high magnification shows that the dark appearance of the quartz is related to the presence of myriads of micropores (Figure 7b). Under crossed-polarized light, the fine-grained colloform quartz shows a mosaic texture consisting of anhedral quartz grains having irregular and interpenetrating grain boundaries (Figure 7c). Microscopy on ultra-thin (15 μm) sections shows that the dark quartz with the micropores is composed of globular aggregates that are 20–30 μm in size (Figure 7b,d). These globular aggregates are composed of fused microspheres that are 1–3 μm in size. The micropores between these microspheres and between the globular aggregates have sichel-like shapes or are irregular and interconnected with concave boundaries.

In addition to the colloform quartz, chalcedonic quartz occurs in the vein samples. The chalcedonic quartz forms spherical, botryoidal, reniform, or mammillary layers that are texturally not unlike the colloform quartz composed of compacted microspheres. However, chalcedonic quartz is composed of radiating and sheaf-like bundles of microfibers. The chalcedonic layers show a fibrous extinction in crossed-polarized light. Based on the observed interference color, the chalcedony fibers are both length-fast and length-slow. The chalcedonic quartz is transparent in thin section and lacks the abundant micropores that are characteristic of the colloform quartz layers.

Moss and comb quartz represent other common primary growth texture present in the veins from the Buckskin National deposit. The moss quartz consists of groups of spheres that are 0.1–1 mm in size. The moss-like aggregates include radiating and concentric patterns of fine-grained quartz. The moss-like quartz is commonly coated by multiple thin colloform quartz layers. Comb quartz is composed of 0.5–1 mm large euhedral crystals. The quartz crystals form parallel or subparallel clusters and can exhibit radial patterns. Near the center of the veins, the euhedral quartz crystals can project into open space. Layers of comb quartz are commonly overgrown by colloform quartz layers.

The crustiform bands in the vein samples from the Buckskin National deposit contain a range of textures interpreted to be related to the recrystallization or replacement of primary textures. All samples collected contain mosaic quartz, which consists of anhedral grains that have irregular and sutured grain boundaries. The texture can be best identified in crossed-polarized light. The anhedral grains are 0.01–0.1 mm in size. Sericite and kaolinite locally are present along the grain boundaries. Mosaic quartz overprints primary growth textures, including colloform and moss quartz which can be recognized in plane-polarized light despite the recrystallization. Other recrystallization textures identified include feathery and flamboyant quartz. In addition, replacement textures recognized in the quartz veins from the Buckskin National deposit include lattice-bladed, ghost-bladed, parallel-bladed, and pseudo-acicular textures. These textures are interpreted to have formed through complete replacement of calcite by quartz.

Precious metal minerals in the vein samples at the Buckskin National deposit primarily occur within the colloform bands. Most abundant are naumannite dendrites that point toward the center of the veins. Electrum is less abundant and typically occurs as inclusions in naumannite. Pyrargyrite locally forms part of the naumannite dendrites [44–46].

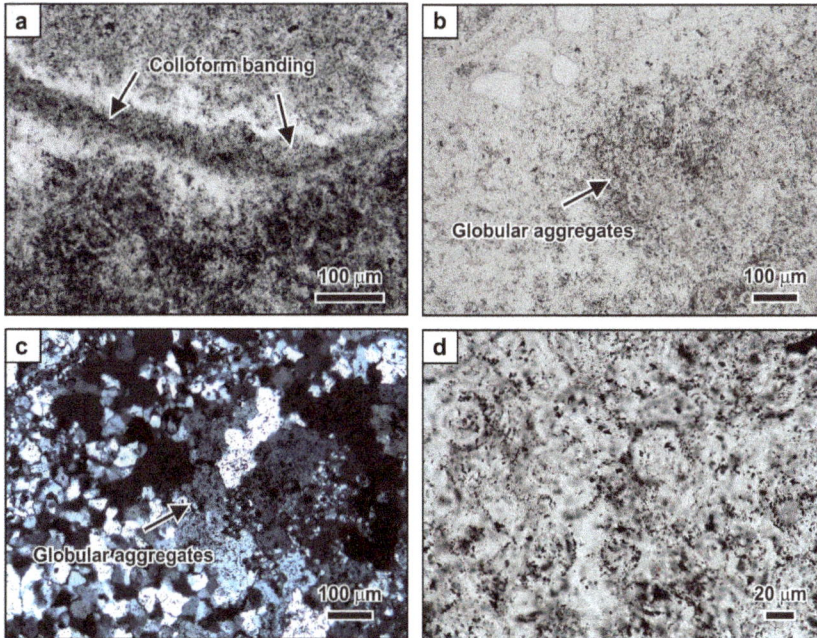

Figure 7. Transmitted-light photomicrographs of colloform banding in bonanza-type quartz veins from the Buckskin National deposit, Nevada. (**a**) Low-magnification image showing a colloform quartz layer. (**b**) Low-magnification image of a colloform quartz layer consisting globular aggregates. Picture was taken on an ultra-thin section. (**c**) Same low-magnification image in crossed-polarized light. The quartz shows a mosaic texture. (**d**) High-magnification image of globular aggregates consisting of fused microspheres. The large number of micropores located between the microspheres gives the quartz a dark appearance.

4. Discussion

4.1. Formation and Recrystallization of Noncrystalline Silica in Modern Sinter Deposits

The sinter studied from the Wairakei geothermal power plant in New Zealand represents a young and highly immature silica deposit that is entirely composed of opal-A. Opal-A represents a noncrystalline hydrated silica phase. Opal-A is noncrystalline as it lacks long-range order [47]. Water is present as absorbed water or forms internal or surface silanol groups [47–49].

Young silica sinters from geothermal areas worldwide consist primarily of opal-A and have similar textural characteristics to those observed in the samples investigated here [50–56]. Smith et al. [57] studied silica deposits sampled from the discharge drain of the Wairakei geothermal power station and the silica sinter terrace of the Orakei Korako geothermal field in New Zealand. Opal-A textures observed included densely packed silica filaments, tangled chains of coalesced spheres forming closely packed mats of silica filaments, and twisted, helical strands of silica. Similar to the present study, opal-A from the Geysir geothermal area in Iceland occurs as featureless low-porosity opal-A laminae that alternate with high-porosity laminae composed of vertical or near-vertical silicified filamentous microbes [58]. The low-porosity laminae were formed of polymerized opal-A microspheres, with porosity being controlled by sphere packing and the amount of opal-A cement. Filamentous microbes were found to have outer mammillary surfaces that are smooth and featureless [58].

Previous studies on silica sinters have shown that the thermodynamically unstable opal-A matures and transforms over time, forming thermodynamically more stable paracrystalline [47] opal-CT, then opal-C, and ultimately blocky, microcrystalline quartz [51,54,55,59–63]. In the case of the Taupo Volcanic Zone in New Zealand, microcrystalline quartz becomes a common phase in silica sinters older than ~20,000 years [51]. Temperature, groundwater interaction, water chemistry, and other environmental factors represent important variables controlling the maturation time of silica sinters [58,64].

The transformation from opal-A to opal-CT is typically accompanied by textural changes [51,54,55,58,59], although the mineralogical changes may outpace textural maturation [63,65]. Opal-CT commonly forms lepispheres that are similar in size to the opal-A microspheres [51,54–56,58,59]. For instance, opal-CT at the Geysir geothermal area in Iceland forms <1-µm lepispheres that are composed of arrays of loosely packed intersecting thin plates or clusters of tightly packed plates. The plates commonly have hexagonal shapes. In addition to lepispheres, opal-CT sometimes also forms complex three-dimensional spindle and barrel frameworks [58]. The transition from opal-A to opal-CT most likely occurs through a dissolution–reprecipitation process [58,62,63].

Continued maturation to opal-C and quartz is commonly associated with another major change in microtextures [51,55,59,62]. At the Roosevelt Hot Springs in Utah, this transition involved a reorganization from blades into elongate, randomly oriented nanorods or blocky aggregates. Diagenetic quartz forms small euhedral crystals oriented parallel to the sinter surface [62]. In addition to opal-C and quartz, moganite can occur in mature sinters [60].

The textural characteristics of opal-CT and opal-C have not been observed in the silica sinter samples investigated from the Wairakei geothermal power plant, confirming that textural maturation of the young deposits has not yet commenced. Smith et al. [57] also showed that silica deposits of up to 2 years in age sampled at the discharge drain of the Wairakei power plant still consist of opal-A, although aging could be demonstrated based on the width of the opal-A band measured in XRD patterns.

4.2. Preservation of Textural Features in Fossil Silica Sinters

Despite textural changes associated with the transition from thermodynamically unstable opal-A to thermodynamically stable quartz with time, fossil sinters preserve a wide range of textural characteristics. Fossil sinters have been recognized throughout geological time, with the oldest being Archean in age [66].

Devonian sinter deposits in Aberdeenshire in Scotland [67] showing massive, vuggy, laminated, lenticular, nodular, and brecciated textures contain abundant silicified plant material. Devonian–Carboniferous sinters of the Drummond Basin in Australia exhibit a wide range of microfacies ranging from high-temperature apparently abiotic geyserites through various forms of stromatolitic sinters to ambient temperature marsh deposits. The sinters contain well-preserved microfossils including cyanobacterial sheaths [68,69]. A detailed study on Jurassic sinters in the Deseado Massif of Argentina showed that these deposits contain small-scale stromatolitic columnar structures, molds of stems and roots of plants, and desiccation cracks [70].

Upper Miocene to Pliocene Waitaia sinter on the east Coromandel Peninsula is the oldest known sinter deposit in New Zealand. The sinter exhibits plant-rich and plant-poor facies as well as detrital-rich vitreous and detrital-rich brecciated facies. In places, the silica sinter is interbedded with swamp deposits. Snails trapped in the sinter have become silicified [55]. The lower to mid-Pliocene Whenuaroa sinter of the Puhipuhi geothermal field in Northland, New Zealand exhibits stromatolitic facies including columnar structures and palisade mats. Quartz and moganite occur as microspheres that represent pseudomorphs of a noncrystalline precursor phase [55,61].

Preservation of delicate macroscopic and microscopic textures in fossil sinters enables their identification in ancient volcanic successions. Information on the presence and location of fossil sinters is used in mineral exploration to locate low-sulfidation epithermal vein deposits and to constrain the location of the paleowater table at the time of mineralization [7,71]. The presence of relic microspherical textures proves that these deposits formed by processes analogous to modern sinters associated with active geothermal systems, which originally involved the precipitation of opal-A that subsequently recrystallized to quartz through intermediate, metastable silica phases such as opal-CT.

4.3. Formation and Recrystallization of Noncrystalline Silica in Epithermal Veins

This study shows that some colloform quartz bands in the vein samples from the Buckskin National deposit in Nevada are composed of mosaic quartz that locally preserves densely packed microspheres texturally resembling those observed in the sinter samples from the Wairakei geothermal power plant in New Zealand. Based on the textural similarity, these microspheres are interpreted to have been originally composed of a noncrystalline silica phase.

This interpretation is consistent with previous works focusing on the ore mineralogy of deposits in the same district. Lindgren [39] described the occurrence of gold at the National deposit and noted that the gold forms elongated rod-like or club-like aggregates that are up to 3 mm in length. The gold aggregates commonly resemble dendrites. He suggested that the gold formed in a yielding medium, implying that the surrounding silica originally was gelatinous mass that slowly crystallized to a fine-grained quartz aggregate subsequent to gold deposition. Lindgren [39] also pointed out that the dendritic nature of the gold must represent a primary growth texture and that the gold and the surrounding silica mass must have been deposited at the same time. Detailed petrographic investigations by Saunders et al. [44,45] and Saunders [46] showed that naumannite in bonanza vein samples from the nearby Buckskin National deposit forms dendritic aggregates pointing towards the center of the veins. The naumannite dendrites probably formed by a similar process as the gold dendrites described by Lindgren [39].

The noncrystalline silica precursor to the colloform bands may have been similar in nature to gel-like silica deposits recovered after a hydrothermal eruption at Porkchop Geyser in Yellowstone [72,73]. Porkchop Geyser was the site of a small eruption in 1989. Ejected blocks were coated by a siliceous gel-like material that was up to 1-cm thick and showed botryoidal textures. Within several days of the eruption, the gelatinous material hardened and became no longer pliable [72,73].

Although the colloform bands in the vein samples from Buckskin National may originally have been composed of noncrystalline opal-A, they today consist entirely of quartz, as confirmed by the XRD experiments. In the vein material investigated, areas showing densely packed microspheres can only locally be recognized through optical microscopy. Under crossed-polarized light, the microspheres are not isotropic, confirming that they are no longer composed of a noncrystalline silica phase. The silica in the vein samples from Buckskin National has fully matured. This maturation process may have involved the formation of intermediate, metastable silica phases such as opal-CT. Investigations by Saunders [19] showed that silica in colloform bands in bonanza-grade samples from the Sleeper deposit in Nevada are virtually isotropic and XRD investigations confirmed the presence of opal-CT. The silica in the vein material from this low-sulfidation epithermal deposit has not been fully transformed to quartz. The degree of maturation may perhaps be related to the evolution of the hydrothermal system following silica deposition. Laboratory studies show that transformation of opal-A to quartz under hydrothermal conditions may occur within days to months [21–24].

In the samples investigated, quartz formed through recrystallization of the noncrystalline precursor phase in the colloform bands is mostly characterized by a mosaic texture, which can be easily recognized under crossed-polarized light. This texture is characterized by highly irregular and interpenetrating grain boundaries. In chert, mosaic textures are known to develop as a result of recrystallization from a noncrystalline silica precursor [74]. A similar origin has been inferred for mosaic quartz in epithermal veins [14,19,20,75].

4.4. Implications for Ore-Forming Processes

The observation that colloform banding exhibiting relic microspheres in the veins at the Buckskin National was originally composed of compacted and merged sphere-like aggregates of a noncrystalline silica precursor phase is in agreement with previous observations by Saunders [19,20] on bonanza-type vein material from the Sleeper deposit in Nevada. Based on careful textural observations, Saunders [19,20] showed that mineralized colloform bands in this deposit formed from coagulated silica. In contrast to this colloform quartz, bands composed of other textural types of quartz are barren. Similarly, Sherlock and Lehrman [76] demonstrated that colloform bands in crustiform veins from the McLaughlin deposit in California consist of compacted microspheres of quartz that formed through recrystallization of a noncrystalline silica precursor phase. The bands contain gold as dendrites or as particles that are concentrated in the interstitial space between the microspheres. At Guanajuato in Mexico, gold grades correlate with quartz vein textures, with the highest gold grades occurring in samples containing abundant colloform bands [14]. At the Koryu deposit in Japan, Shimizu [17] also demonstrated that colloform bands are the main host to precious metal minerals. The evidence available from these deposits collectively suggests that the formation of colloform bands originally composed of noncrystalline silica is directly related to the process of precious metal deposition in epithermal vein deposits.

In epithermal systems, silica supersaturation with respect to quartz leading to the deposition of opal-A can be accomplished as a result of fluid immiscibility [77,78]. Fluid inclusion evidence from Buckskin National in Nevada [40,42], Sleeper in Nevada [79], McLaughlin in California [76,80], Guanajuato in Mexico [14,81], and Koryu in Japan [17] is interpreted to indicate that phase separation of the hydrothermal liquids occurred during vein formation, even though the exact conditions of noncrystalline silica deposition cannot be ascertained due to the lack of primary fluid inclusions in the colloform bands composed of relic microspheres. Additional evidence for the occurrence of boiling at these deposits includes the presence of platy calcite which is replaced by quartz pseudomorphs (cf. [11]).

Moncada et al. [14] suggested that two end-member types of fluid immiscibility can be distinguished in epithermal systems based on the "intensity" of vapor production. During "gentle" boiling, a small proportion of the hydrothermal liquid is converted to vapor as the ascending hydrothermal liquid intersects the liquid plus vapor coexistence boundary. The small amount of vapor produced this way rises slowly through the fracture network. The remaining liquid cools as a result of boiling and continues to rise in the presence of vapor. During "violent" boiling, referred to as flashing, vapor is produced due to near-instantaneous vaporization of a large amount of hydrothermal liquid. This process may occur in response to a seismic event or dike-induced faulting [8]. Propagation of vaporstatic conditions in the fracture will cause any liquid present at depth or within the surrounding wall rock to flash to vapor [82]. Flashing may be associated with the formation of extensive zones of brecciation at depth. Hydrothermal eruption craters may develop at surface [83–86].

Flashing of the hydrothermal liquids would result in the near-instantaneous deposition of silica, as silica solubility in the vapor phase is significantly lower than in the liquid. The colloform bands composed of compacted silica microspheres may record such events of transient fluid flashing as extreme silica supersaturation with respect to quartz could be easily achieved by this process [82]. This conclusion is consistent with the observation that fluid inclusion evidence for gentle boiling can be recognized in a range of different textural types of quartz, many of which are not directly associated with ore minerals [14].

Deposition of noncrystalline silica in the veins may be concomitant with the formation of the precious metal minerals because flashing results in the preferential partitioning of H_2S into the vapor phase, reducing the amount of H_2S in solution in the coexisting liquid [9]. This process of metal deposition is observed in modern geothermal systems where sharp decreases in pressure occur such as on back-pressure plates in surface pipes of geothermal power plants [9]. Sanchez-Alfaro et al. [87] showed that flashing is a more effective mechanism of gold precipitation than gentle boiling.

4.5. Exploration Implications

The textural observations on crustiform quartz vein samples from the Buckskin National deposit in Nevada suggest that the precious metal minerals are primarily contained in colloform bands containing relic silica microspheres. Delicate intergrowth between the ore minerals and the microspherical silica in these bands [45–47] strongly supports the hypothesis that deposition of both were caused by the same process and flashing of the hydrothermal fluids represents the most likely process allowing rapid codeposition. Therefore, textural analysis on epithermal veins could be used to identify hydrothermal systems that underwent transient periods of fluid flashing. These systems are likely to be associated with veins having high gold grades, as fluid flashing represents the most efficient process of gold deposition in the epithermal environment [9,87]. Conversely, hydrothermal systems in which fluid ascent is only accompanied by gentle boiling or cooling (cf. [13,75]) are more likely to form lower-grade precious metal vein deposits.

The conclusion that gold precipitation in bonanza-type low-sulfidation epithermal deposits is linked to fluid flashing also has implications for the depth at which economic ore zones can be expected to occur below the water table. In systems experiencing transient fluid flashing, the depth at which the first gentle boiling occurs does not represent the main control on the location of the ore zone (cf. [12,88]). The ore zone will occur at the depth at which fluid flashing has occurred causing gold supersaturation or, alternatively, at a shallower depth if colloidal precious metals are mechanically transported upward during the flashing event. This has significant implications for the design of drilling programs aimed at finding high-grade ore zones in low-sulfidation epithermal vein systems.

5. Conclusions

Textural comparison between silica sinters from the Wairakei geothermal power plant in New Zealand and bonanza vein samples from the Buckskin National deposit in Nevada revealed that the precipitation of noncrystalline silica in hydrothermal systems can occur over a range of temperatures and pressures. Precipitation in the low-temperature surface environment occurred in response to rapid cooling (<100 °C) of the hydrothermal liquid flashed to atmospheric pressure. Deposition of noncrystalline silica in hydrothermal veins took place as a result of rapid pressure changes causing near-instantaneous vaporization of a large amount of hydrothermal liquid at temperatures of 200–250 °C at several hundred meters below surface at subhydrostatic conditions (<15.5–40 bar). In both cases, rapid deposition inhibited quartz precipitation resulting in a high degree of silica supersaturation with respect to quartz in the hydrothermal fluids.

The observation that mineralized bands in bonanza veins from the Buckskin National deposit in Nevada were originally composed of a noncrystalline silica precursor has significant implications for the understanding of ore-forming processes in the epithermal environment and the design of exploration strategies for low-sulfidation epithermal veins. The results of this study suggest that supersaturation of silica and precious metals only occurred episodically, as other texturally distinct quartz layers in the crustiform veins lack ore minerals. Transient flashing of the hydrothermal liquids, which may be seismically induced, represents a key mechanism in the formation of bonanza grades in the hydrothermal veins. Ores may form at the depth of flashing or closer to the water table, as colloidal precious metals may have been mechanically transported upward during flashing.

Author Contributions: T.M. conducted the sampling at Wairakei. T.T. sampled the quartz vein samples at the Buckskin National deposit. The study was conceived jointly by T.T., T.M., and T.J.R. The microscopic work was conducted by T.T. under supervision of T.M. and T.J.R. The results of the study were jointly discussed by all authors. The manuscript was written by T.T. and T.M. T.J.R. edited an earlier version of the manuscript.

Acknowledgments: We thank Abbie Dean and Fabian Sepulveda of Contact Energy for the opportunity to sample the silica sinter at the Wairakei geothermal power plant and comments that helped us improving an earlier version of this manuscript. We acknowledge Contact Energy for granting permission to publish the research results. We are indebted to James Saunders for help provided during sample collection at the Buckskin National deposit. We thank Richard Wendlandt and Reinhard Kleeberg for performing the X-ray diffraction analyses. Susann Stolze is thanked for identifying the pollen present in the sinter from Wairakei. The manuscript benefited from discussions with James Saunders and Lauren Zeeck on the textural characteristics of epithermal quartz veins. Research on the samples from the Buckskin National deposit was financially supported by the Geological Society of Nevada Elko Chapter.

Conflicts of Interest: The authors declare no conflict of interest.

References

1. Lindgren, W. *Mineral Deposits*, 4th ed.; McGraw-Hill: New York, NY, USA, 1933; p. 930.
2. Buchanan, L.J. Precious metal deposits associated with volcanic environments in the southwest. *Geol. Soc. Ariz. Dig.* **1981**, *14*, 237–262.
3. Bodnar, R.J.; Reynolds, T.J.; Kuehn, C.A. Fluid-inclusion systematics in epithermal systems. *Rev. Econ. Geol.* **1985**, *2*, 73–97.
4. White, N.C.; Hedenquist, J.W. Epithermal gold deposits: Styles, characteristics and exploration. *SEG Newsl.* **1995**, *23*, 9–13.
5. Cooke, D.R.; Simmons, S.F. Characteristics and genesis of epithermal gold deposits. *Rev. Econ. Geol.* **2000**, *13*, 221–244.
6. Simmons, S.F.; White, N.C.; John, D.A. Geological characteristics of epithermal precious and base metal deposits. In *Economic Geology*; Hedenquist, J.W., Thompson, J.F.H., Goldfarb, R.J., Richards, J.P., Eds.; Society of Economic Geologists: Littleton, CO, USA, 2005; Volume 100, pp. 485–522.
7. Hedenquist, J.W.; Arribas, A.; Gonzalez-Urien, E. Exploration for epithermal gold deposits. *Rev. Econ. Geol.* **2000**, *13*, 45–47.
8. Rowland, J.V.; Simmons, S.F. Hydrologic, magmatic, and tectonic controls on hydrothermal flow, Taupo Volcanic Zone, New Zealand: Implications for the formation of epithermal vein deposits. *Econ. Geol.* **2012**, *107*, 427–457. [CrossRef]
9. Brown, K.L. Gold deposition from geothermal discharges in New Zealand. *Econ. Geol.* **1986**, *81*, 979–983. [CrossRef]
10. Clark, J.R.; Williams-Jones, A.E. Analogues of epithermal gold-silver deposition in geothermal well scales. *Nature* **1990**, *346*, 644–645. [CrossRef]
11. Simmons, S.F.; Christenson, B.W. Origins of calcite in a boiling geothermal system. *Am. J. Sci.* **1994**, *294*, 361–400. [CrossRef]
12. Simmons, S.F.; Browne, P.R.L. Hydrothermal minerals and precious metals in the Broadlands-Ohaaki geothermal system: Implications for understanding low-sulfidation epithermal environments. *Econ. Geol.* **2000**, *95*, 971–999. [CrossRef]
13. Albinson, T.; Norman, D.I.; Cole, D.; Chomiak, B. Controls on formation of low sulfidation epithermal deposits in Mexico: Constraints from fluid inclusion and stable isotope data. *SEG Spec. Publ.* **2001**, *8*, 1–32.
14. Moncada, D.; Mutchler, S.; Nieto, A.; Reynolds, T.J.; Rimstidt, J.D.; Bodnar, R.J. Mineral textures and fluid inclusion petrography of the epithermal Ag-Au deposits at Guanajuato, Mexico: Application to exploration. *J. Geochem. Explor.* **2012**, *114*, 20–35. [CrossRef]
15. Bobis, R.E. A review of the description, classification and origin of quartz textures in low sulfidation epithermal veins. *J. Geol. Soc. Philipp.* **1994**, *49*, 15–39.
16. Dong, G.; Morrison, G.; Jaireth, S. Quartz textures in epithermal veins, Queensland—Classification, origin, and implication. *Econ. Geol.* **1995**, *90*, 1841–1856. [CrossRef]
17. Shimizu, T. Reinterpretation of quartz textures in terms of hydrothermal fluid evolution at the Koryu Au-Ag deposit, Japan. *Econ. Geol.* **2014**, *109*, 2051–2065. [CrossRef]
18. Sander, M.V.; Black, J.E. Crystallization and recrystallization of growth-zoned vein quartz crystals from epithermal system—Implications for fluid inclusion studies. *Econ. Geol.* **1988**, *83*, 1052–1060. [CrossRef]
19. Saunders, J.A. Colloidal transport of gold and silica in epithermal precious-metal systems: Evidence from the Sleeper deposit, Nevada. *Geology* **1990**, *18*, 757–760. [CrossRef]
20. Saunders, J.A. Silica and gold textures in bonanza ores of the Sleeper deposit, Humboldt County, Nevada: Evidence for colloids and implications for epithermal ore-forming processes. *Econ. Geol.* **1994**, *89*, 628–638. [CrossRef]
21. Ernst, E.G.; Calvert, S.E. An experimental study of the recrystallization of porcelanite and its bearing on the origin of some bedded cherts. *Am. J. Sci.* **1969**, *267*, 114–133.

22. Mitzutani, S. Silica minerals in the early stage of diagenesis. *Sedimentology* **1970**, *15*, 419–436. [CrossRef]

23. Bettermann, P.; Liebau, F. The transformation of amorphous silica to crystalline silica under hydrothermal conditions. *Contrib. Mineral. Petrol.* **1975**, *53*, 25–36. [CrossRef]

24. Oehler, J.H. Hydrothermal crystallization of silica gel. *Geol. Soc. Am. Bull.* **1976**, *87*, 1143–1152. [CrossRef]

25. Mortimer, N. Origin of the Torlesse Terrane and coeval rocks, North Island, New Zealand. *Int. Geol. Rev.* **1994**, *36*, 891–910. [CrossRef]

26. Rogan, M. A geophysical study of the Taupo Volcanic Zone, New Zealand. *J. Geophys. Res.* **1982**, *87*, 4073–4088. [CrossRef]

27. Rowland, J.V.; Sibson, R.H. Extensional fault kinematics within the Taupo Volcanic Zone, New Zealand: Soft-linked segmentation of a continental rift system. *N. Z. J. Geol. Geophys.* **2001**, *44*, 271–284. [CrossRef]

28. Rae, A.J.; Rosenberg, M.D.; Bignall, G.; Kilgour, G.N.; Milicich, S.D. Geological Results of Production Well Drilling in the Western Steamfield, Ohaaki Geothermal System: 2005–2007. In Proceedings of the 29th New Zealand Geothermal Workshop, Auckland, New Zealand, 19–21 November 2007; University of Auckland: Auckland, New Zealand, 2007; p. 6.

29. Bignall, G.; Milicich, S.; Ramirez, E.; Rosenberg, M.; Kilgour, G.; Rae, A. Geology of the Wairakei-Tauhara Geothermal System, New Zealand. In Proceedings of the World Geothermal Congress, Bali, Indonesia, 25–29 April 2010; International Geothermal Association: Bochum, Germany, 2010; p. 8.

30. Darby, D.J.; Hodgkinson, K.M.; Blick, G.H. Geodetic measurement of deformation in the Taupo Volcanic Zone, New Zealand: The North Taupo network revisited. *N. Z. J. Geol. Geophys.* **2000**, *43*, 157–170. [CrossRef]

31. Villamor, P.; Berryman, K. A late Quaternary extension rate in the Taupo Volcanic Zone, New Zealand, derived from fault slip data. *N. Z. J. Geol. Geophys.* **2001**, *44*, 243–269. [CrossRef]

32. Acocella, V.; Spinks, K.; Cole, J.; Nicol, A. Oblique back arc rifting of Taupo Volcanic Zone, New Zealand. *Tectonics* **2003**, *22*, 1045. [CrossRef]

33. Rowland, J.V.; Sibson, R.H. Structural controls on hydrothermal flow in a segmented rift system, Taupo Volcanic Zone, New Zealand. *Geofluids* **2004**, *4*, 259–283. [CrossRef]

34. Rosenberg, M.D.; Bignall, G.; Rae, A.J. The geological framework of the Wairakei-Tauhara geothermal system, New Zealand. *Geothermics* **2009**, *38*, 72–84. [CrossRef]

35. Houghton, B.F.; Wilson, C.J.N.; McWilliams, M.O.; Lanphere, M.A.; Weaver, S.D.; Briggs, R.M.; Pringle, M.S. Chronology and dynamics of a large silicic magmatic system: Central Taupo Volcanic Zone, New Zealand. *Geology* **1995**, *23*, 13–16. [CrossRef]

36. Hunt, T.M.; Bromley, C.J.; Risk, G.F.; Sherburn, S.; Soengkono, S. Geophysical investigations of the Wairakei field. *Geothermics* **2009**, *38*, 85–97. [CrossRef]

37. Browne, P.R.L.; Ellis, A.J. The Ohaki-Broadlands hydrothermal area, New Zealand: Mineralogy and related geochemistry. *Am. J. Sci.* **1970**, *269*, 97–131. [CrossRef]

38. Steiner, A. The Wairakei geothermal area, North Island, New Zealand: Its subsurface, geology, and hydrothermal rock alteration. *N. Z. Geol. Surv. Bull.* **1977**, *90*, 134.

39. Lindgren, W. Geology and mineral deposits of the National Mining district, Nevada. *USGS Bull.* **1915**, *601*, 58.

40. Vikre, P.G. Precious metal vein systems in the National district, Humboldt County, Nevada. *Econ. Geol.* **1985**, *80*, 360–393. [CrossRef]

41. Vikre, P.G. Paleohydrology of Buckskin Mountain, National district, Humboldt County, Nevada. *Econ. Geol.* **1987**, *82*, 934–950. [CrossRef]

42. Vikre, P.G. Sinter-vein correlations at Buckskin Mountain, National district, Humboldt County, Nevada. *Econ. Geol.* **2007**, *102*, 193–224. [CrossRef]

43. Roberts, R.J. Quicksilver deposit at Buckskin Peak National mining district, Humboldt County, Nevada. *USGS Bull.* **1940**, *922*, 115–133.

44. Saunders, J.A.; Unger, D.L.; Kamenov, G.D.; Fayek, M.; Hames, W.E.; Utterback, W.C. Genesis of Middle Miocene Yellowstone hotspot-related bonanza epithermal Au-Ag deposits, Northern Great Basin, USA. *Miner. Deposita* **2008**, *43*, 715–734. [CrossRef]

45. Saunders, J.A.; Beasley, L.; Vikre, P.; Unger, D. Colloidal and physical transport textures exhibited by electrum and naumannite in bonanza epithermal veins from Western USA, and their significance. In *Great Basin Evolution and Metallogeny, Proceedings of Geological Society of Nevada 2010 Symposium, Reno-Sparks, NV, USA, 14–22 May 2010*; Steininger, R., Pennell, B., Eds.; Geological Society of Nevada: Reno-Sparks, NV, USA, 2011; pp. 825–832.

46. Saunders, J.A. Textural evidence of episodic introduction of metallic nanoparticles into bonanza epithermal ores. *Minerals* **2012**, *2*, 228–243. [CrossRef]

47. Smith, D.K. Opal, cristobalite, and tridymite: Noncrystallinity versus crystallinity, nomenclature of the silica minerals and bibliography. *Powder Diffr.* **1998**, *13*, 2–19. [CrossRef]

48. Jones, B.; Renaut, R.W. Water content of opal-A: Implications for the origin of laminae in geyserite and sinter. *J. Sediment. Res.* **2004**, *74*, 117–128. [CrossRef]

49. Day, R.; Jones, B. Variations in water content in opal-A and opal-CT from Geyser discharge aprons. *J. Sediment. Res.* **2008**, *78*, 301–315. [CrossRef]

50. Jones, B.; Renaut, R.W.; Rosen, M.R. Biogenicity of silica precipitation around geysers and hot-spring vents, North Island, New Zealand. *J. Sediment. Res.* **1997**, *67*, 88–104.

51. Herdianita, N.R.; Browne, P.R.L.; Rodgers, K.A.; Campbell, K.A. Mineralogical and textural changes accompanying ageing of silica sinter. *Miner. Deposita* **2000**, *35*, 48–62. [CrossRef]

52. Campbell, K.A.; Rodgers, K.A.; Brotheridge, J.M.A.; Browne, P.R.L. An unusual modern silica-carbonate sinter from Pavlova spring, Ngatamariki, New Zealand. *Sedimentology* **2002**, *49*, 835–854. [CrossRef]

53. Guidry, S.A.; Chafetz, H.S. Anatomy of siliceous hot springs: Examples from Yellowstone National Park, Wyoming, USA. *Sediment. Geol.* **2003**, *157*, 71–106. [CrossRef]

54. Lynne, B.Y.; Campbell, K.A. Morphologic and mineralogic transitions from opal-A to opal-CT in low-temperature siliceous sinter diagenesis, Taupo Volcanic Zone, New Zealand. *J. Sediment. Res.* **2004**, *74*, 561–579. [CrossRef]

55. Rodgers, K.A.; Browne, P.R.L.; Buddle, T.F.; Cook, K.L.; Greatrex, R.A.; Hampton, W.A.; Herdianita, N.R.; Holland, G.R.; Lynne, B.Y.; Martin, R.; et al. Silica phases in sinters and residues from geothermal fields of New Zealand. *Earth Sci. Rev.* **2004**, *66*, 1–61. [CrossRef]

56. Fernandez-Turiel, J.L.; Garcia-Valles, M.; Gimeno-Torrente, D.; Saavedra-Alonso, J.; Martinez-Manent, S. The hot spring and geyser sinters of El Tatio, northern Chile. *Sediment. Geol.* **2005**, *180*, 125–147. [CrossRef]

57. Smith, B.Y.; Turner, S.J.; Rodgers, K.A. Opal-A and associated microbes from Wairakei, New Zealand: The first 300 days. *Mineral. Mag.* **2003**, *67*, 563–579. [CrossRef]

58. Jones, B.; Renaut, R.W. Microstructural changes accompanying the opal-A to opal-CT transition: New evidence from the siliceous sinters of Geysir, Haukadalur, Iceland. *Sedimentology* **2007**, *54*, 921–948. [CrossRef]

59. Campbell, K.A.; Sannazzaro, K.; Rodgers, K.A.; Herdianita, N.R.; Browne, P.R.L. Sedimentary facies and mineralogy of the late Pleistocene Umukuri silica sinter, Taupo Volcanic Zone, New Zealand. *J. Sediment. Res.* **2001**, *71*, 727–746. [CrossRef]

60. Rodgers, K.A.; Cressey, G. The occurrence, detection and significance of moganite (SiO_2) among some silica sinters. *Mineral. Mag.* **2001**, *65*, 157–167. [CrossRef]

61. Rodgers, K.A.; Hampton, W.A. Laser Raman identification of silica phases comprising microtextural components of sinters. *Mineral. Mag.* **2003**, *67*, 1–13. [CrossRef]

62. Lynne, B.Y.; Campbell, K.A.; Moore, J.N.; Browne, P.R.L. Diagenesis of 1900-year-old siliceous sinter (opal-A to quartz) at Opal Mound, Roosevelt Hot Springs, Utah, U.S.A. *Sediment. Geol.* **2005**, *179*, 249–278. [CrossRef]

63. Lynne, B.Y.; Cambell, K.A.; James, B.J.; Browne, P.R.L.; Moore, J. Tracking crystallinity in siliceous hot-spring deposits. *Am. J. Sci.* **2007**, *307*, 612–641. [CrossRef]

64. Lynne, B.Y.; Cambell, K.A.; Perry, R.S.; Browne, P.R.L.; Moore, J.N. Acceleration of sinter diagenesis in an active fumarole, Taupo Volcanic Zone, New Zealand. *Geology* **2006**, *34*, 749–752. [CrossRef]

65. Liesegang, M.; Milke, R.; Berthold, C. Amorphous silica maturation in chemically weathered clastic sediments. *Sediment. Geol.* **2018**, *365*, 54–61. [CrossRef]

66. Djokic, T.; Van, K.; Martin, J.; Campbell, K.A.; Walter, M.R.; Ward, C.R. Earliest signs of life on land preserved in ca. 3.5 Ga hot spring deposits. *Nat. Commun.* **2017**, *8*, 15263. [CrossRef] [PubMed]

67. Trewin, N.H. Depositional environment and preservation of biota in the Lower Devonian hot-springs of Rhynie, Aberdeenshire, Scotland. *Trans. R. Soc. Edinb. Earth. Sci.* **1993**, *84*, 433–442. [CrossRef]

68. White, N.C.; Wood, D.G.; Lee, M.C. Epithermal sinters of Paleozoic age in north Queensland, Australia. *Geology* **1989**, *17*, 718–722. [CrossRef]

69. Walter, M.R.; Desmarais, D.; Farmer, J.D.; Hinman, N.W. Lithofacies and biofacies of mid-Paleozoic thermal spring deposits in the Drummond Basin, Queensland, Australia. *Palaios* **1996**, *11*, 497–518. [CrossRef] [PubMed]

70. Guido, D.; de Barrio, R.; Schalamuk, I. La Marciana Jurassic sinter—Implications for exploration for epithermal precious-metal deposits in Deseado Massif, southern Patagonia, Argentina. *Appl. Earth Sci.* **2002**, *111*, 106–113. [CrossRef]

71. Sillitoe, R.H. Epithermal paleosurfaces. *Miner. Deposita* **2015**, *50*, 767–793. [CrossRef]

72. Fournier, R.O.; Thompson, J.M.; Cunningham, C.G.; Hutchinson, R.A. Conditions leading to a recent small hydrothermal explosion at Yellowstone National Park. *Geol. Soc. Am. Bull.* **1991**, *103*, 1114–1120. [CrossRef]

73. Keith, T.E.C. 1992 A look at silica phases in evolving hydrothermal systems. In *Water-Rock Interaction, Proceedings of the 7th International Symposium on Water-Rock Interaction, Park City, UT, USA, 13–18 July 1992*; Kharaka, Y.K., Maest, A.S., Eds.; Balkema: Rotterdam, The Netherlands, 1992; pp. 1423–1426.

74. Lovering, T.G. *Jasperoid in the United States—Its Characteristics, Origin, and Economic Significance*; USGS Prof. Prep; United States Government Publishing Office: Washington, DC, USA, 1972; Volume 710, p. 164.

75. Camprubí, A.; Albinson, T. Epithermal deposits in México—Update of current knowledge, and an empirical reclassification. *Geol. Soc. Am. Spec. Pap.* **2007**, *422*, 377–415.

76. Sherlock, R.L.; Lehrman, N.J. Occurrences of dendritic gold at the McLaughlin mine hot-spring gold deposit. *Miner. Deposita* **1995**, *30*, 323–327. [CrossRef]

77. Drummond, S.E.; Ohmoto, H. Chemical evolution and mineral deposition in boiling hydrothermal systems. *Econ. Geol.* **1985**, *80*, 126–147. [CrossRef]

78. Fournier, R.O. The behavior of silica in hydrothermal solutions. *Rev. Econ. Geol.* **1985**, *2*, 45–61.

79. Saunders, J.A.; Schoenly, P.A. Boiling, colloid nucleation and aggregation, and the genesis of bonanza Au-Ag ores of the Sleeper deposit, Nevada. *Miner. Deposita* **1995**, *30*, 199–210. [CrossRef]

80. Sherlock, R.L.; Tosdal, R.M.; Lehrman, N.J.; Graney, J.R.; Losh, S.; Jowett, E.C.; Kesler, S.E. Origin of the McLaughlin mine sheeted vein complex: Metal zoning, fluid inclusion, and isotopic evidence. *Econ. Geol.* **1995**, *90*, 2156–2181. [CrossRef]

81. Buchanan, L.J. The Las Torres Mine, Guanajuato, Mexico: Ore Controls of a Fossil Geothermal System. Ph.D. Dissertation, Colorado School of Mines, Golden, CO, USA, 1979.

82. Henley, R.W.; Hughes, G.O. Underground fumaroles; "Excess heat" effects in vein formation. *Econ. Geol.* **2000**, *95*, 453–466. [CrossRef]

83. Muffler, L.J.P.; White, D.E.; Truesdell, A.H. Hydrothermal explosion craters in Yellowstone National Park. *Geol. Soc. Am. Bull.* **1971**, *82*, 723–740. [CrossRef]

84. Hedenquist, J.W.; Henley, R.W. Hydrothermal eruptions in the Waiotapu geothermal system, New Zealand: Their origin, associated breccias, and relation to precious metal mineralization. *Econ. Geol.* **1985**, *80*, 1640–1668. [CrossRef]

85. Browne, P.R.L.; Lawless, J.V. Characteristics of hydrothermal eruptions, with examples from New Zealand and elsewhere. *Earth Sci. Rev.* **2001**, *52*, 299–331. [CrossRef]

86. Monecke, T.; Petersen, S.; Hannington, M.D.; Anzidei, M.; Esposito, A.; Giordano, G.; Garbe-Schönberg, D.; Augustin, N.; Melchert, B.; Hocking, M. Explosion craters associated with shallow submarine gas venting off Panarea island, Italy. *Bull. Volcanol.* **2012**, *74*, 1937–1944. [CrossRef]

87. Sanchez-Alfaro, P.; Reich, M.; Driesner, T.; Cembrano, J.; Arancibia, G.; Pérez-Flores, P.; Heinrich, C.A.; Rowland, J.; Tardani, D.; Lange, D.; et al. The optimal windows for seismically-enhanced gold precipitation in the epithermal environment. *Ore Geol. Rev.* **2016**, *79*, 463–473. [CrossRef]

88. Cline, J.S.; Bodnar, R.J.; Rimstidt, J.D. Numerical simulation of fluid flow and silica transport and deposition in boiling hydrothermal solutions: Application to epithermal gold deposits. *J. Geophys. Res.* **1992**, *97*, 9085–9103. [CrossRef]

minerals MDPI

Article

Multi-Stage Evolution of Gold-Bearing Hydrothermal Quartz Veins at the Mokrsko Gold Deposit (Czech Republic) Based on Cathodoluminescence, Spectroscopic, and Trace Elements Analyses

Vojtěch Wertich [1],*, Jaromír Leichmann [1], Marek Dosbaba [2] and Jens Götze [3]

[1] Department of Geological Sciences, Masaryk University, Kotlářská267/2, 611 37 Brno, Czech Republic; leichman@sci.muni.cz

[2] TESCAN ORSAY HOLDING, a.s., Libušina tř. 21, 623 00 Brno, Czech Republic; marek.dosbaba@tescan.com

[3] Institute of Mineralogy, TU Bergakademie Freiberg, Brennhausgasse 14, 09599 Freiberg, Germany; jens.goetze@mineral.tu-freiberg.de

* Correspondence: wertich@sci.muni.cz; Tel.: +42-073-267-9084

Received: 13 June 2018; Accepted: 2 August 2018; Published: 4 August 2018

Abstract: We performed a detailed analysis of hydrothermal quartz at the Mokrsko gold deposit (Čelina, Mokrsko-East, and Mokrsko-West deposits). Twenty-one samples were studied by scanning electron microscopy cathodoluminescence (CL) imagining, CL emission spectra and trace elements were measured on six selected samples. Four quartz growth generations Q1 to Q4 were described. Homogeneous early blue CL Q1 with initial emission spectra at 380 and 500 nm was observed at the Čelina deposit with typical titanium concentrations in the range of 20–50 ppm. Hydrothermal quartz at Mokrsko-West, which also includes early Q1, late subhedral faces of yellow CL Q2, and microfissures of greenish CL Q3 (both 570 nm), is characterized by titanium depletion. The titanium concentration is comparable to previous studies of crystallization temperatures proving titanium concentration in quartz as a good geothermal indicator. Q4, developed in microfissures only at Čelina, has no visual CL effect. Mokrsko-West is specific in comparison to Mokrsko-East and Čelina by germanium enrichments in hydrothermal quartz (up to 17 ppm) and the presence of fluorite. Tectonic (sheeted veinlets system, regional tectonic setting) and geochemical (germanium in quartz, the presence of fluorite) characteristics of the quartz veins link the late mineralization stages at the Mokrsko-West deposit to the temporally related Blatná intrusive suite.

Keywords: hydrothermal quartz; gold deposit; cathodoluminescence (CL); trace elements; titanium; germanium; intrusion-related

1. Introduction

Quartz is a very common gangue mineral in hydrothermal metallic ore veins [1]. Though gangue minerals are defined as "non-valuable" [2], quartz can preserve information about physicochemical conditions of a vein genesis that can be valuable in order to understand mineral deposit formations [3].

Despite the rather simple structure of hydrothermal quartz (SiO_4 tetrahedrons), it often contains point defects, line defects, and three-dimensional defects. Such structural defects reflect changes in conditions during different geological processes such as primary crystallization, metamorphism, changes in crystallization temperatures, alterations, secondary dissolution-precipitation, etc. [1,3–6].

Defects in quartz crystal structure are considered activators of cathodoluminescence (CL) [7,8]. CL phenomena can be described as the emission of photons of characteristic wavelengths evoked by "bombarding" the material with an accelerated beam of high-energy electrons [7–9]. These emissions can be captured by a CL detector, providing an image of the structure of the analysed material unobservable by any other method. When CL is combined with other analytical methods such as, e.g., the analysis of chemical composition, stable isotopes, or fluid inclusions, the obtained data can clarify issues of the material's origin [10–12]. At least five characteristic emission bands of different wavelengths in quartz and their activation centres have been described [13–16].

According to Rusk [3], hydrothermal quartz displays more variable CL textures than other types of quartz due to many different geological processes that occur during the genesis of hydrothermal veins and mineral deposits. Thanks to CL, it is possible to detect and study different generations of hydrothermal quartz and their mutual relationship or relation to ore mineralization. Thus, CL analyses are frequently utilized in mineral deposit studies [17–20].

Mokrsko, with its ~100 t of gold resources (~3.2 Moz), ranks among the largest gold deposits in Europe [21]. Despite extensive exploration [21,22] and several scientific studies [23–25], the origin of the deposit is still not fully understood. There are several types of gold-bearing hydrothermal quartz veins crystallized under different temperature and pressure conditions [22,23].

This paper aims to identify individual quartz growth generations, its succession, and mutual relationship using a high-resolution CL imaging of up to 1 cm diagonally large sectors of quartz veins accompanied by the analysis of CL emission spectra. The geochemistry of trace elements in hydrothermal quartz veins was also studied by low detection limit laser ablation inductively coupled plasma mass spectrometry (LA-ICP-MS) analyses. Detailed mapping of CL textures of hydrothermal quartz and its trace elements chemistry will provide insight into the structure and composition of veins at the gold deposit. Such information will help to better understand the quartz vein genesis at the Mokrsko gold deposit.

2. Mokrsko Gold Deposit

2.1. Geological Setting

The main gold metallogenetic phases in the Bohemian Massif, including the formation of the Mokrsko gold deposit, were temporally and spatially connected with the Variscan orogeny [26,27], which formed the main tectonic units in Western and Central Europe during the Carboniferous time (Figure 1, [28]). Many of the gold deposits in the Bohemian Massif were formed in conjunction with a large granitoid mass of the Variscan Central Bohemian Plutonic Complex (~360–330 Ma [29,30]).

Two of three main ore bodies of the Mokrsko gold deposit, the Čelina and the Mokrsko-East deposits (Figure 1), are hosted by volcanic rocks of acidic to basic composition belonging to the Jílové Belt, part of the Neoproterozoic Teplá–Barrandian Unit [31,32]. Toward the west, the Jílové Belt was affected by contact metamorphism caused by the emplacement of the Sázava suite. The Variscan Sázava suite (Figure 1) is composed mostly of calc-alkaline biotite-amphibole tonalite, granodiorite, and quartz diorite. This contact zone hosts the largest ore body of the ore district—the Mokrsko-West deposit (Figure 1). U–Pb zircon radiometric dating of the Sázava suite gave 354.1 ± 3.5 Ma [30]. The other Variscan intrusive suite in the vicinity of the ore district is the Blatná suite, the largest suite of the Central Bohemian Plutonic Complex. The Blatná suite is characterised by high-potassic calc-alkaline biotite granodiorite and granite [33]. The Blatná suite is slightly younger then the Sázava suite; U–Pb zircon radiometric dating has yielded crystallization ages from 340 to 350 Ma [29,33,34]. Re–Os radiometric dating of molybdenite from hydrothermal quartz veins in the gold deposit gave ages from 340 to 345 Ma [27,35].

Both intrusive suites are distinct in regional tectonic setting as well as in magma origin. The Sázava suite is a geochemically rather primitive intrusion with magma originated from a depleted mantle wedge [30,33] that was emplaced in the upper crustal Teplá–Barrandian Unit during regional shortening/transpression [36]. Differently, the Blatná suite represents a geochemically more evolved intrusion that originated from the mixing between acidic (granodioritic) magma and a basic component derived from a moderately enriched mantle source [33]. The tectonic setting is also different with the

Blatná suite, indicating emplacement during the switch from transpression to the onset of exhumation of the middle to lower-crustal Moldanubian Unit [37].

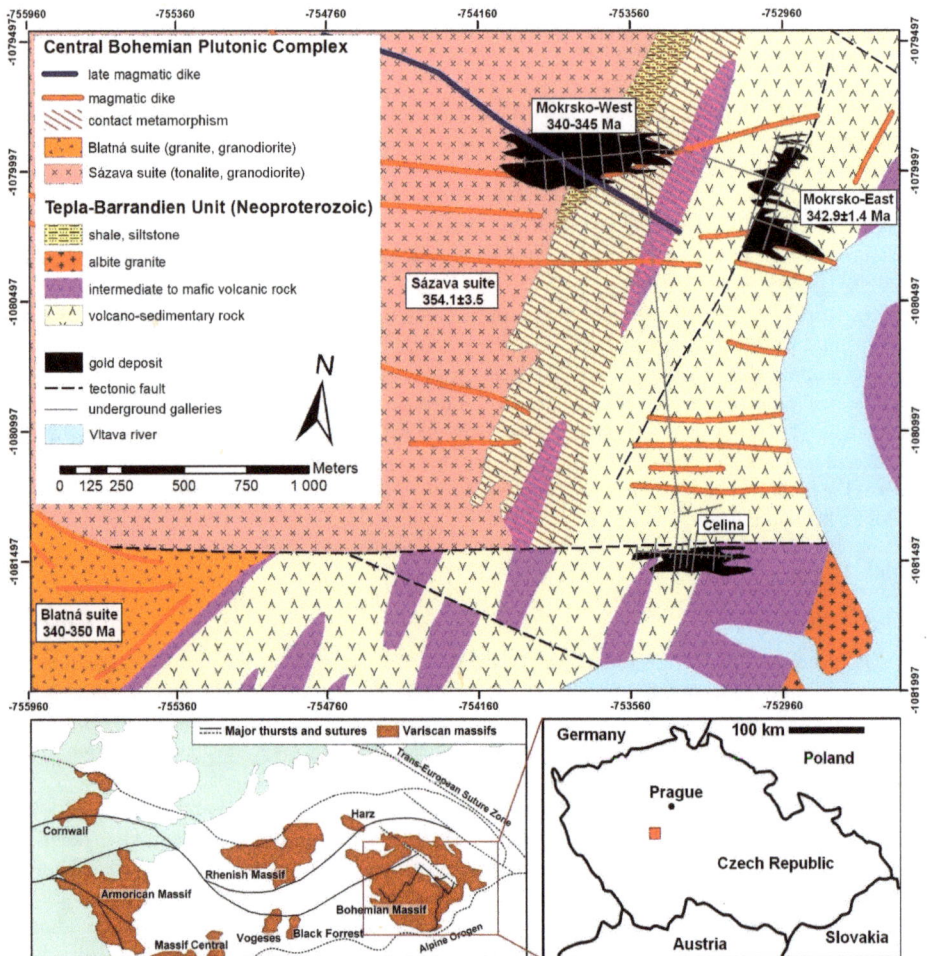

Figure 1. The geological map surrounding the Mokrsko gold deposit edited after Morávek [22]. The ages of Mokrsko-West and Mokrsko-East deposits came from Re–Os molybdenite dating [27,35]. The notice area of magmatic rocks in the west belonging to two Variscan intrusive suites—the Sázava suite and the Blatná suite. Dating is based on U–Pb the zircon ages [29,30,33,34]. The eastern side is a domain of volcanic and volcano-sedimentary rocks of Neoproterozoic Jílové Belt, part of the Teplá–Barrandian Unit. The bottom left is a projection of the Bohemian Massif position in Central Europe; edited after Franke [28]. The red square is an area of the geological map, situated about 50 km south of Prague, Czech Republic.

2.2. Characteristics of the Mokrsko Gold Deposit

Quartz veins of all three ore bodies (Figure 1) were emplaced in a fissure system that formed in an east-west direction due to the north-south extension in both the Jílové Belt and the Central Bohemian Plutonic Complex. The extension allowed the emplacement of both sub-vertical up to one-meter thick quartz veins and a sheeted veinlet system [22]. Morávek et al. [21], Boiron et al. [23],

Zachariáš et al. [24] and Zachariáš [25] divided the mineralization into three phases based on fluid inclusions, stable O-isotope composition, as well as mineralogical composition and vein structure.

Zachariáš et al. [24] marked the earliest phase as the quartz stage consisting of massive quartz mineralization accompanied by minor sulphide mineralization. Morávek et al. [22] reported the pressure and temperature conditions of ~270 MPa and 450 °C for this early stage. Zachariáš et al. [24] calculated the P–T window for early quartz veins within the range of 450–550 °C and 250–400 MPa.

The distinctive mineralizing event of the Mokrsko gold deposit—the formation of a sheeted veinlet system—took place during the second phase, the quartz-sulphide stage, according to Zachariáš et al. [24]. The veins are 0.1 mm to 50 mm wide and their frequency is highest in the western part of the Mokrsko-West deposit, with up to more than one hundred veins per meter [21,22]. According to Zachariáš et al. [24], most of the quartz gangue, as well as sulphides, was precipitated during this phase. Arsenopyrite is the most common gold-bearing sulphide at Mokrsko-West and Mokrsko-East, whereas pyrite and pyrrhotine prevail at the Čelina deposit [22]. For this phase, Zachariáš [24] estimated temperatures of ~395 °C to ~475 °C based on arsenopyrite thermometry. Boiron et al. [23] described this phase as the early ore stage and presented a fluid inclusion homogenization temperature of 300–380 °C and pressure of 170–270 MPa. The sheeted veinlet system is developed only at the Mokrsko-West deposit. It is missing on both the Čelina and the Mokrsko-East deposits which makes it difficult to distinguish between the different generations of veins for these two deposits [25].

The last phase is the precipitation of quartz; ore minerals with significant gold content, calcite, and chlorite in irregular micro-fissures developed in older quartz veins [21,23]. Morávek et al. [22], Boiron et al. [23] and Zachariáš et al. [24] concluded that both temperature and pressure were continuously decreasing toward this late mineralization phase, with temperatures in the range of 200–300 °C.

Results from studies conducted during exploration presented by Morávek et al. [21], Morávek [22], and detailed fluid inclusion analyses conducted by Boiron et al. [23] suggested a hydrothermal-metamorphic origin of gold-bearing quartz veins nowadays classified as orogenic gold deposits, for example Groves et al. [38] and Goldfarb et al. [39]. Both Morávek et al. [21] and Boiron et al. [23] assumed a very limited input of magmatic fluids. In their point of view, intrusion plays the role of a heat source, causing fluid circulation, rather than the role of the main source of fluids.

Zachariáš et al. [24] describe the features typical for orogenic gold deposits and for so-called intrusive-related gold systems, which include, especially, the presence of a sheeted veinlet system. According Hart and Goldfarb [40] and Hart [41] these two deposit models are difficult to distinguish because they have many common features, such as low-salinity CO_2 fluids; Au accompanied with W, Bi, and Te; and also temporal and/or spatial relationship to intrusion. Such features are present also at the Mokrsko gold deposit [22–24].

3. Samples and Analytical Methods

Sampling was done in situ from quartz veins and veinlets at the Mokrsko gold deposit. Five samples come from a one-meter wide sub-vertical vein from the Čelina deposit and eight samples from different places at the Mokrsko-West deposit. Mokrsko-East is inaccessible, eight samples were obtained from archives of the Czech Geological Survey and from the personal collection of Petr Morávek.

Polished and carbon-coated thin sections that were 30 μm and 100 μm thick were used for CL and LA-ICP-MS.

The most representative samples, demonstrating relationships between individual CL textures well, were chosen for the acquisition of high-resolution coloured CL images of up to one centimetre diagonally large sectors of thin sections. One CL image for Čelina, two images for Mokrsko-East and three images for Mokrsko-West. High-resolution CL images were obtained using a scanning electron microscope (SEM) TESCAN MIRA3 (TESCAN ORSAY HOLDING a.s., Brno, Czech Republic) with a field emission electron source. The system is equipped with a rainbow colour CL detector using a compact polymethyl methacrylate light guide, transparent in the range of 350 to 650 nm. The working distance was 20 mm; the accelerating voltage and current were set to 25 kV and 6 nA, respectively. Scanning speed ranged from 32 to 1000 ms/pixel, depending on the CL emissivity of the imaged area. Backscatter electron (BSE) images were obtained under identical conditions but with higher scanning speeds.

Selected samples based on high-resolution CL images were subjected to additional CL analyses and measuring of CL emission spectra using a "hot cathode" CL microscope HC1-LM (LUMIC, Bochum, Germany) [42] operated at 14 kV accelerating voltage and a current of 0.2 mA (current density of about 10 $\mu A/mm^2$). Luminescence images were captured "on-line" during CL operations using a Peltier cooled digital video-camera (OLYMPUS DP72). CL spectra in wavelengths ranging from 380 to 1000 nm were recorded with an Acton Research SP-2356 digital triple-grating spectrograph with a Princeton Spec-10 CCD detector attached to the CL microscope by a silica-glass fibre guide.

CL emission spectra were measured under standardized conditions with wavelength calibration by a Hg-halogen lamp, spot width 30 μm, and a measuring time of 5 s. For characterization of the transient CL behaviour of the hydrothermal quartz samples, two representative CL spectra were taken. The first initially and the second after 180 s of electron irradiation.

Trace elements analyses were conducted on the selected samples with high-resolution CL images. We performed 69 analyses at different places in the sample from the up-to-one-meter thick quartz vein at the Čelina deposit, 55 analyses on the two samples from the Mokrsko-East deposit, and 60 analyses on the three samples from the Mokrsko-West deposit. Analyses were done by LA-ICP-MS with the setup consisting of a NewWave Research UP213 laser ablation system emitting a wavelength of 213 nm coupled to an Agilent 7500ce (Agilent Technologies, Hachioji, Japan) quadrupole ICP-MS. The analyses of quartz were performed by using a laser beam diameter of 65 μm, a repetition rate of 10 Hz, laser beam fluence of 8 J/cm^{-2}, and duration of ablation of 60 s. The ablated material was carried out by helium (1.0 L/min). Prior to the ICP-MS, argon (0.6 L/min) was admixed into the helium flow. The certified reference material NIST612 was used for quantification purposes. The isotope ^{28}Si was utilized as an internal standard for suppressing different ablation rates between NIST612 and the quartz samples.

Al, Li, Ti, Ge, Ga Na, K, Ca, Ge, Ga, Fe, and As were chosen for analyses according to their common presence in quartz [1,6,12]. Na, K, and Ca, as well as Fe and As, are elements which are preferably concentrated in fluid and mineral microinclusions in quartz vein deposits with sulphides [1,6,12]. To minimalize the risk of contamination by microinclusions, the measured signal was continuously observed. Ablation spots with signals showing order-scale deviations against the prevailing quartz composition were omitted and assumed to be affected by inclusions. Secondly, the analyses with Na > 800 ppm, K > 800 ppm, Ca > 500 ppm, Fe > 200 ppm, and As > 100 ppm were excluded from the dataset; these threshold values were set based on the average abundance of the trace elements in hydrothermal quartz [1,3,6,12,18].

Laser ablation craters 65 μm wide and similarly deep did not allow for the clear measurement of individual quartz growth generations because CL analyses show only the surface of the sample and some quartz generations formed zones only a few μm wide, especially in young microfissure-forming quartz. Therefore, laser ablation spots were measured with the same samples with high-resolution CL images, but we were unable to target individual quartz growth generations defined by CL analyses.

4. Results

4.1. Cathodoluminescence Textures and CL Spectral Analyses

The Čelina deposit shows a rather homogenous to slightly mottled blue CL texture of Q1 (Figure 2). The spectral analysis reveals two initial emission bands. One, only visible as a shoulder at ca. 380 nm and the second at ca. 500 nm (Figure 2) that had strongly decreased after 180 s of electron irradiation. Two additional emission bands were detected at ca. 450 and 650 nm in the spectra, particularly after 180 s of electron irradiation. The secondary quartz growth generation at Čelina, Q4 precipitated in the microfissures formed during a brittle tectonic phase. This quartz is seen black in the SEM-CL image (Figure 2).

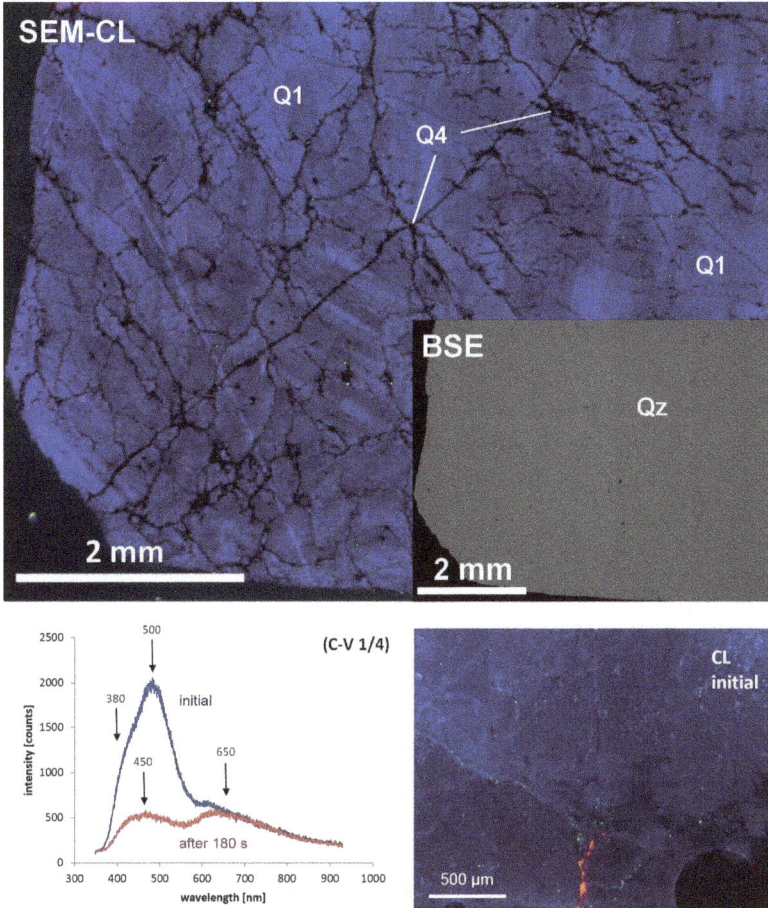

Figure 2. The scanning electron microscopy cathodoluminescence (SEM-CL) image showing part of an up to 1 m wide quartz vein from the eastern part of the Čelina deposit (sample C-V 1/4). Notice that small irregular fissures filled up by the younger Q4 cannot be seen at the BSE image of the same area as is the SEM-CL image. Cathodoluminescence (CL) emission spectra of the same sample were acquired initially and after 180 s of electron irradiation. The hot cathode CL image was acquired together with the initial blue spectrum.

Mokrsko-East is predominated by blue CL quartz Q1 (Figure 3), although it is less homogeneous than at the Čelina deposit. Q1 is intersected by younger quartz which precipitated in microfissures. The fissure quartz Q3 (Figure 3) is developed in a dense network of irregular veins, penetrating older blue CL quartz Q1. Different from the fissure quartz at Čelina, Q3 has a yellow to greenish CL colour.

Due to a dense network of quartz Q3 fissures, the emission spectra represent a mixture of two quartz growth generations with distinct CL colours—blue Q1 and yellow to greenish Q3. The broad emission band consists of at least three bands at ca. 450 nm, 570 nm, and 650 nm (Figure 3).

At the Mokrsko-West deposit, CL reveals that both older thick quartz veins and younger sheeted veinlet system consist of at least three quartz growth generations (Figures 4 and 5). The oldest quartz growth generation is Q1 with blue CL, which is predominant also at the Mokrsko-West deposit (Figure 5). Blue CL Q1 is overgrown by younger Q2 (observed only at Mokrsko-West) and was penetrated by the youngest Q3.

The Q2 quartz generation with bright yellow CL typically exhibits preserved subhedral shapes and growth zones, although in some cases, it has irregular shapes. The bright yellow CL shows a broad emission band with a maximum at ca. 570 nm (Figure 4).

The Q3 quartz growth generation with greenish CL developed as a dense network or cobweb of irregular veinlets (Figure 4). This quartz fills the microfissures (Figure 5), which cut both the previous quartz generations (Figure 4). The greenish CL quartz produced spectra similar to the yellow CL Q2, only with significantly lower intensity compared to Q2, resulting in a less intense CL shade.

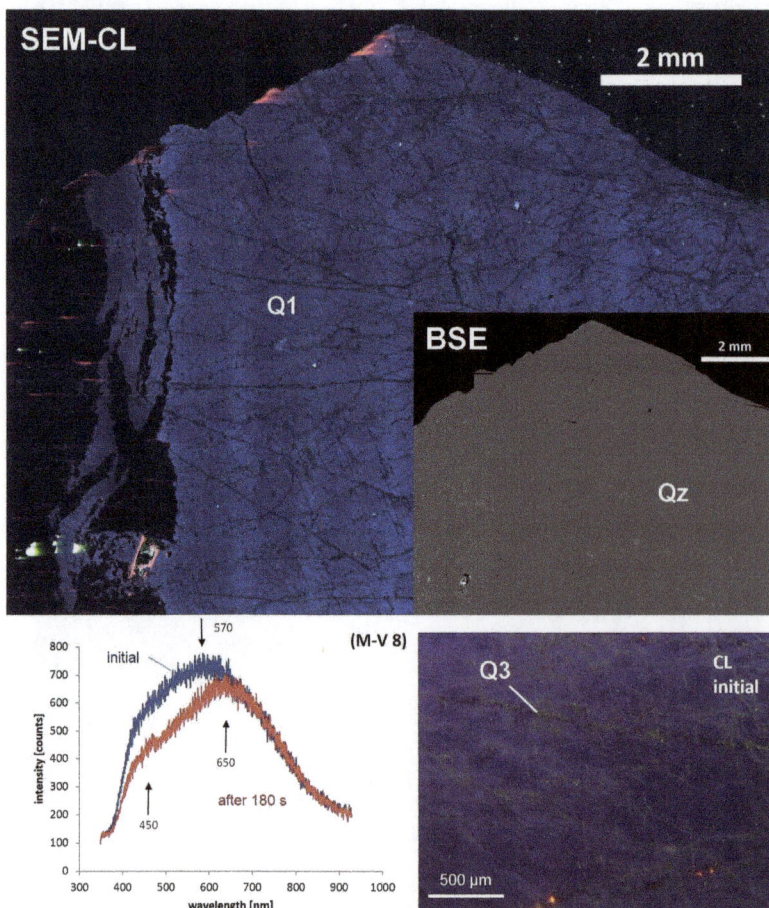

Figure 3. The SEM-CL and back-scattered electron (BSE) image showing part of a quartz vein from the Mokrsko-East deposit (sample M-V 8). CL emission spectra from a mixture of blue CL Q1 and yellow-greenish CL Q3. The hot cathode CL image was acquired together with the initial spectrum. Qz: quartz.

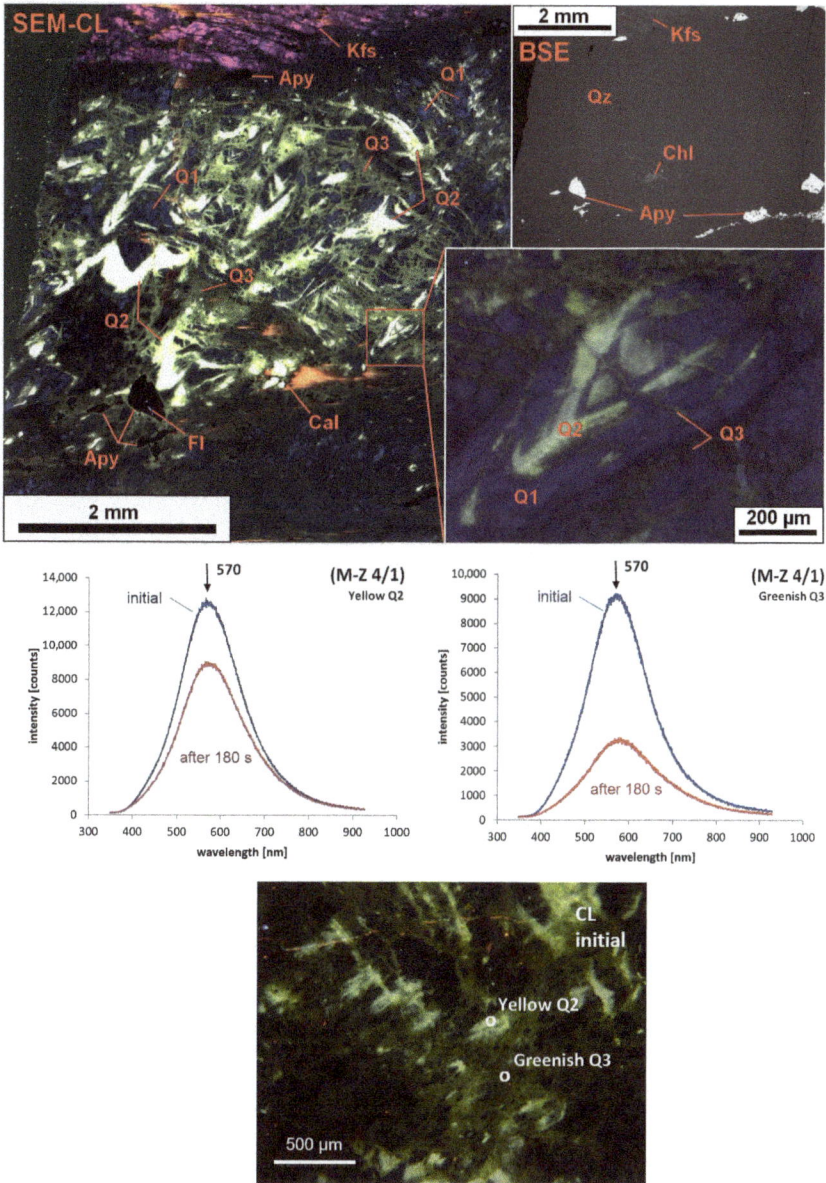

Figure 4. The SEM-CL and BSE images showing a quartz vein from the Mokrsko-West deposit (sample M-Z 4/1). The quartz vein has a complex CL texture consisting of 3 quartz grow th generations. Detail shows the oldest blue CL quartz (Q1) overgrown by younger quartz (Q2) with intense yellow CL and partly preserved growth zonation. The youngest yellow-greenish CL quartz (Q3) is present as a dense network of tiny irregular veinlets. Notice a few μm large inclusions of fluorite in arsenopyrite. The CL emission spectra of yellow CL Q2, greenish Q3, and hot cathode CL image of an area of the initial spectra. Qz: quartz; Kfs: potassium feldspar; Apy: arsenopyrite; Cal: calcite; Fl: fluorite.

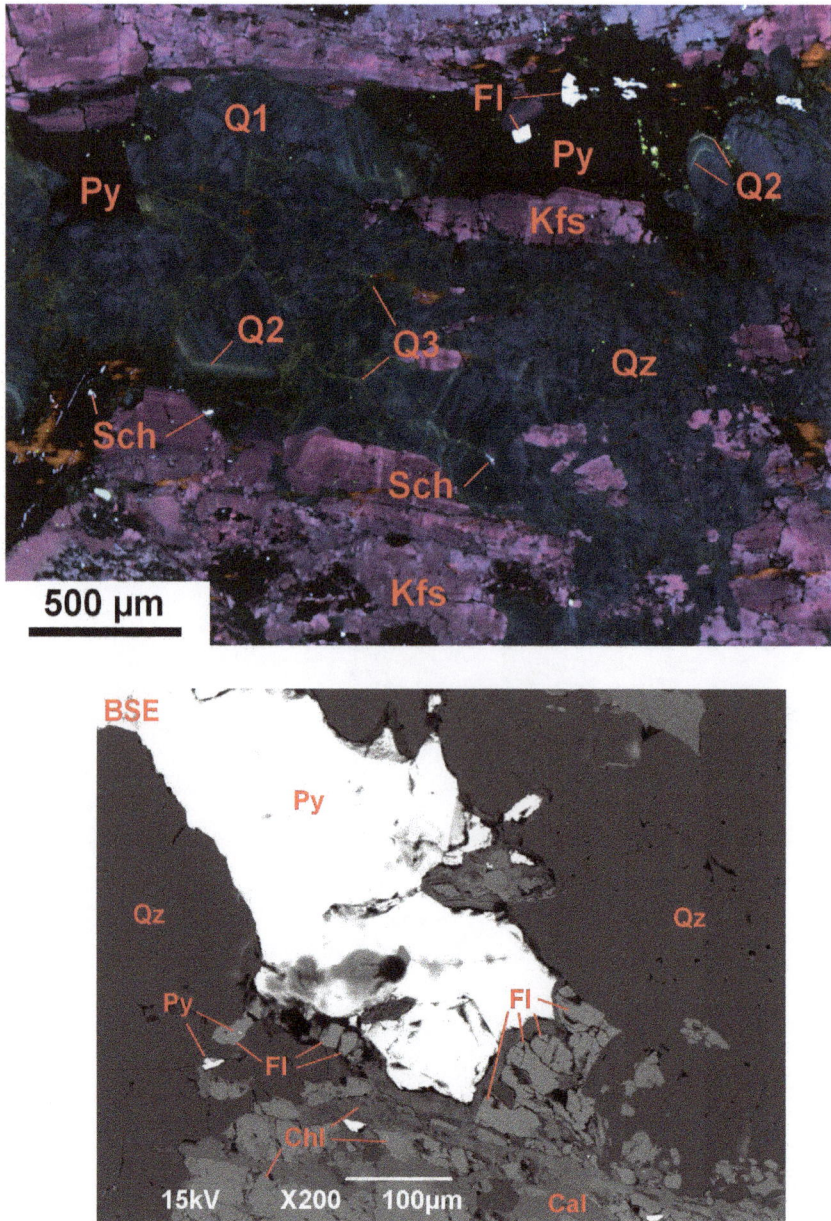

Figure 5. The detailed images of two quartz veins from a sheeted veinlet system at the Mokrsko-West deposit (sample M-Z 6/1). The SEM-CL image shows the spatial relationship of fluorite and pyrite. Notice also the scheelite mineralization at the vein edge and wall rock. Scheelite has a more bluish shade compared to the CL of fluorite. The BSE image from the same sample shows grains of fluorite overgrowing pyrite and also micro-inclusions of pyrite in fluorite; Py: pyrite; Qz: quartz; Fl: fluorite; Chl: chlorite; Cal: calcite; Kfs: potassium feldspar; Sch: scheelite.

SEM-CL imaging reveals a systematic presence of up to 50 μm large grains of fluorite with a strong CL effect (Figure 5). The common occurrence of fluorite at Mokrsko-West was usually not recorded by previous studies. In many cases, fluorite overgrows or is present as inclusions in sulphides (Figures 4 and 5).

The Q1 and Q2 growth generations, as well as the youngest Q3, do not penetrate gold-bearing arsenopyrite (Figure 4) or pyrite (Figure 5).

4.2. Trace Element Analyses

The Čelina deposit quartz has aluminium concentrations from tens to several hundred ppm. The values exceeding 200 ppm are very rare (Figure 6). The titanium concentration is 15 to 74 ppm, with 20–50 ppm being most common. The concentrations of lithium, which reached values from 11 to 36 ppm, are distinct. The concentrations of germanium are low at the Čelina deposit with values up to a maximum of 3 ppm (Figure 6). The microfissures of late Q4 at the Čelina deposit do not develop such dense network as compared to the microfissures of Q3 at Mokrsko-East and Mokrsko-West. This allowed us to avoid hitting the microfissures during the LA-ICP-MS analysis to obtain the trace element composition of Q1.

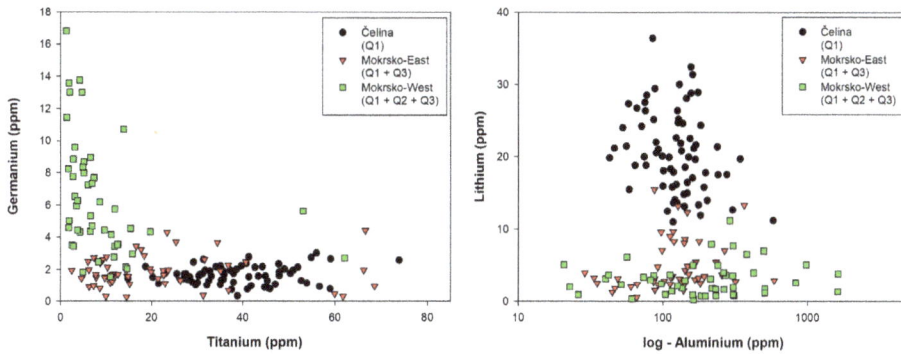

Figure 6. The trace element concentrations for hydrothermal quartz veins—Čelina (69 analyses), Mokrsko-East (55 analyses), and Mokrsko-West (60 analyses). Analyses should be taken as the total geochemical composition of hydrothermal quartz at individual deposits. Nevertheless, quartz generations which are present at these deposits are labelled. Several measurements of lithium yield concentrations below the limits of detection (7 at Mokrsko-East and 11 at Mokrsko-West), thus, these pairs with aluminium are not displayed in the Al-Li diagram. Data used for the diagrams, as well as other measured data, are available in the supplementary materials (Tables S1 and S2).

At Mokrsko-East, the concentration of aluminium, titanium, and germanium are similar to the Čelina deposit (Figure 7). The aluminium concentration is 16 ppm to several hundred ppm and titanium is from 2 to 69 ppm, however, in contrast to Čelina, the predominant values are lower, mostly ranging from 2 to 20 ppm. Germanium was relatively depleted at the Mokrsko-East deposit, with values mostly of a few ppm. The lithium concentration was also relatively depleted as compared to the Čelina deposit with concentrations only exceptionally exceeding 10 ppm.

Mokrsko-West is similar to Mokrsko-East in low lithium concentration with maximum values of 11 ppm. Titanium only exceptionally exceeds 15 ppm and germanium is high, mostly with concentrations of 2–14 ppm (Figure 6). The aluminium concentrations have a much larger range than at the other deposits with 20 to several hundreds of ppm and with a few analyses exceeding 1000 ppm (Figure 7).

The concentration of gallium, a common pair to germanium, was depleted at all three deposits; values are in many cases below the limit of detection (72% measurements at Čelina, 75% at Mokrsko-East, and 39% at Mokrsko-West). Additionally, the values of iron and arsenic are very low with maxima of a few ppm or, in many cases, below the detection limit at the Čelina and the Mokrsko-East deposits. The iron and arsenic concentrations at the Mokrsko-West deposit are from 3 to threshold values of 200 ppm and 2–64 ppm, respectively. The other measured elements (Na, K, and Ca) did not show any distinctive variations between the different deposits and concentrations are from below the detection limit up to the threshold values (Table S2).

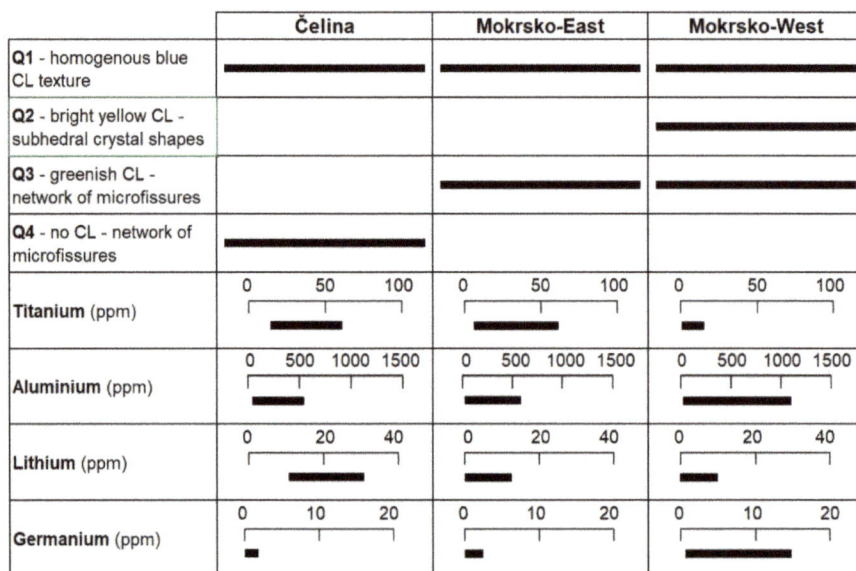

	Čelina	Mokrsko-East	Mokrsko-West
Q1 - homogenous blue CL texture	▬▬▬▬▬▬	▬▬▬▬▬▬	▬▬▬▬▬
Q2 - bright yellow CL - subhedral crystal shapes			▬▬▬▬▬▬
Q3 - greenish CL - network of microfissures		▬▬▬▬▬▬	▬▬▬▬▬▬
Q4 - no CL - network of microfissures	▬▬▬▬▬▬		
Titanium (ppm)	0 50 100	0 50 100	0 50 100
Aluminium (ppm)	0 500 1000 1500	0 500 1000 1500	0 500 1000 1500
Lithium (ppm)	0 20 40	0 20 40	0 20 40
Germanium (ppm)	0 10 20	0 10 20	0 10 20

Figure 7. The graphical comparison of individual quartz growth generations, its CL textures, and concentrations of trace elements at the Čelina, the Mokrsko-East, and the Mokrsko-West deposits.

5. Discussion

CL emission spectra of homogenously blue CL Q1, measured at the Čelina deposit, are characteristic for hydrothermal quartz [12,13] and for quartz from pegmatites [15]. The CL activation centre of these two emission bands is assumed to be a substitution of Si^{4+}, the most commonly trivalent aluminium accompanied by monovalent compensation ion, with Li^+ being the most common [7,10,12,13]. High aluminium concentrations in quartz at all three deposits suggest that this element is the main substituent for Si^{4+}. Elevated lithium concentrations at the Čelina deposit might suggest lithium as a compensation ion to aluminium. On the other hand, the very low lithium concentrations at both Mokrsko-East and Mokrsko-West do not support this substitution mechanism. A different element will be contributed to the charge balancing of trivalent aluminium. It could possibly be H^+, which has been proven to exist in this role [43], but a different analytical method (infrared spectroscopy) is needed for the analyses of the hydrogen concentration.

Distinctive emission spectra of Q2 and Q3, which had overgrown and penetrated Q1 at Mokrsko-East and Mokrsko-West, have a different origin. The dense penetration of Q1 caused the mixing of spectra of individual quartz generations as is most pronounced in mixed emission spectra obtained at Mokrsko-East. Similarly, the dense penetration of Q1 combined with the relatively large (65 μm) spot size of the LA-ICP-MS analysis also indicates that the geochemical composition of Q1 is locally mixed with those of Q2 and Q3 at Mokrsko-East and Mokrsko-West.

Only at the Čelina deposit, where Q1 is not densely penetrated by other quartz generations, is it possible to link the geochemical signature of hydrothermal quartz to blue CL Q1. The analysed hydrothermal quartz vein at Čelina can be considered as a product of an early quartz stage, for which the crystallization temperature in the range of 450–500 °C has been described [21,24]. The relatively high temperature seems to match with the elevated titanium concentrations at the Čelina deposit. Many authors, e.g., Wark and Watson [44] Thomas et al. [45] or Huang and Audétat [46] have suggested that the titanium concentration is controlled by the quartz crystallization temperature. The titanium concentration in quartz at Čelina of >20 ppm is in accordance with the datasets of Rusk et al. [47] from different deposits relating titanium concentrations above 10 ppm to crystalization temperatures above 400 °C.

Subhedral shapes of yellow CL Q2 and late microfissures of yellow-greenish CL Q3 with an emission band at 570 nm can be related to the fast crystallization at temperatures below 250 °C [16]. This agrees with temperature estimation of 200–300 °C, suggested for late quartz developed in the irregular microfissures at Mokrsko-West [21,23,24].

We assume that the measured geochemical signature of quartz from Mokrsko-West and Mokrsko-East is, in most cases, composed of more than one quartz growth generation because the microfissures of Q3 strongly penetrate the older quartz generations and the laser ablation beam was unable to avoid it. In spite of that, the very low titanium concentrations at Mokrsko-West are in accordance with the mineral deposits database of Rusk et al. [47], that is, in assigning concentrations of titanium under 10 ppm to crystallization temperatures below 350 °C.

The elevated concentration of germanium in the hydrothermal quartz points to a different origin of at least one quartz growth generation at Mokrsko-West. The enrichment of germanium in quartz is typical for late stages of magmatic fractionation of granitic melts [48,49], indicating a possible magmatic origin of the Ge-enriched quartz phase at Mokrsko-West. Breiter et al. [48] also described the relative increase of germanium concentration from the early to late intrusive suites of the Central Bohemian Plutonic Complex, which suggests an affiliation of the Ge-enriched quartz phase to late magmatic processes.

Gold-bearing sulphides unaffected by quartz generations suggest later or simultaneous crystallization with the latest quartz generation. Fluorite, which often overgrows sulphides, is also related to late mineralization. The presence of fluorite can possibly support a late magmatic derivation of the assemblage as fluorine is another element typical for late crystallization stages of evolved granitic rocks [50,51].

On the other hand, the homogeneous to slightly mottled texture of the oldest blue CL Q1 that is visible at all three deposits, is very common in hydrothermal quartz from orogenic gold deposits with fluids of metamorphic origin [3,52]. This is in accordance with studies of Morávek et al. [21] and Boiron et al. [23] suggesting a hydrothermal-metamorphic origin of the quartz veins. However, the CL textures of partly preserved growth zones of Q2, later the "spiderweb" textures of Q3 and Q4, are more typical for porphyry deposits [17,52]. This suggests a magmatic influence of the hydrothermal system. Moreover, steep hydrothermal quartz veins formed in the east-west trending tectonic zone with an extensive regime [22], are a characteristic tectonic setting of the intrusion-related deposits [38,41] in contrast to orogenic gold, where there is the typically compressional strain [39,40].

The Sázava suite was emplaced during the compressional (shortening) and transpressional regime [36], whereas the emplacement of the Blatná suite indicates a switch from transpression to the onset of exhumation of the crustal unit, which led to the formation of extensional shear zones [37]. Dating of the molybdenite at Mokrsko-West [35] and Mokrsko-East [27] shows ages from 340 to 345 Ma, which is about 10 Ma younger than the zircon crystallization ages of Sázava suite—the host rock of the Mokrsko-West deposit. However, it is possible that the Blatná suite, which is younger (340–350 Ma [33,34]) than the Sázava suite, extends to larger depths under the Sázava suite and the gold deposits.

Another indicator of the prolonged presence of magmatic processes during the origin of the mineral deposits are several pre-ore mineralization granitic porphyry dikes at Mokrsko-West and Mokrsko-East and also presence of one mafic lamprophyre dike at the Mokrsko-West deposit, which postdates the ore mineralization [22].

6. Conclusions

Hydrothermal quartz veins at the Mokrsko-West, Mokrsko-East, and Čelina deposit have distinctive CL textures and spectra, as well as trace element compositions, which allow us to distinguish the quartz veins from each deposit. Four different quartz growth generations were characterized based on SEM-CL mapping of hydrothermal quartz veins.

High concentrations of titanium in early phases, represented by the homogenous blue CL Q1 at Čelina deposit, correspond to high-T stages identified by previous studies using other methods. This supports the titanium concentration in quartz being a good indicator of the relative crystallization temperature in one system, where decreasing temperature can be correlated with lowering titanium concentration in hydrothermal quartz.

Mokrsko-East looks similar to Čelina, but the microfissures formed during the brittle tectonic phase were precipitated by quartz with different signatures. Microfissures at Čelina were filled by Q4 which had no CL effect, whereas yellow to greenish CL Q3 precipitated at Mokrsko-East. The presence of Q3 and the more frequent values of titanium below 20 ppm links Mokrsko-East to the processes which took place at the Mokrsko-West deposit.

Despite the mixing of the geochemical signatures of three quartz growth generations at Mokrsko-West, the overall depleted titanium concentration, as well as the presence of low-temperature yellow CL Q2, are consistent with the decrease of the crystallization temperature in the late mineralization stages. Hydrothermal quartz at Mokrsko-West shows an enrichment in germanium and the presence of yellow Q2, which distinguishes it from the other two deposits, suggesting a different source of late quartz growth generations.

The enrichment of germanium in hydrothermal quartz, as well as the presence of fluorite, indicate a magmatic source as they are both characteristics of the late stages of the magmatic differentiation of granitic melts. The affiliation of late quartz growth generations to magmatic processes is further supported by the temporal relationship to the intrusion of the Blatná suite, the presence of the sheeted veinlet system, and the observed CL textures of Q2, Q3, and Q4, which are typical for gold deposits related to intrusion.

Supplementary Materials: The following are available online at http://www.mdpi.com/2075-163X/8/8/335/s1, Table S1: Table of trace elements analyses (Al, Ti, Li, Ge) from all three deposits; Table S2: Table of other measured trace elements (Na, K, Fe, As) from all three deposits.

Author Contributions: V.W. is the main author of the paper; J.L. evaluated the data, was consulted for the results, and supervised the main author. M.D. performed the high-resolution CL imaging in the laboratories at TESCAN ORSAY HOLDING, was consulted for the results, and contributed to the Samples and Analytical Methods chapter; J.G. conducted and described the results from hot-cathode CL imaging and CL emission spectroscopy analyses. His contributions also appear in the Samples and Analytical Methods chapter, as well as in the Results and Discussion chapter.

Funding: The research was supported by project Geodyn2018 (MUNI/A/1088/2017).

Acknowledgments: Research was supported by the project MUNI/A/1088/2017. Special thanks belong to the TESCAN ORSAY HOLDING for cooperation in large-scale SEM-CL imaging. We are grateful to Peter Morávek for consultation, guidance at the Mokrsko gold deposit, and for providing samples from the inaccessible Mokrsko-East deposit. We are very thankful to Carita Augustsson and two anonymous reviewers for their comments and suggestions which substantially improved the paper. I would like also to express my thanks to Jan Cempírek for useful consultations.

Conflicts of Interest: The authors declare no conflict of interest. The founding sponsors had no role in the design of the study; in the collection, analyses, or interpretation of data; in the writing of the manuscript, and in the decision to publish the results.

References

1. Götze, J. Chemistry, textures and physical properties of quartz—Geological interpretation and technical application. *Mineral. Mag.* **2009**, *73*, 645–671. [CrossRef]

2.   Fairbanks, E.E. Gangue minerals. In *The Encyclopedia of Mineralogy*; Frye, K., Ed.; Hutchinson Ross Pub. Co.: Stroudsburg, PA, USA, 1981; p. 163.

3.   Rusk, B. Quartz cathodoluminescence: Textures, Trace Elements, and Geological Applications. In *Cathodoluminescence and its Application to Geoscience*; Coulson, I.M., Ed.; Mineralogical Association of Canada: Québec City, QC, Canada, 2014; pp. 127–141. ISBN 978-0-921294-55-9.

4.   Götte, T.; Ramseyer, K. Trace Element Characteristics, Luminescence Properties and Real Structure of Quartz. In *Quartz: Deposits, Mineralogy and Analytics*; Götze, J., Möckel, R., Eds.; Springer: Berlin/Heidelberg, Germany, 2012; pp. 256–285.

5.   Weil, J.A. A review of the EPR spectroscopy of the point defects in a-quartz: The decade 1982–1992. In *Physics and Chemistry of SiO$_2$ and the Si-SiO$_2$ Interface 2*; Deal, B.E., Helms, C.R., Eds.; Plenum Press: New York, NY, USA, 1993; pp. 131–144, ISBN 978-1-4899-0776-9.

6.   Gerler, J. Geochemische Untersuchungen an hydrothermalen, metamorphen, granitischen und pegmatitischen Quarzen und deren Flüssigkeitseinschlüssen. Ph.D. Thesis, Göttingen University, Göttingen, Germany, 1990.

7.   Götze, J.; Plötze, M.; Habermann, D. Origin, spectral characteristics and practical applications of the cathodoluminescence (CL) of quartz—A review. *Mineral. Pet.* **2001**, *71*, 225–250. [CrossRef]

8.   Henry, D. Cathodoluminescence Theory. The Science Education Research Centre at Carleton College. 2012. Available online: https://serc.carleton.edu/research_education/geochemsheets/CLTheory.html (accessed on 1 December 2017).

9.   Mason, R. The Physics and Chemistry of Cathodoluminescence. In *Cathodoluminescence and its Application to Geoscience*; Coulson, I.M., Ed.; Mineralogical Association of Canada: Québec City, QC, Canada, 2014; pp. 1–10, ISBN 978-0-921294-55-9.

10. Botis, S.; Nokhrin, S.M.; Pan, Y.; Xu, Y.; Bonli, T.; Sopuck, V. Natural Radiation-Induced Damage in Quartz. I. Correlations between Cathodoluminence Colors and Paramagnetic Defects. *Can. Mineral.* **2005**, *43*, 1565–1580. [CrossRef]

11. Jourdan, A.L.; Vennemann, T.W.; Mullis, J.; Ramseyer, K. Oxygen isotope sector zoning in natural hydrothermal quartz. *Mineral. Mag.* **2009**, *73*, 615–632. [CrossRef]

12. Götze, J.; Plötze, M.; Graupner, T.; Hallbauer, D.K.; Bray, C.J. Trace element incorporation into quartz: A combined study by ICP-MS, electron spin resonance, cathodoluminescence, capillary ion analysis, and gas chromatography. *Geochim. Cosmochim. Acta* **2004**, *68*, 3741–3759. [CrossRef]

13. Perny, B.; Eberhardt, P.; Ramseyer, K.; Pankarth, R. Microdistribution of Al, Li, and Na in α quartz: Possible causes and correlation with short-lived cathodoluminescence. *Am. Mineral.* **1992**, *77*, 534–544.

14. Ramseyer, K.; Baumann, J.; Matter, A.; Mullis, J. Cathodoluminescence colours of a-quartz. *Mineral. Mag.* **1988**, *52*, 669–677. [CrossRef]

15. Götze, J.; Plötze, M.; Trautmann, T. Structure and luminescence characteristics of quartz from pegmatites. *Am. Mineral.* **2005**, *90*, 13–21. [CrossRef]

16. Götze, J.; Pan, Y.; Stevens-Kalceff, M.; Kempe, U.; Müller, A. Origin and significance of the yellow cathodoluminescence (CL) of quartz. *Am. Mineral.* **2015**, *100*, 1469–1482. [CrossRef]

17. Rusk, B.; Reed, M. Scanning electron microscope–cathodoluminescence analysis of quartz reveals complex growth histories in veins from the Butte porphyry copper deposit, Montana. *Geology* **2002**, *30*, 727–730. [CrossRef]

18. Monecke, T.; Kempe, U.; Götze, J. Genetic significance of the trace element content in metamorphic and hydrothermal quartz: A reconnaissance study. *Earth Planet. Sci. Lett.* **2002**, *202*, 709–724. [CrossRef]

19. Müller, A.; Herrington, R.; Armstrong, R.; Seltmann, R.; Kirwin, D.J.; Stenina, N.G.; Kronz, A. Trace elements and cathodoluminescence of quartz in stockwork veins of Mongolian porphyry-style deposits. *Mineral. Depos.* **2010**, *45*, 707–727. [CrossRef]

20. Frelinger, S.N.; Ledvina, M.D.; Kyle, J.R.; Zhao, D. Scanning electron microscopy cathodoluminescence of quartz: Principles, techniques and applications in ore geology. *Ore Geol. Rev.* **2015**, *65*, 840–852. [CrossRef]

21. Morávek, P.; Janatka, J.; Pertoldová, J.; Straka, J.; Ďurišová, E.; Pudilová, M. The Mokrsko Gold Deposit—The Largest Gold Deposit in the Bohemian Massif, Czechoslovakia. *Econ. Geol. Monogr. Ser.* **1988**, *1989*, 252–259. [CrossRef]

22. Morávek, P. The Mokrsko gold deposit. In *Gold Deposits in Bohemia*, 2nd ed.; Morávek, P., Ed.; Czech Geological Survey: Prague, Czech Republic, 1996; pp. 31–56, ISBN 80-7075-202-5.

23. Boiron, M.C.; Barakat, A.; Cathelineau, M.; Banks, D.A.; Durisová, J.; Morávek, P. Geometry and *p*–V–T–X conditions of microfissural ore fluid migration: The Mokrsko gold deposit (Bohemia). *Chem. Geol.* **2001**, *173*, 207–225. [CrossRef]

24. Zachariáš, J.; Morávek, P.; Gadas, P.; Pertoldová, J. The Mokrsko-West gold deposit, Bohemian Massif, Czech Republic: Mineralogy, deposit setting and classification. *Ore Geol. Rev.* **2014**, 238–263. [CrossRef]

25. Zachariáš, J. Structural evolution of the Mokrsko-West, Mokrsko-East and Čelina gold deposits, Bohemian Massif, Czech Republic: Role of fluid overpressure. *Ore Geol. Rev.* **2016**, *74*, 170–195. [CrossRef]

26. Moravek, P.; Pouba, Z. Precambrian and Phanerozoic history of gold mineralization in the Bohemian Massif. *Econ. Geol.* **1987**, *82*, 2098–2114. [CrossRef]

27. Zachariáš, J.; Stein, H. Re-Os Ages of Variscan Hydrothermal Gold Mineralizations, Central Bohemian Metallogenic Zone, Czech Republic. In *Mineral Deposits at the Beginning of the 21st Century*; Piestrzyński, A., Ed.; Swets & Zeitlinger Publishers: Lisse, The Netherlands, 2001; pp. 851–854, ISBN 9789026518461.

28. Franke, W. The Variscan orogen in Central Europe: Construction and collapse. *Geol. Soc. Lond. Mem.* **2006**, *32*, 333–343. [CrossRef]

29. Holub, F.V.; Cocherie, A.; Rossi, P. Radiometric dating of granitic rocks from the Central Bohemian Plutonic Complex (Czech Republic): Constraints on the chronology of thermal and tectonic events along the Moldanubian-Barrandian boundary. *Comptes Rendus De L'Académie Des Sci.* **1997**, *325*, 19–26. [CrossRef]

30. Janoušek, V.; Braithwaite, C.J.R.; Bowes, D.R.; Gerdes, A. Magma-mixing in the genesis of Hercynian calc-alkaline granitoids: An integrated petrographic and geochemical study of the Sázava intrusion, Central Bohemian Pluton, Czech Republic. *Lithos* **2004**, *78*, 15–26. [CrossRef]

31. Hajná, J.; Žák, J.; Kachlík, V. Structure and stratigraphy of the Teplá–Barrandian Neoproterozoic, Bohemian Massif: A new plate-tectonic reinterpretation. *Gondwana Res.* **2011**, *19*, 495–508. [CrossRef]

32. Waldhauserová, J. Proterozoic volcanites and intrusive rocks of the Jílové zone in Central Bohemia. *Krystalinikum* **1984**, *17*, 77–97.

33. Janoušek, V.; Wiegand, B.A.; Žák, J. Dating the onset of Variscan crustal exhumation in the core of the Bohemian Massif: New U-Pb single zircon ages from the high-K calc-alkaline granodiorites of the Blatna suite, Central Bohemian Plutonic Complex. *J. Geol. Soc.* **2010**, *167*, 347–360. [CrossRef]

34. Dörr, W.; Zulauf, G. Elevator tectonics and orogenic collapse of a Tibetan-style plateau in the European Variscides: The role of the Bohemian shear zone. *Int. J. Earth Sci.* **2010**, *99*, 299–325. [CrossRef]

35. Ackerman, L.; Haluzová, E.; Creaser, R.A.; Pašava, J.; Veselovský, F.; Breiter, K.; Erban, V.; Drábek, M. Temporal evolution of mineralization events in the Bohemian Massif inferred from the Re–Os geochronology of molybdenite. *Miner. Depos.* **2017**, *52*, 651–662. [CrossRef]

36. Žák, J.; Schulmann, K.; Hrouda, F. Multiple magmatic fabrics in the Sázava pluton (Bohemian Massif, Czech Republic): A result of superposition of wrench-dominated regional transpression on final emplacement. *J. Struct. Geol.* **2005**, *27*, 805–822. [CrossRef]

37. Žák, J.; Holub, F.V.; Verner, K. Tectonic evolution of a continental magmatic arc from transpression in the upper crust to exhumation of mid-crustal orogenic root recorded by episodically emplaced plutons: The Central Bohemian Plutonic Complex (Bohemian Massif). *Int. J. Earth Sci.* **2005**, *94*, 385–400. [CrossRef]

38. Groves, D.I.; Goldfarb, R.J.; Robert, F.; Hart, C.J.R. Gold Deposits in Metamorphic Belts: Overview of Current Understanding, Outstanding Problems, Future Research, and Exploration Significance. *Econ. Geol.* **2003**, *98*, 1–29. [CrossRef]

39. Goldfarb, R.J.; Bakker, T.; Dubé, B.; Groves, D.I.; Hart, C.J.R.; Gosselin, P. Distribution, Character, and Genesis of Gold Deposits in Metamorphic Terranes. In *100th Anniversary Volume*; Hedenquist, J.W., Thompson, J.F.H., Goldfarb, R.J., Richards, J.P., Eds.; Society of Economic Geologists: Littleton, CO, USA, 2005; pp. 407–450, ISBN 978-1-887483-01-8.

40. Hart, C.; Goldfarb, R.J. Distinguishing intrusion-related from orogenic gold systems. In *Proceedings of the New Zealand Minerals Conference: Realising New Zealand's Mineral Potential, Auckland, New Zealand, 13–16 November 2005*; Crown Minerals, Ministry of Economic Development and Australasian Institute of Mining and Metallurgy, New Zealand Branch: Wellington, New Zealand, 2005; pp. 125–133, ISBN 0478284551.

41. Hart, C.J.R. Reduced Intrusion-related Gold system. In *Mineral Deposits of Canada: A Synthesis of Major Deposit Types, District Metallogeny, the Evolution of Geological Provinces and Exploration Methods*; Goodfellow, W.D., Ed.; Geological Association of Canada—Mineral Deposits Division: St. John's, NL, Canada, 2007; Volume 102, pp. 95–112.

42. Neuser, R.D.; Bruhn, F.; Götze, J.; Habermann, D.; Richter, D.K. Kathodolumineszenz: Methodik und Anwendung. *Zentralblatt für Geologie und Paläontologie Teil I H* **1995**, *1/2*, 287–306.

43. Miyoshi, N. Successive zoning of Al and H in hydrothermal vein quartz. *Am. Mineral.* **2005**, *90*, 310–315. [CrossRef]

44. Wark, D.A.; Watson, E.B. TitaniQ: A titanium-in-quartz geothermometer. *Contrib. Mineral. Pet.* **2006**, *152*, 743–754. [CrossRef]

45. Thomas, J.B.; Bruce Watson, E.; Spear, F.S.; Shemella, P.T.; Nayak, S.K.; Lanzirotti, A. TitaniQ under pressure: The effect of pressure and temperature on the solubility of Ti in quartz. *Contrib. Mineral. Pet.* **2010**, *160*, 743–759. [CrossRef]

46. Huang, R.; Audétat, A. The titanium-in-quartz (TitaniQ) thermobarometer: A critical examination and re-calibration. *Geochim. Cosmochim. Acta* **2012**, *84*, 75–89. [CrossRef]

47. Rusk, B.G.; Lowers, H.A.; Reed, M.H. Trace elements in hydrothermal quartz: Relationships to cathodoluminescent textures and insights into vein formation. *Geology* **2008**, *36*, 547–550. [CrossRef]

48. Breiter, K.; Gardenová, N.; Kanický, V.; Vaculovič, T. Gallium and germanium geochemistry during magmatic fractionation and post-magmatic alteration in different types of granitoids: A case study from the Bohemian Massif (Czech Republic). *Geol. Carpath.* **2013**, *64*, 171–180. [CrossRef]

49. Höll, R.; Kling, M.; Schroll, E. Metallogenesis of germanium—A review. *Ore Geol. Rev.* **2007**, *30*, 145–180. [CrossRef]

50. Bailey, J.C. Fluorine in granitic rocks and melts: A review. *Chem. Geol.* **1977**, *19*, 1–42. [CrossRef]

51. Yang, X.-M.; Lentz, D.R. Chemical composition of rock-forming minerals in gold-related granitoid intrusions, southwestern New Brunswick, Canada: Implications for crystallization conditions, volatile exsolution, and fluorine-chlorine activity. *Contrib. Mineral. Petrol.* **2005**, *150*, 287–305. [CrossRef]

52. Rusk, B. Cathodoluminescent Textures and Trace Elements in Hydrothermal Quartz. In *Quartz: Deposits, Mineralogy and Analytics*; Götze, J., Möckel, R., Eds.; Springer: Berlin/Heidelberg, Germany, 2012; pp. 307–329.

minerals

Article

Investigation of Fluids in Macrocrystalline and Microcrystalline Quartz in Agate Using Thermogravimetry-Mass-Spectrometry

Julia Richter-Feig [1], Robert Möckel [2], Jens Götze [1] and Gerhard Heide [1,*]

[1] TU Bergakademie Freiberg, Institute of Mineralogy, Brennhausgasse 14, 09596 Freiberg, Germany; richterjulia1989@gmail.com (J.R.-F.); jens.goetze@mineral.tu-freiberg.de (J.G.)

[2] Helmholtz-Zentrum Dresden-Rossendorf, Helmoltz Institute Freiberg for Resource Technology, Chemnitzer Str. 40, 09599 Freiberg, Germany; r.moeckel@hzdr.de

* Correspondence: richterjulia1989@gmail.com

Received: 3 November 2017; Accepted: 14 February 2018; Published: 17 February 2018

Abstract: Gaseous and liquid fluids in agates (banded chalcedony—SiO_2) of different localities were investigated systematically by thermogravimetry-mass-spectrometry within a temperature range from 25 to 1450 °C, for the first time. Chalcedony and macrocrystalline quartz from twelve agate samples were investigated, from Germany (Schlottwitz, St. Egidien, Chemnitz and Zwickau), Brazil (Rio Grande do Sul), Scotland (Ayrshire) and the USA (Montana). They originate from mafic and felsic volcanic rocks as well as hydrothermal and sedimentary environments. The results were evaluated regarding compounds of hydrogen with fluorine, chlorine, nitrogen, carbon and sulphur. Additionally, oxygen compounds were recognized with hydrogen, fluorine, nitrogen, sulphur and carbon. The nature of the compounds was identified based on their mass-charge-ratio and the intensity ratios of the associated fragments. Due to interferences of different compounds with the same mass-charge-ratio, only H_2O, HF, NO, S, SO, CO_3—as well as several hydrocarbon compounds (for example CO_3^{2-} or CO)—could be properly identified. The main degassing temperatures were detected at around 500 and 1000 °C. Generally, a difference between quartz and chalcedony regarding the composition of their fluids could not be found. The results indicate a silica source for the agate formation from aqueous solutions but also a possible role of fluorine compounds. Additionally, CO_2 and other fluids were involved in the alteration of volcanic rocks and the mobilization and transport of SiO_2.

Keywords: agate; quartz; chalcedony; thermogravimetry-mass-spectrometry; EGA

1. Introduction

Agates (banded chalcedony—SiO_2) are spectacular products of nature, which have been investigated for decades regarding the conditions of their formation (e.g., [1–10]). In detail, agates have a very complex composition consisting of certain SiO_2 polymorphs and morphological quartz varieties (e.g., [4,11–13]). For instance, quartzine, opal-A, opal-CT and/or moganite can be intergrown or intercalated with chalcedony layers and macrocrystalline quartz in agate. Moreover, agates can contain considerable amounts of water (molecular water and/or silanol groups) and mineral inclusions, which are often responsible for the different colouration of agates [4,7,12–14].

The process of agate formation is as complex as its composition and may differ depending on the type of parent rocks and formation environment. Most agates occur in volcanic host rocks.

The chemical and mineralogical composition of agates in volcanic host rocks as well as their association with certain mineral products of alteration processes (e.g., zeolites, clay minerals, iron oxides) led to the conclusion that the formation of these agates is closely connected to late-

and post-volcanic alteration or weathering of the parent rocks (e.g., [6–11]).The complex processes lead to the accumulation of silica in cavities, so that agates represent a mixture of certain SiO_2 polymorphs and morphological quartz varieties (e.g., [4,7,11,12]).

Transport and accumulation of silica in cavities of the host rocks is predominantly realized by diffusion processes. In basic volcanics, vesicular cavities form during the solidification of the lava, whereas in acidic volcanics so-called lithophysae (high-temperature crystallization domains [15]) are formed at first, due to the devitrification and degassing of the volcanic glass/melt.

The formation of vein agates is related to fissures and veins within different types of crystalline rocks, which enable a free movement of silica-bearing mineralizing fluids through a system of cracks. Here, SiO_2 is accumulated by hydrothermal-magmatic solutions, whereas silica in sedimentary agates preferentially derives from SiO_2-rich pore solutions [16].

The accumulation and condensation of silicic acid result in the formation of silica sols and amorphous silica as precursors for the development of the typical agate structures. It is assumed that the formation of the typical agate microstructure is governed by processes of self-organization, starting with the spherulitic growth of chalcedony and continuing into chalcedony fibres [10,12]. Macrocrystalline quartz crystallizes when the SiO_2 concentration in the mineralizing fluid is low. The estimation of the temperature of agate formation using oxygen isotopes, Al concentrations or homogenization temperatures of fluid inclusions provided a temperature range between ca. 20 and 200 °C [7].

Although much geochemical and mineralogical data of agates exist, there are still open questions and controversial discussions, especially regarding the transport of the enormous amounts of silica necessary for the formation of agates. In general, it is assumed that diffusion of monomeric silicic acid in pore fluids could be the main transport process (e.g., [1,7,17,18]). However, geochemical data indicate that a transport of elements and chemical compounds in aqueous fluids cannot be the only process involved in agate formation. In certain agates (especially those of acidic volcanic rocks), elevated concentrations of Ge (>10 ppm), U (>15 ppm) and B (>30 ppm) as well as the occurrence of paragenetic calcite and fluorite indicate that other fluids can play a role in the alteration of volcanic rocks and the mobilization and transport of SiO_2 and other chemical compounds [11,19]. In consequence, additional chemical transport reactions (CTR) of gases and liquids by stable fluorine (and chlorine) compounds such as SiF_4, BF_3, GeF_4 and UO_2F_2 could explain the processes during agate formation better than exclusive element transport by silicic acid in aqueous solutions (e.g., [20,21]). The analysis of the fluids in the different agates provides information about the chemical composition of mineral-forming fluids and for the reconstruction of the agate formation processes.

The sample material includes agates from basic and acidic volcanic rocks, hydrothermal vein agates and agates of sedimentary origin to enable a comparison of these different formation environments. In the present study, a systematic study of gaseous and liquid fluids in agates from different localities and of different origin was performed by evolved gas analysis (in this case and further called thermogravimetry-mass-spectrometry). Moreover, microcrystalline chalcedony and macro-crystalline quartz were analysed separately to reveal possible differences in the fluid composition. Because of the lack of visible fluid inclusions within the micro-crystalline agate matrix, conventional techniques for the characterization of the fluids such as microscopy or Raman spectroscopy could not be applied.

2. Materials and Methods

2.1. Sample Material

In the present study, agates from eight different localities and different geological environments were analysed (Table 1, Figure 1), five from Saxony/Germany (Chemnitz-Furth, Chemnitz-Altendorf, Schlottwitz, St. Egidien and Zwickau) and each one from Brazil, Scotland and the USA, respectively. The material includes samples from basic, intermediate and acidic volcanic host rocks as well as hydrothermal vein agate and agate of sedimentary origin. Twenty-two measurements in total were made including multiple measurements on some samples.

Table 1. Compilation of the investigated agate samples and their genetic types.

Location	Genetic Type	Age of Host Rock	Measurements Per Sample	Reference
Soledade (Rio Grande do Sul, Brazil)	Basic volcanic rock (basalt)	~135 Ma	2	[22]
Heads of Ayr (Dunure area, Scotland)	Basic volcanic rock (basalt)	~412 Ma	2	[23,24]
Zwickau- Planitz (Saxony, Germany)	Acidic volcanic rock (pitchstone)	~290 Ma	2	[19]
St.Egidien (Saxony, Germany)	Acidic volcanic rock (ignimbrite)	~290 Ma	6	[19]
Chemnitz-Furth (Saxony, Germany)	Acidic volcanic rock (ignimbrite)	~290 Ma	4	[19]
Chemnitz-Altendorf (Saxony, Germany)	Vein agate in altered Pitchstone	~290 Ma	2	[19]
Schlottwitz (Saxony, Germany)	Hydrothermal vein agate	~270 Ma	2	[25]
Dryhead area, Prior Mountains (Montana, USA)	Sedimentary agate (clay shale and siltstone)	250–300 Ma	2	[26]

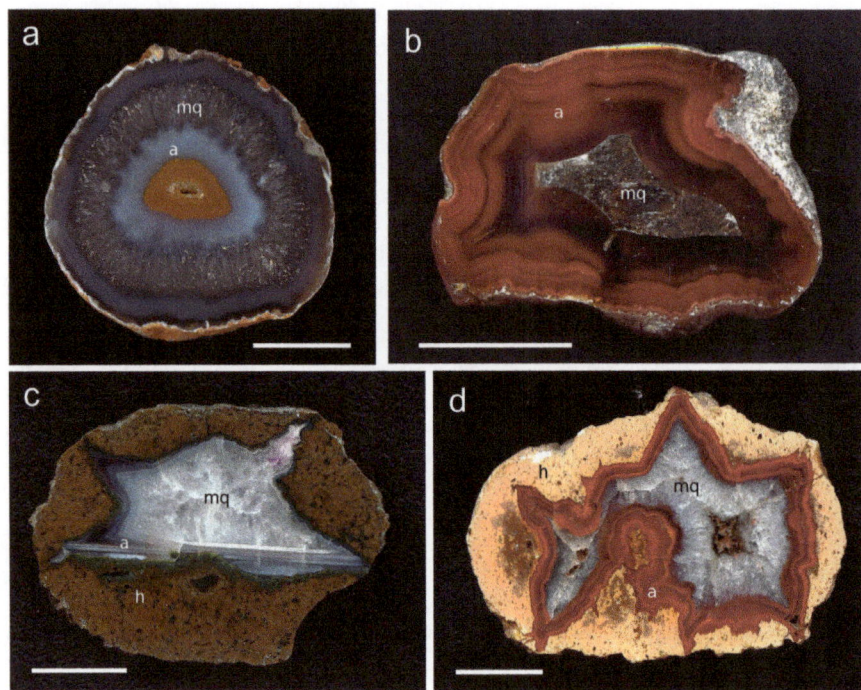

Figure 1. *Cont.*

Figure 1. Investigated agate samples of the present study with a = agate, mq = macrocrystalline quartz, h = host rock; (**a**) Soledade, Rio Grande do Sul (Brazil), (**b**) Heads of Ayr, Dunure area (Scotland), (**c**) Zwickau-Planitz, Saxony (Germany), (**d**) St. Egidien, Saxony (Germany), (**e**) Chemnitz-Furth, Saxony (Germany), (**f**) Chemnitz-Altendorf, Saxony (Germany), (**g**) Schlottwitz, Saxony (Germany), (**h**) Dryhead area, Montana (USA); scale bar is 2 cm.

The sample material was crushed with a small steel hammer to a size between 0.4 and 1 mm and then handpicked under a binocular microscope in order to separate the quartz and chalcedony parts of the samples and to avoid impurities and mineral inclusions.

2.2. Analytical Method

The method of thermogravimetry-mass-spectrometry is based on temperature dependent changes of the physical properties of a substance [27]. Fluids and gases, which are included in a solid, can be released during heating. In order to analyse escaping fluids, the thermo-balance (with evacuated oven chamber) is directly coupled with a mass spectrometer [27].

This analytical arrangement enables a faster analysis and prevents reactions of the released gases as the distance between sample and mass-spectrometer is kept as short as possible [28].

The gases are ionized in an ionization chamber and then separated and measured according to their mass-charge ratio [29]. The equipment used in this study was a thermo-analytic system NETZSCH STA 409 (Netsch, Selb, Germany), directly coupled to a QMS 403/5 quadrupole mass spectrometer (Pfeiffer Vacuum, Aßlar, Germany) (Figure 2). A background measurement under similar conditions was taken prior to a set of three to five sample measurements.

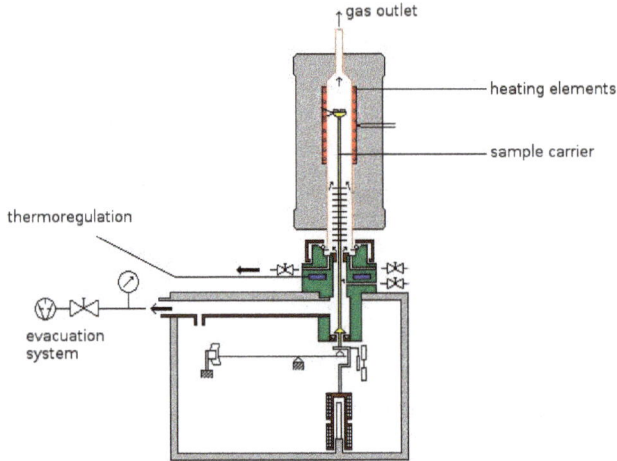

Figure 2. Schematic diagram of the apparatus NETZSCH STA 409. The gas outlet is directly coupled to the quadrupole mass-spectrometer.

Thirty milligram sample material were heated under vacuum ($<10^{-5}$ mbar) up to a temperature of 1450 °C with a constant heating-rate of 10 K/min. For correct temperature measurements, the thermo-element is located directly beneath the sample holder.

3. Results

The focus of this investigation was the characterization of the compounds that escaped during heating at various temperatures. They were identified according to their mass-charge-ratio and the corresponding intensity relations of the associated fragments. The detected and measured fluids consisted of compounds of fluorine, chlorine, sulphur, carbon and nitrogen with oxygen and/or hydrogen.

By default, the thermographic curves were also logged, revealing total mass losses during heating to 1450 °C between 0.5% and 1.5% for chalcedony and <0.5% for the quartz samples. Some similarities are found between the samples: chalcedony tends to have a rather unspecific mass loss during heating, with two main loss ranges between 100–500 °C and above 1000 °C (e.g., Chemnitz-Furth, Chemnitz-Altendorf, St. Egidien, Zwickau). Quartz revealed even more unspecific characteristics, except for the samples from Chemnitz-Furth and St. Egidien, both showing a step at around 300 °C with mass loss of 0.1%–0.2%.

3.1. Detected Compounds

3.1.1. Hydrogen

Hydrogen does not occur separately but mainly results from splitting off from a compound after ionization. The remaining hydrogen either appears as H or H_2 on the m/q = 1 and 2 (Figure 3), which also represent the mass-charge-ratios that identify hydrogen. Both H and H_2 mainly appear around 500–550 °C though H_2 often appears around 900–950 °C as well.

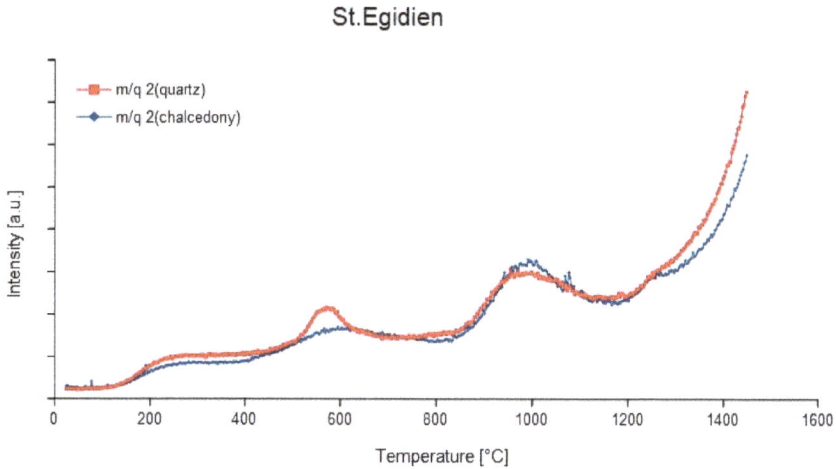

Figure 3. Temperature dependent degassing curves of the mass-charge ratio 2 ($^1H_2$) of quartz and chalcedony in agate from St. Egidien, Germany.

3.1.2. Chlorine

Chlorine occurs on the mass-charge-ratios 35 and 37. The main part (75.8%) belongs to ^{35}Cl, which is consistent with the mass-charge-ratio 35. The rest is associated with ^{37}Cl (24.2%) [30]. The Cl_2 molecule appears on the mass-charge-ratios 70, 72 (for ^{36}Cl which is not stable or $^{35}Cl^{37}Cl$) and 74, respectively. Together with H, chlorine forms HCl, which is detectable on the mass-charge-ratios 36 (Figure 4), 37 and 38 (Figure 5). In chalcedony, all mass-charge-ratios for chlorine could be identified except for the mass-charge-ratio = 35, which only occurred in the Rio Grande do Sul sample. However, in quartz all mass-charge-ratios could be identified, although m/q = 35 only occurs in the samples Chemnitz-Furth, St. Egidien and Rio Grande do Sul. The measurements showed that all compounds degas between 300–500 °C and around 1000 °C. The identification is difficult due to the interference with other compounds. The mass-charge-ratios 35 and 36 can be interfered by H_2S, 37 by $^{12}C_3^1H$ and 38 by $^{19}F_2$ and $^{12}C_3^1H_2$. All chlorine molecules show interference with hydrocarbon compounds and with SO_2.

Rio Grande do Sul

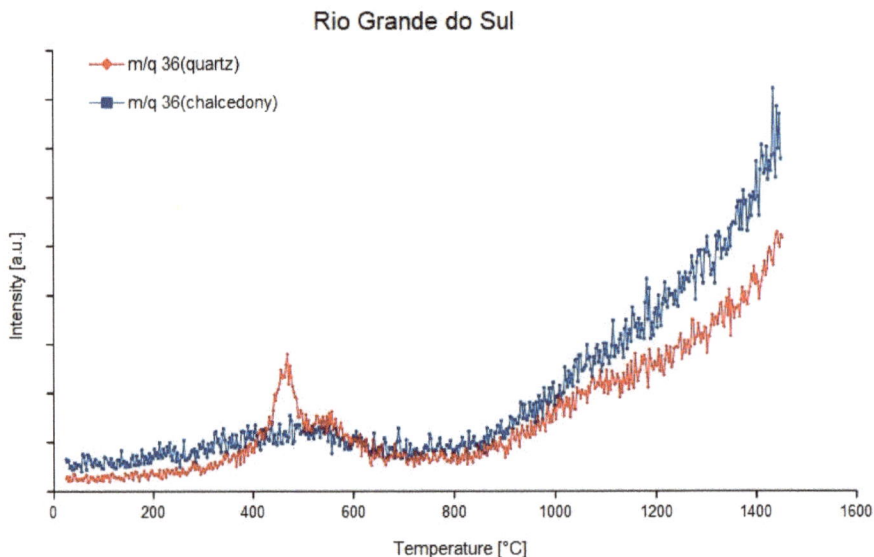

Figure 4. Degassing curves of the mass-charge ratio 36 ($^1H^{35}Cl$) of quartz and chalcedony in agate from Rio Grande do Sul, Brazil.

Montana

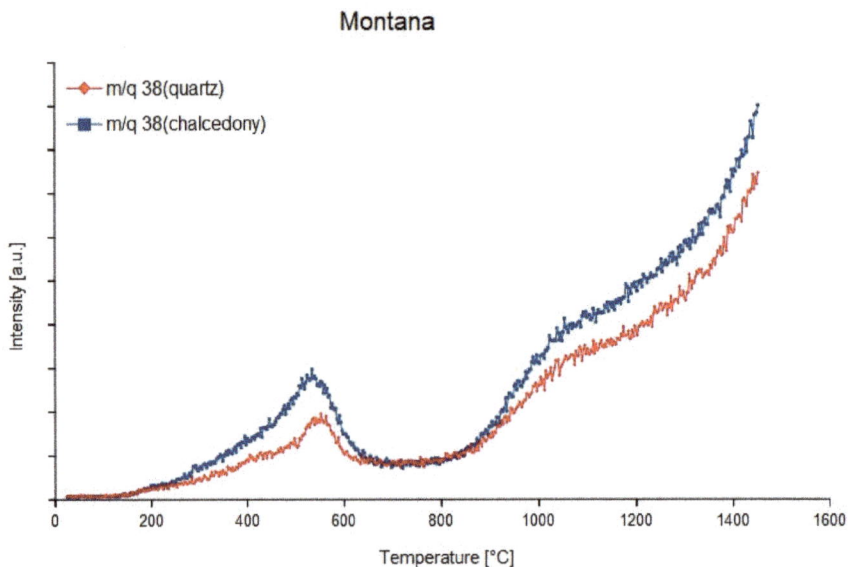

Figure 5. Degassing curves of the mass-charge ratio 38 ($^1H^{37}Cl$) of quartz and chalcedony in sedimentary agate from Montana, USA.

3.1.3. Nitrogen

Nitrogen occurs as ^{14}N and ^{15}N, though ^{15}N only makes 0.4% of the natural nitrogen. The nitrogen molecule $^{14}N_2$ can be identified on the mass-charge-ratio 28 which is more abundant besides 29. NO, N_2O and NO_2 can be formed with oxygen and can be identified by the mass-charge-ratios 30

($^{14}N^{16}O$), 44 and 45 (N_2O) and 46 ($^{14}N^{16}O_2$). N_2O and associated fragments show dominant peaks at m/q = 44, 30, 16 and 14 (Figure 6), whereas in NO_2 appears at m/q = 30, 46, 16 and 14. In combination with hydrogen, $^{14}N^1H_3$ is formed with m/q = 14–17, with the peak at m/q = 17 as the dominant. The main temperatures of degassing peak around 500 °C and around 900 °C. The nitrogen compounds show a variety of peak developments. In some samples (Zwickau, St. Egidien—see Figure 7) no peak was formed but a continuous curve occurs, especially at the mass-charge ratios 28, 30 and 44. Sometimes, strongly increasing curves appear (e.g., m/q = 44, 45 and 46 in Figure 8) and only a few samples exhibit well developed peaks (e.g., quartz in sample Chemnitz Furth and Rio Grande do Sul, respectively). A problem when identifying these compounds is the fact, that all of them (except m/q = 31 for $^{15}N^{16}O$) interfere with compounds of carbon.

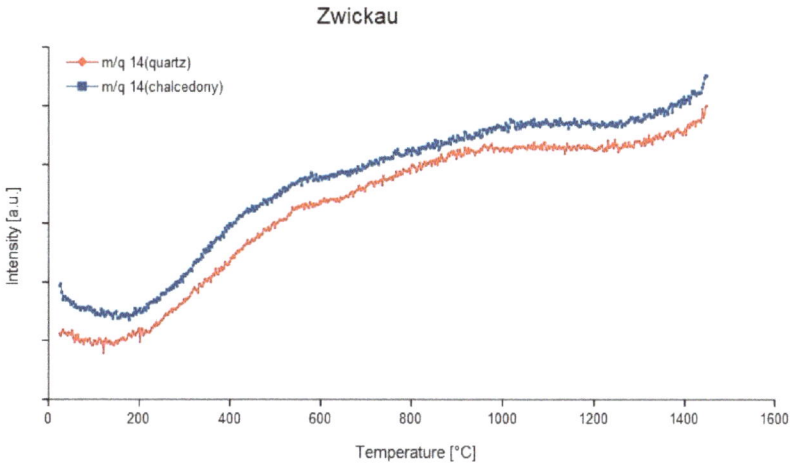

Figure 6. Continuous degassing curves of mass-charge ratio 14 (N) of the quartz and chalcedony parts in the agate from Zwickau, Germany.

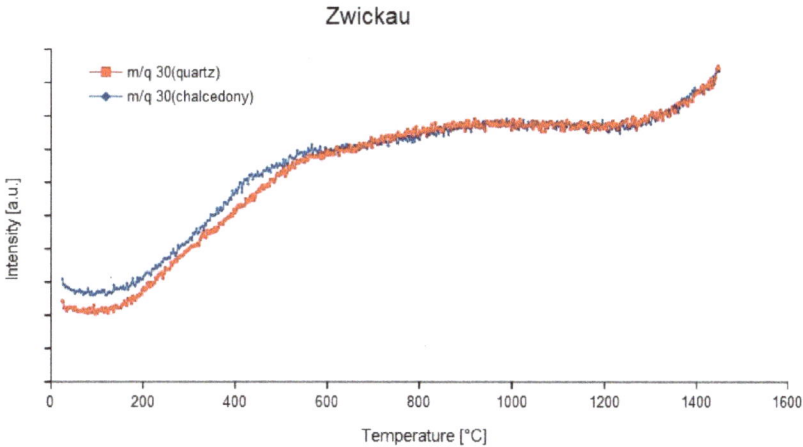

Figure 7. Continuous degassing curves of mass-charge ratio 30 ($^{14}N^{16}O$) of the quartz and chalcedony parts in the agate from Zwickau, Germany.

Chemnitz- Furth

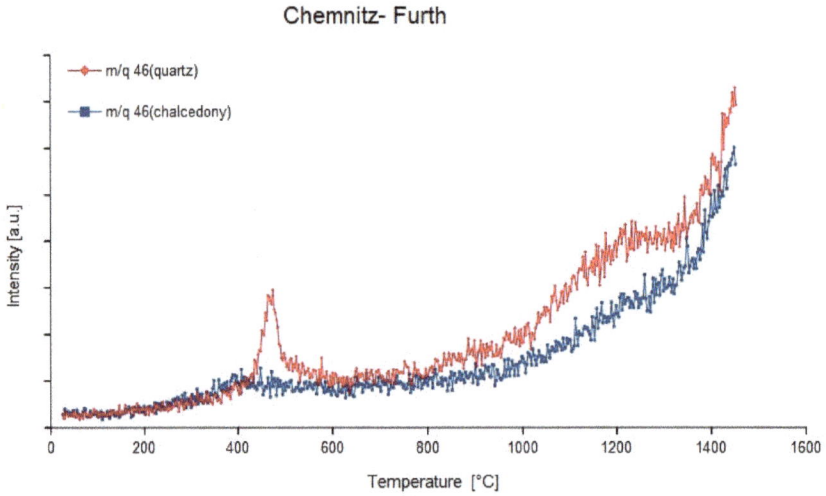

Figure 8. Degassing curves of the mass-charge ratio 46 ($^{14}N^{16}O_2$) of quartz and chalcedony in the agate sample from Chemnitz- Furth, Germany.

3.1.4. Fluorine

Fluorine compounds are associated with the mass-charge-ratios 19 to 21 (F = 19, HF = 20 and 21). The fluorine molecule $^{19}F_2$ appears at the mass-charge ratio 38. The degassing temperature was found mainly around 500 °C but also rises up to 1000 °C. Results of [20] with degassing temperatures around 900 °C confirm the recent measurements. Due to their overlap with water (m/z =18–20) the curves for ^{19}F and $^{1}H^{19}F$ do not show one clear peak but either several peaks (e.g., Rio Grande do Sul, St. Egidien in Figure 9) or a continuous curve of degassing. The interference of water with ^{19}F and $^{1}H^{19}F$ might be low but cannot be ignored. $^{19}F_2$ interferes with $^{1}H^{37}Cl$ and $^{12}C_3^{1}H_2$.

St.Egidien

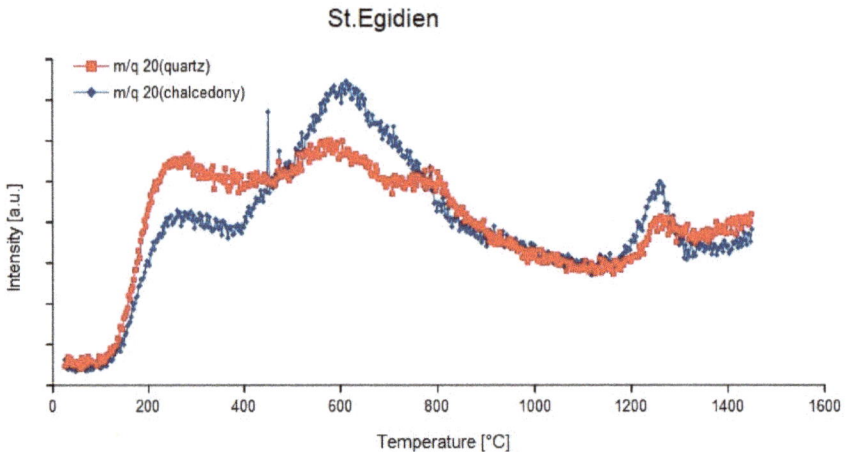

Figure 9. Degassing curves of the mass-charge ratio 20 ($^{1}H^{19}F$) of quartz and chalcedony in the agate sample from St. Egidien, Germany.

3.1.5. Carbon

Carbon has the mass-charge ratios 12 and 13. The intensity on $m/q = 12$ should be significantly higher than that on $m/q = 13$. Several hydrocarbon compounds are formed in combination with H as well as CO and CO_2 with oxygen. They can be identified via the mass-charge ratios 28 (Figure 10) and 44 which should show the highest intensity in comparison with other mass-charge ratios such as $m/q = 12$ (^{12}C), 16 (^{16}O), 29 ($^{13}C^{16}O$), 45 ($^{13}C^{16}O_2$) or 46 ($^{12}C^{17}O_2$). Furthermore, carbonic acid ($^1H_2{}^{12}C^{16}O_3$) is another compound formed with hydrogen and oxygen. It is detectable on the mass-charge-ratio 62 and the dissociation products on 61($^1H^{12}C^{16}O_3{}^-$) and 60 ($^{12}C^{16}O_3{}^{2-}$—Figure 11).

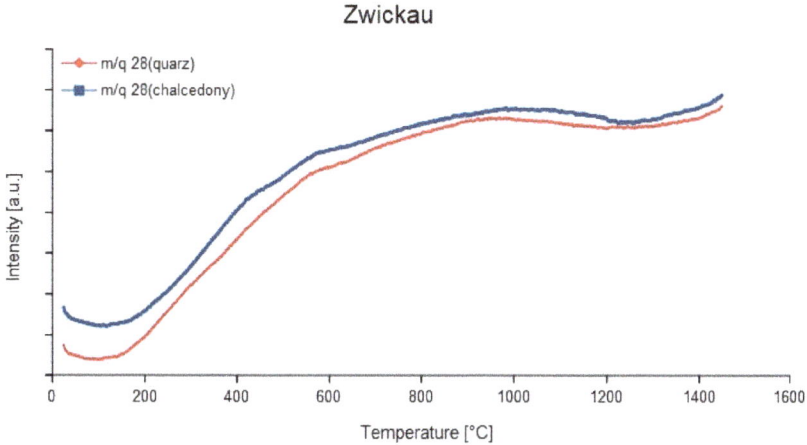

Figure 10. Continuous degassing curves of mass-charge ratio 28 ($^{12}C^{16}O$) of the quartz and chalcedony parts in the agate from Zwickau, Germany.

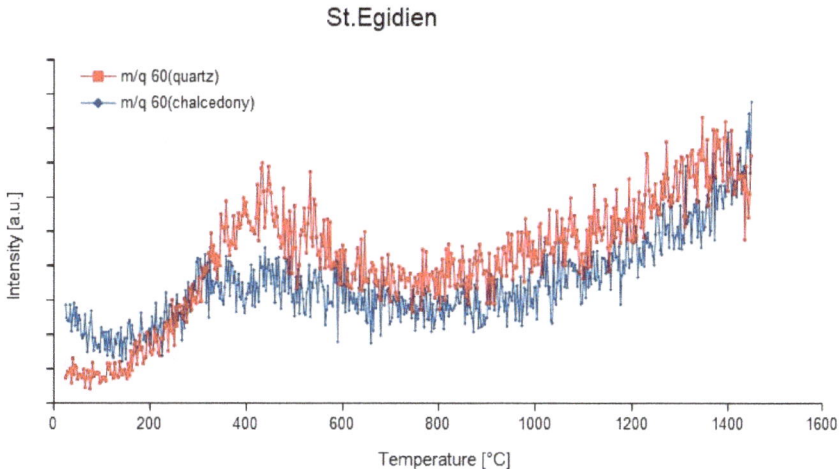

Figure 11. Degassing curves of $m/q = 60$ ($^{12}C^{16}O_3{}^{2-}$) of quartz and chalcedony in agate from St. Egidien, Germany.

Hydrocarbon compounds can also be present in hydrothermal and volcanic fluids. Studies of [31] detected hydrocarbon in agates of acidic volcanic rocks from Novy Kosciol (Poland). The most important hydrocarbon compounds are methane ($^{12}C^1H_4$, m/q = 12), ethane ($^{12}C_2^1H_6$, m/q = 30), propane ($^{12}C_3^1H_8$, m/q = 44), butane ($^{12}C_4^1H_{10}$, m/q = 58), heptane ($^{12}C_5^1H_{12}$, m/q = 72), hexane ($^{12}C_6^1H_{14}$, m/q = 86) and heptane ($^{12}C_7^1H_{16}$, m/q = 100). During ionization hydrogen splits up from the compounds and several dissociation products can form.

Looking at the degassing curves, either distinct peaks or exponential-like increasing curves can occur. The exponential increase is characteristic for the mass-charge ratios 12 (^{12}C, Figure 12), 28 ($^{12}C^{16}O$) and 44 ($^{12}C^{16}O_2$) in most samples except the agate from Chemnitz Furth (m/q = 12 and 44). All other compounds degas at temperatures around 500 °C with slight differences between quartz and chalcedony. An exception is m/q = 60 ($^{12}C^{16}O_3^{2-}$), which was only detected in chalcedony of the samples from Montana, Rio Grande do Sul, Zwickau, St. Egidien and Schlottwitz at 300–400 °C and in the quartz fraction of all samples around 400 °C. At higher temperatures (1000 °C) only hydrocarbon compounds degas.

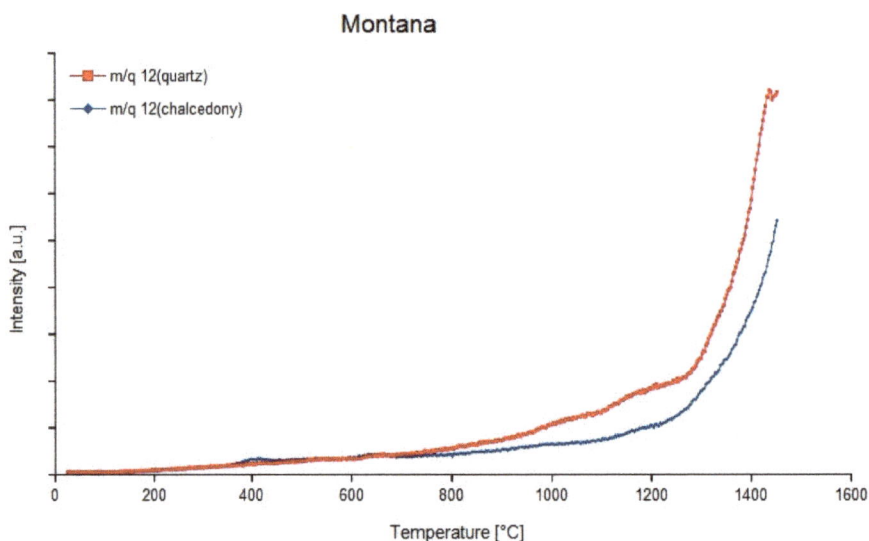

Figure 12. Degassing curves m/q = 12 (C) of quartz and chalcedony from the Montana agate showing exponential shape.

Due to their complexity, carbon compounds interfere with many other compounds. For instance, $^{12}C^{16}O$ and $^{12}C^{16}O_2$ interfere with the nitrogen compounds $^{14}N_2$, $^{14}N_2^{16}O$ and $^{14}N^{16}O_2$, heptane with chlorine (m/q = 70, 72), or sulphur compounds at high mass-charge ratios with other hydrocarbon compounds. Furthermore, different carbon compounds can interfere among each other such as $^{12}C^{16}O$ and $^{12}C_2^1H_4$ (m/q = 28—Figure 9) or $^{12}C^{16}O_2$ and $^{12}C_3^1H_8$ (m/q = 44).

3.1.6. Sulphur

Natural sulphur consists of 95% ^{32}S, 4.2% ^{33}S and 0.75% ^{34}S [30]. It can form compounds with hydrogen and oxygen, such as SO_2, H_2S, H_2SO_4 and their decomposition products. SO_2 can be identified with the mass-charge ratios 64 ($^{32}S^{16}O_2$, Figure 13), 48 ($^{32}S^{16}O$) and 32 (^{32}S), which are the most important besides 65, 50, 49, 34 and 33. H_2S should show the highest intensities on the mass-charge ratios 34 ($^1H_2^{32}S$, Figure 14), 35($^1H_2^{33}S$) and 32 (^{32}S), whereas m/q = 35 and 36 have lower intensities. H_2SO_4 is detectable on the mass-charge ratios 80 (Figure 15), 81, 82, 98, 64 and 65, which are important for identification.

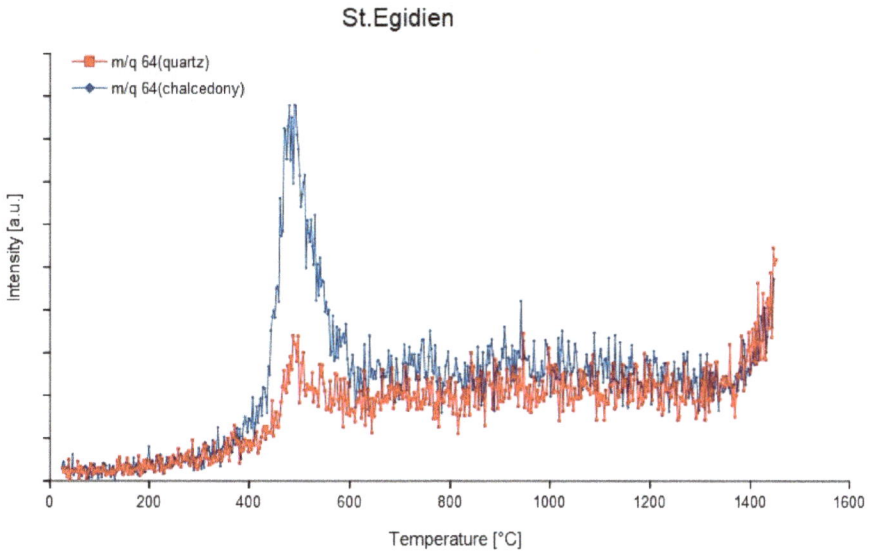

Figure 13. Degassing curves of quartz and chalcedony from the agate sample St. Egidien, Germany at mass-charge ratio 64 ($^{32}S^{16}O_2$).

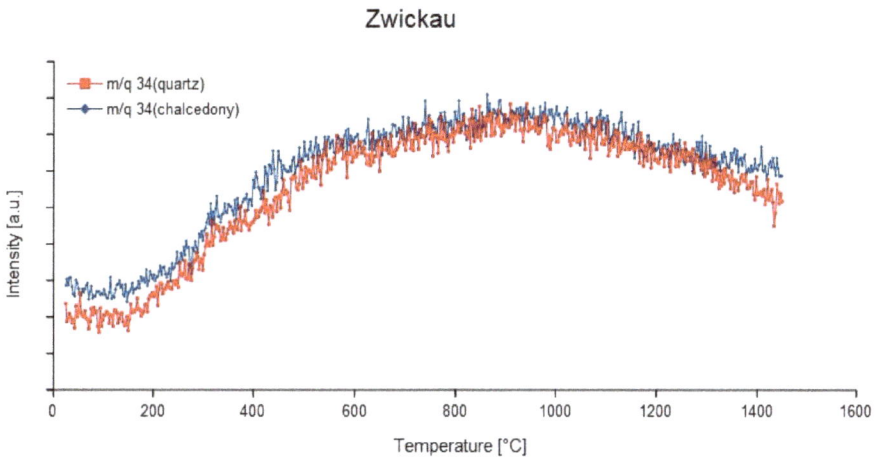

Figure 14. Continuous degassing curves of quartz and chalcedony in the agate from Zwickau, Germany at mass-charge ratio 34 ($^1H_2{}^{32}S$).

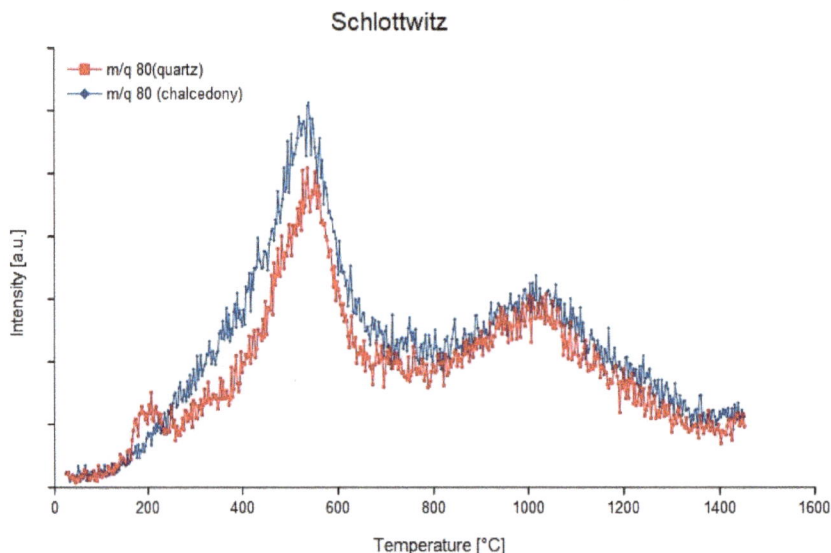

Figure 15. Degassing curves of quartz and chalcedony from the hydrothermal vein agate of Schlottwitz, Germany at mass-charge ratio 80 ($^1H_2{}^{32}S^{16}O_4$).

Most sulphur compounds degas at low temperatures between 300 and 350 °C such as $^{32}S^{16}O$ (m/q = 48), $^1H_2{}^{32}S$ (m/q = 34) and S (m/q = 33). The mass-charge ratio 64 often shows no distinct degassing peak but a strongly increasing curve, except for the agate from St.Egidien which shows a strong peak around 500 °C. For the m/q = 32, which only occurs in the chalcedony samples Rio Grande do Sul, St. Egidien and Zwickau, a curve with concave shape to the underground appears. In the quartz samples, this mass-charge ratio is not detectable in the agate from Chemnitz Furth.

Sulphur compounds show interferences with chlorine compounds, e.g., on the mass-charge ratios 35 ($^1H_2{}^{33}S$ and ^{35}Cl) and 36 ($^1H_2{}^{34}S$ and $^1H^{35}Cl$) and with hydrocarbon compounds (except ^{33}S m/q = 33 and $^{32}S^{16}O$ m/q = 48).

3.2. Volatiles Compounds in Agates of Different Origins

3.2.1. Agates from Mafic Volcanic Rocks

The agates from Rio Grande do Sul (Brazil) and Heads of Ayr (Scotland) originate from mafic volcanic rocks (basalts). The quartz part of the Rio Grande agate degasses between 250 and 550 °C with no degasification at a higher temperature range, whereas chalcedony of these samples shows degassing between 850 and 1300 °C. In contrast, both quartz and chalcedony from Heads of Ayre degas at low and high temperatures. The lower temperature range is between 150 and 600 °C with a maximum of degassing at 500 °C. At higher temperatures between 950 and 1200 °C only few volatiles could be detected (mainly hydrocarbon compounds). Both samples contain various sulphur-, chlorine- and hydrocarbon compounds as well as HF only in chalcedony.

3.2.2. Agates from Acidic and Intermediate Volcanic Rocks

Agates originating from acidic to intermediate volcanic rocks are represented by the samples from Zwickau, Chemnitz-Furth and St. Egidien, Germany. Two main ranges of degasification were detected at 400–500 °C and 900–1000 °C with some additional compounds degassing between. Chalcedony tends to degas at slightly higher temperatures (up to 1250 °C, Zwickau), whereas the escape temperatures of volatiles in macrocrystalline quartz are mostly below 1000 °C.

All samples contain compounds of nitrogen, chlorine, sulphur and carbon. The fluorine compounds ^{19}F and $^{19}F_2$ were detected in all samples, whereas $^1H^{19}F$ was only found in the chalcedony separates. Carbonates were detected as well, preferentially in quartz. For instance, H_2CO_3 and associated fragmentary compounds were measured during degassing of macrocrystalline quartz in the agate from Chemnitz Furth.

3.2.3. Vein Agates

Vein agates are represented by the samples from Chemnitz Altendorf and Schlottwitz. Both investigated agates contain sulphur and chlorine compounds as well as $^1H_2{}^{12}C^{16}O_3$. In addition, ^{19}F and $^{19}F_2$ were detected in the Chemnitz-Altendorf sample and $^{12}C^{16}O_3{}^{2-}$, $^1H^{12}C^{16}O_3$ as well as nitrogen compounds in the agate from Schlottwitz.

The agate from Chemnitz Altendorf showed a narrow temperature range of degasification. Both chalcedony and quartz mainly degas at 500 °C, quartz shows an additional escape of volatiles at a temperature of ca. 1000 °C (up to 1200 °C). The thermal behaviour of the agate from Schlottwitz is quite different. Volatiles in macrocrystalline quartz show narrow degassing temperature ranges around 250, 550 and 1000 °C, whereas chalcedony starts to degas in a wider range between 300 and 500 °C up to 950–1250 °C with some mass-charge-ratios showing additional escaping fluids in the temperature range between.

3.2.4. Sedimentary Agates

The agate from Montana, USA is also known as Dryhead agate which is formed in sedimentary rocks [26]. In this agate type compounds of nitrogen, chlorine, fluorine, sulphur and carbonate were detected. Chalcedony seems to contain more nitrogen compounds ($^{14}N^1H_3$ and $^{14}N^1H_2$), whereas fluorine (^{19}F) is more abundant in macrocrystalline quartz.

Between 200–600 °C and 900–1050 °C, maxima of degassing were detected in both chalcedony and macrocrystalline quartz, although the escape of fluids from quartz occurred in a narrower range with a maximum around 500 °C. Chalcedony on the other hand shows varying maxima of degassing. Both show additional slight degasification between or above these temperature ranges on some mass-charge-ratios.

4. Discussion

The results of the experiments illustrate that there is a number of fluids present in chalcedony and macrocrystalline quartz of agates of different genetic types and from different occurrences worldwide. The main compounds together with the corresponding degassing temperatures are summarized in Table 2. It has to be considered that the chemical compounds listed are not identical with the primary fluids included in the agate samples. During degassing and ionization of the fluids at elevated temperature, processes of dissociation can change the original chemical composition of the compounds. Therefore, the detected compounds provide only indications regarding the chemical composition of the included fluids.

Table 2. Detected compounds and corresponding temperatures of degassing in the investigated agate samples.

Sample	Material	Compounds	Temperature (°C)
Rio Grande do Sul	Quartz	$^{14}N^{16}O$, SO, S, $^{12}C^{16}O$, $^{12}C^1H$, $^1H^{19}F$	400–450
	Chalcedony	$^{14}N^{16}O$, SO, S, $^{12}C^{16}O_3{}^{2-}$, $^{12}C^1H$	150–500; 900–1050
Heads of Ayre	Quartz	$^{14}N^{16}O$, SO, $^{12}C^{16}O_3{}^{2-}$, $^{12}C^1H$	400; 600
	Chalcedony *		400; 500
Zwickau	Quartz	$^{14}N^{16}O$, SO, $^{12}C^{16}O_3{}^{2-}$	400–500; 900–1000
	Chalcedony	$^{14}N^{16}O$, $^{12}C^{16}O_3{}^{2-}$	400–500; 900–1000
Chemnitz Furth	Quartz	$^{14}N^{16}O$, $^{14}N^{16}O_2$, SO, S, $^{12}C^{16}O_3{}^{2-}$, $^{12}C^1H$, ^{12}C, $^1H^{19}F$	500; 1000
	Chalcedony	SO, $^{12}C^1H$	500; 1000
St. Egidien	Quartz	$^{14}N^{16}O$, SO, $^{12}C^{16}O_3{}^{2-}$, $^{12}C^1H$, $^1H^{19}F$	400–500; 900–1000
	Chalcedony	$^{14}N^{16}O$, SO, $^{12}C^{16}O_3{}^{2-}$, $^{12}C^1H$	400–500; 900–1000
Montana	Quartz	$^{14}N^{16}O$, SO, S, $^{12}C^{16}O_3{}^{2-}$, $^{12}C^1H$,	200–400; 500; 1000
	Chalcedony	$^{14}N^{16}O$, SO, $^{12}C^{16}O_3{}^{2-}$, $^{12}C^1H$	300–500; 1000
Chemnitz Altendorf	Quartz	$^{14}N^{16}O$, SO, S, $^{12}C^{16}O_3{}^{2-}$, $^{12}C^1H$, $^1H^{19}F$,	500; 1000
	Chalcedony	$^{14}N^{16}O$, SO, S, $^{12}C^{16}O$, $^{12}C^1H$, $^1H^{19}F$	500
Schlottwitz	Quartz	$^{14}N^{16}O$, SO, S, $^{12}C^{16}O_3{}^{2-}$, $^{12}C^1H$	500; 1000
	Chalcedony	$^{14}N^{16}O$, SO, $^{12}C^{16}O_3{}^{2-}$, $^{12}C^1H$	300–350; 450–500; 950–1000

* The chalcedony separated from the Heads of Ayre sample did not provide properly identifiable compounds due to interferences.

The degassing curves (Figures 3–15) illustrate that the escape of fluids can occur in two different ways: abruptly or as diffusion process. Sharp peaks (spikes) are interpreted to be explosion-like decrepitation effects due to the rupture of inclusions, caused by the increasing pressure in the inclusions and resulting in abrupt degassing [32]. On the other hand, a steeply rising curve is assumed to be correlated with the opening of cracks (around 800 °C). Whereas the spikes were detected at varying temperatures, in our samples the crack-correlated increase of the signals occurred around 500 and 900 °C respectively.

Diffusion, on the other hand, results in broad maxima because of a rather slow and continuous process of volatile release. It occurs when structurally bound molecules diffuse out of the quartz structure at a certain temperature. The simultaneous occurrence of crack opening and diffusion processes causes defined peaks with a broad foot, which are found for most mass-charge ratios [33,34].

The chemical composition of fluids found in nearly all samples is characterized by carbonates, nitrogen-, sulphur- and fluorine compounds as indicated in Table 2. Furthermore, chlorine was found though it shows interference with other compounds. Its presence, especially the mass-charge ratio 35, was verified by Götze et al. [19], who found chlorine as a salt in the inclusions of the samples Chemnitz-Furth and St. Egidien.

Although there was no marked difference in abundance of measured fluid compounds between quartz and chalcedony the mass charge ratios measured in quartz occurred at more temperature ranges. In addition, carbonates seem to appear more commonly in macrocrystalline quartz, whereas fluorine compounds were more frequently detected in chalcedony.

Except for the sample Heads of Ayr, the investigated samples, show two ranges of degasification for both chalcedony and macrocrystalline quartz. Most compounds escape in the lower temperature range around 500 °C. At higher temperatures between 950–1050 °C mainly hydrocarbon compounds, carbonates and sulphur compounds were detected. Chalcedony showed additional degassing between 150 to 350 °C and 1100 to 1250 °C, which is interpreted as degassing from liquids from grain boundaries and enclosed inclusions.

Differences were detected for agate samples from different genetic environments, in particular when comparing the temperatures of degassing. In general, agates from mafic volcanic rocks tend to have a broad temperature range of degassing, whereas the agates of felsic origin show a narrower one. The first release of volatiles happens around 500 °C, the second degassing around 1000 °C is more pronounced in the agates from felsic host rocks. Agates from a sedimentary host rock show two temperature ranges of degassing with the lower one between 200 and 500 °C. These different degassing temperatures show that differences in the agate micro-structure may influence the thermal release of fluids. A looser structure results in degassing at lower temperatures.

In general, the results from a recent study [35] are confirmed by this investigation. Moxon [35] differentiated between different types of water with distinct degassing temperatures (e.g., molecular water <190 °C and silanol water >1000 °C). Although the focus in the present study was more on the different compounds that evolve during heating, distinct temperature ranges were also found for both the mass loss and the detected components (see Table 2). Differences in the temperatures that were described in detail in [35] and the presented data may be found in the different design and objective of the experiments.

An interesting result is the possible role of chlorine and fluorine compounds in the transport and accumulation of silica and other elements during agate formation [20]. Both compounds could be detected, although the presence of chlorine is uncertain due to the strong interference with other compounds. Nevertheless, the results indicate the possible transport of silica in other forms than as diffusing silicic acid in aqueous pore solutions. Such a scenario is emphasized by Schrön [21], who proposed the volatile transport of silica as SiF_4.

The results in Table 2 emphasize that fluorine compounds were in particular detected in agates from volcanic environments. This can be explained by the preferred occurrence of F-bearing volatiles in volcanic processes, which can act as transport media for certain chemical compounds. Götze et al. [20] detected high concentrations of Ge, U and B in agates of acidic volcanic rocks and concluded that other fluids than aqueous ones can play a role in the alteration of volcanic rocks and the mobilization and transport of SiO_2 and other chemical compounds. Chemical transport reactions by stable fluorine compounds such as SiF_4, BF_3, GeF_4 and UO_2F_2 could be responsible for the accumulation of these elements and additional phases such as calcite or fluorite besides SiO_2 modifications in volcanic agates. Accordingly, element transport for agate formation in volcanic environments is not exclusively related to aqueous solutions.

The results of the present study also provided some more information regarding the participation of different fluids during the formation of the agates. The occurrence of hydrocarbon compounds in agates of volcanic origin was proven, which was first suggested by Dumańska-Słowik et al. [31] and Götze et al. [20]. In the case of the agates from Nowy Kościół (Poland), Dumańska-Słowik et al. [31] detected solid bitumen, which they related to algal or a mixed algal-humic origin based on the stable carbon isotope composition.

In the agates investigated in this study, certain hydrocarbon compounds as well as carbonic acid could be detected. The incorporation of hydro-carbon compounds as inclusions into the SiO_2 microstructure points to co-precipitation of both materials, probably from the same source. Hydrothermal methane and/or higher-molecular hydrocarbon compounds in the volatiles participating in the silica accumulation could have served as precursors for the detected organic fluids in the agate.

5. Conclusions

Separated chalcedony and macrocrystalline quartz from eight agates of different origin and localities were analysed by thermoanalysis directly coupled with a mass spectrometer. The temperature ranges at which degassing occurs were mainly found around 500 and 1000 °C. The analysed volatiles consisted of compounds of C, N, S, F and Cl with H and/or O. They could be identified according to their specific mass-charge ratio and the corresponding temperature of degassing. Due to interferences of different compounds with similar mass-charge ratios not all compounds could be definitely identified. Nevertheless, the results gave indications regarding the sources and transport of silica for the agate formation. Besides the transport and accumulation of silica in aqueous solutions, the possible role of fluorine compounds, CO_2 and other fluids in the alteration of rocks and the mobilization and transport of SiO_2 have to be taken into account.

The method of thermogravimetry-mass-spectrometry used is suitable for the investigation of fluid phases in minerals because of the limited requirements for preparation and the possibility for direct measurements of escaping gases. However, limitations in the proper verification of chemical compounds are given by interfering mass-charge-ratios due to the limited resolution of the mass-spectrometer. Moreover, the measured chemical compounds are not always identical with the primary included fluids, since ionization and dissociation of the volatiles at elevated temperatures can change the original chemical composition of the compounds. Therefore, the detected compounds often represent fragments of the primary volatiles.

Acknowledgments: We are grateful to Terry Moxon (Doncaster, UK) for providing samples of Scottish agates. The reviews of T. Moxon, an anonymous reviewer, improved the quality of the paper significantly.

Author Contributions: Julia Richter-Feig and Robert Möckel prepared the samples and performed the measurements. Gerhard Heide provided the analytical equipment for the analysis and Jens Götze provided the samples. The results were analysed by Julia Richter-Feig and further evaluated by Robert Möckel, Jens Götze and Gerhard Heide. Julia Richter-Feig wrote the first draft if the paper and Jens Götze and Robert Möckel substantially revised it.

Conflicts of Interest: The authors declare no conflict of interest.

References

1. Landmesser, M. Das Problem der Achatgenese. *Mitteilungen Pollichia* **1984**, *72*, 5–137. (In German)
2. Godovikov, A.A.; Ripinen, O.I.; Motorin, S.G. *Agaty*; Moskva, Nedra: Moscow, Russia, 1987; p. 368.
3. Blankenburg, H.-J. *Achat*; VEB Deutscher Verlag für Grundstoffindustrie: Leipzig, Germany, 1988; p. 203.
4. Moxon, T.; Rios, S. Moganite and water content as a function of age in agate: An XRD and thermogravimetric study. *Eur. J. Mineral.* **2004**, *4*, 693–706. [CrossRef]
5. Holzhey, G. Vorkommen und Genese der Achate und Paragensemineralien in Rhyolithkugeln aus Rotliegendvulkaniten des Thüringer Waldes. Ph.D. Thesis, TU Bergakademie Freiberg, Freiberg, Germany, 1993. (In German)
6. Pabian, R.K.; Zarins, A. *Banded Agates—Origins and Inclusions*; Educational Circular No. 12; University of Bebraska: Lincoln, NE, USA, 1994; p. 32.
7. Götze, J. Agate—Fascination between legend and science. In *Agates III*; Zenz, J., Ed.; Bode-Verlag: Salzhemmendorf, Germany, 2011; pp. 19–133.
8. Moxon, T.; Petrone, C.M.; Reed, S.J.B. Characterization and genesis of horizontal banding in Brazilian agate: An X-ray diffraction, thermogravimetric and electron microprobe study. *Mineral. Mag.* **2013**, *77*, 227–248. [CrossRef]
9. Moxon, T.; Reed, S.J.B. Agate and chalcedony from igneous and sedimentary hosts aged from 13 to 3480 Ma: A cathodoluminescence study. *Mineral. Mag.* **2006**, *70*, 485–498. [CrossRef]
10. Merino, E.; Wang, Y.; Deloule, E. Genesis of agates in flood basalts: twisting of chalcedony fibers and trace-element geochemistry. *Am. J. Sci.* **1995**, *295*, 1156–1176. [CrossRef]
11. Heaney, P.J. A proposed mechanism for the growth of chalcedony. *Contrib. Mineral. Petrol.* **1993**, *115*, 66–74. [CrossRef]
12. Graetsch, H. Structural characteristics of opaline and microcrystalline silica minerals. *Rev. Mineral. Geochem.* **1994**, *29*, 209–232.
13. Götze, J.; Nasdala, L.; Kleeberg, R.; Wenzel, M. Occurrence and distribution of "moganite" in agate/chalcedony: A combined micro-Raman, Rietveld and cathodoluminescence study. *Contrib. Mineral. Petrol.* **1998**, *133*, 96–105. [CrossRef]
14. Flörke, O.W.; Graetsch, H.; Martin, B.; Röller, K.; Wirth, R. Nomenclature of Micro- and Non-Crystalline Silica Minerals, Based on Structure and Microstructure. *Neues Jahrbuch Mineralogie Abhandlungen* **1991**, *163*, 19–42.
15. Breitkreuz, C. Spherulites and lithophysae—200 years of investigation on high-temperature crystallization domains in silica-rich volcanic rocks. *Bull. Volcanol.* **2013**, *75*, 705–720. [CrossRef]
16. Petránek, J. Sedimentäre Achate. *Der Aufschluss* **2009**, *60*, 291–302.
17. Holzhey, G. Herkunft und Akkumulation des SiO_2 in Rhyolithkugeln aus Rotliegendvulkaniten des Thüringer Waldes. *Geowissenschaftliche Mitteilungen von Thüringen* **1995**, *3*, 31–59. (In German)

18. Moxon, T. On the origin of agate with particular reference to fortification agate found in the Midland Valley, Scotland. *Chemie der Erde* **1991**, *51*, 251–260.

19. Götze, J.; Möckel, R.; Vennemann, T.; Müller, A. Origin and geochemistry of agates from Permian volcanic rocks of the Sub-Erzgebirge basin (Saxony, Germany). *Chem. Geol.* **2016**, *428*, 77–91.

20. Götze, J.; Schrön, W.; Möckel, R.; Heide, K. The role of fluids in the formation of agates. *Chemie der Erde* **2012**, *72*, 283–286. [CrossRef]

21. Schrön, W. Chemical fluid transport (CFT)—A window into Earth and its development. *Chemie der Erde* **2013**, *73*, 421–428. [CrossRef]

22. Gilg, H.A.; Morteani, G.; Kostitsyn, Y.; Preinfalk, C.; Gatter, I.; Strieder, A.J. Genesis of amethyst geodes in basaltic rocks of the Serra Geral formation (Ametista do Sul, Rio Grande do Sul, Brazil): A fluid inclusion, REE, oxygen, carbon and Sr isotope study on basalt, quarz and calcite. *Miner. Depos.* **2003**, *38*, 1009–1025. [CrossRef]

23. Phillips, E.R. *Petrology of the Igneous Rocks Exposed in the Ayr District (Sheet 14 W) of the Southern Midland Valley, Scotland*; British Geological Survey: Nottingham, UK, 1999.

24. Thirlwell, M.F. Geochronology of Late Caledonian magmatism in northern Britain. *J. Geol. Soc. Lond.* **1988**, *145*, 951–967. [CrossRef]

25. Haake, R.; Fischer, J.; Reissmann, R. Über die Achat - und Amethyst-Vorkommen von Schlottwitz im Osterzgebirge. *Mineralien-Welt* **1991**, *1*, 20–24.

26. Götze, J.; Möckel, R.; Kempe, U.; Kapitonov, I.; Vennemann, T. Characteristics and origin of agates in sedimentary rocks from the the Dryhead area, Montana, USA. *Mineral. Mag.* **2009**, *73*, 673–690. [CrossRef]

27. Heide, K. 1. Aufl. Leipzig. In *Dynamische Thermische Analysenmethoden*; Deutscher Verlag für Grundstoffindustrie: Leipzig, Germany, 1979.

28. Schöps, D.; Schmidt, C.M.; Heide, K. Quantitative EGA analysis of H_2O in silicate glasses. *J. Therm. Anal. Calorim.* **2005**, *80*, 749–752. [CrossRef]

29. Otto, M. 3. Aufl. Weinheim. In *Analytische Chemie*; WILEY-VCH Verlag GmbH & Co. KGaA: Weinheim, Germany, 2006.

30. Faure, G.; Mensing, T.M. *Isotopes Principles and Application*; John Wiley & Sons: Hoboken, NJ, USA, 2005.

31. Dumańska-Słowik, M.; Natkaniec-Nowak, L.; Kotarba, M.J.; Sikorska, M.; Rzymełka, J.A.; Łoboda, A.; Gaweł, A. Mineralogical and geochemical characterization of the "bituminous" agates from Nowy Kościół (Lower Silesia, Poland). *N. Jb. Miner. Abh.* **2008**, *184*, 255–268. [CrossRef]

32. Barker, C.; Smith, M.P. Mass spectrometric determination of gases in individual fluid inclusions in natural minerals. *Anal. Chem.* **1986**, *58*, 1330–1333. [CrossRef]

33. Roedder, E. *Fluid Inclusions*; Reviews in Mineralogy Volume 12; Mineralogical Society of America: Chantilly, VA, USA, 1984; p. 646.

34. Lambrecht, G.; Diamond, L.W. Morphological ripening of fluid inclusions and coupled zone-refining in quartz crystals revealed by cathodoluminescence imaging: Implications for CL-petrography, fluid inclusion analysis and trace-element geothermometry. *Geochim. Cosmochim. Acta* **2014**, *141*, 381–406. [CrossRef]

35. Moxon, T. A re-examination of water in agate and its bearing on the agate genesis enigma. *Mineral. Mag.* **2017**, *81*, 1223–1244. [CrossRef]

![minerals logo] *minerals*

MDPI

Article

Cathodoluminescence (CL) Characteristics of Quartz from Different Metamorphic Rocks within the Kaoko Belt (Namibia)

Jonathan Sittner * and Jens Götze

Institute of Mineralogy, TU Bergakademie Freiberg, Brennhausgasse 14, 09599 Freiberg, Germany;
jens.goetze@mineral.tu-freiberg.de
* Correspondence: sittner@mailserver.tu-freiberg.de; Tel.: +49-3731-392638

Received: 11 November 2017; Accepted: 28 March 2018; Published: 1 May 2018

Abstract: Quartz of metamorphic rocks from the Kaoko belt (Namibia) representing metamorphic zones from greenshist to granulite facies were investigated by cathodoluminescence (CL) microscopy and spectroscopy to characterize their CL properties. The samples cover P-T conditions from the garnet zone (500 ± 30 °C, 9 ± 1 kbar) up to the garnet-cordierite-sillimanite-K-feldspar zone (750 ± 30 °C, 4.0–5.5 kbar). Quartz from 10 different localities and metamorphic environments exclusively exhibits blue CL. The observed CL colors and spectra seem to be more or less independent of the metamorphic grade of the host rocks, but are determined by the regional geological conditions. Quartz from different localities of the garnet-cordierite-sillimanite-K-feldspar zone shows a dominant 450 nm emission band similar to quartz from igneous rocks, which might be related to recrystallization processes. In contrast, quartz from different metamorphic zones in the western part of the central Kaoko zone (garnet, staurolite, kyanite, and kyanite-sillimanite-muscovite zone) is characterized by a heterogeneous blue-green CL and a dominant 500 nm emission band that strongly decreases in intensity under electron irradiation. Such CL characteristics are typical for quartz of pegmatitic and/or hydrothermal origin and indicate the participation of fluids during neoformation of quartz during metamorphism.

Keywords: quartz; metamorphic rocks; cathodoluminescence; Kaoko belt

---

1. Introduction

Quartz (trigonal alpha-quartz) is one of the most important constituents of the Earth's crust and the most frequent silica mineral. It occurs in magmatic, metamorphic, and sedimentary rocks and thus its properties are used as an indicator for specific conditions of formation and for the reconstruction of geological processes [1,2]. In particular, cathodoluminescence (CL) properties are useful, since numerous studies of the luminescence behavior of quartz have shown highly variable characteristics depending on the specific P,T,X conditions during quartz formation [3–5].

Investigations of natural and synthetic quartz specimens showed various luminescence emission bands, which cause the visible luminescence colors under the electron beam [6–11]. The visible CL of natural quartz mainly consists of two broad emission bands centered at ~450 nm (blue emission) and 620–650 nm (red emission). The blue band is usually very broad and consists of up to four overlapping component bands centered at 390, 420, 450, and 500 nm [12]. The most frequent 450 nm emission is due to the recombination of self-trapped excitons, which involves an irradiation-induced oxygen Frenkel pair consisting of an oxygen vacancy and a peroxy linkage ($\equiv$Si–O–O–Si$\equiv$) [6]. The orange to red emission band at about 620–650 nm has been detected in almost all synthetic and natural quartz types. This emission is attributed to the recombination of electrons in the non-bridging oxygen band-gap state with holes in the valence band [13].

Zinkernagel [3] established one of the first classification schemes of quartz based on its CL colors, which he used for the provenance analysis of detrital quartz in sandstones. Over the past decades a considerable number of such methodological provenance studies have been published based both on CL colors and the spectral characteristics of quartz [13–21]. The recent state of a general classification of quartz from different environments based on investigations of a wide spectrum of quartz-bearing rocks showed that quartz of igneous, volcanic, hydrothermal, pegmatitic, and sedimentary origin can mostly be recognized due to their specific CL colors, spectra, and textures, e.g., [4,14,16,17,21,22]. In contrast, the knowledge about the CL of metamorphic quartz, in particular the development and possible transformation of CL characteristics during ongoing metamorphic processes, is still incomplete.

Zinkernagel [3] distinguished brown luminescent quartz (major peak around 620 nm and minor peak near 450 nm) of low-grade metamorphic rocks or slowly cooled high-grade metamorphic rocks from violet or blue luminescent quartz (major peaks at 450 and 620 nm) of high-grade metamorphic rocks, which have undergone relatively fast cooling. Similar observations were made by Sprunt et al. [23], who found a relationship between CL color of quartz and metamorphic grade of naturally deformed quartzite. The originally different CL colors in the undeformed parent rock material change toward a uniform brownish-red color through high-grade levels of metamorphism. Owen [24] suggested that quartz CL turns to a uniform reddish-brown color above the garnet zone during high-grade metamorphism. Augustsson and Reker [21] indicated that the change from quartz with brown/dark blue CL to brighter blue CL takes place at ca. 500–600 °C in amphibolite facies. So, metamorphic quartz that recrystallizes at high temperatures (e.g., granulite) reverts to a blue CL color comparable to that of plutonic quartz [4,5,21]. In addition, diffusion during metamorphic processes as well as fluid-controlled recovery seems to influence the CL characteristics of quartz in high-grade metamorphic rocks. For instance, heterogeneous internal textures in quartz and a correlation of the blue CL with the Ti content have been observed [25,26].

The literature data show that CL of quartz in metamorphic rocks is relatively complex and not all factors influencing the CL behaviour are known. The present study aimed to enhance knowledge about the CL characteristics of quartz in different metamorphic rocks and possible variations' dependence on the metamorphic conditions. The precondition for the detection of changes in CL properties of metamorphic quartz is the systematic investigation of a metamorphic profile covering distinct metamorphic grades in a defined geological frame. Therefore, the study was based on samples from a profile along the Gomatum and Hoarusib valleys within the Kaoko belt (Namibia), made up of rocks of greenshist to granulite facies. The mineralogical composition and thermobarometric development of these rocks are well known [27], so the CL data can be discussed in the established petrogenetic context. Combined analyses by polarizing microscopy and CL microscopy and spectroscopy provided new data concerning the CL characteristics of quartz in different grades of metamorphic rocks and revealed possible changes in the development of the CL properties through ongoing metamorphism.

2. Geological Background and Sample Material

The Kaoko belt is located in northwestern Namibia and represents a Neoproterozoic orogen, which was generated during the Pan-African-Brasiliano orogenic cycle in West Gondwana [28]. Simultaneously to this orogenesis, a collision between the Kongo, Kalahari, Rio-de-la-Plata, and São-Francisco cratons took place, forming these belt systems with the Kaoko, Damara and Gariep belt on the African side and the Dom Feliciano and Ribeira belts on the South American side [29].

The Kaoko belt has been subdivided by Miller [30] into three tectono-stratigraphic zones, Eastern, Central and Western. The NNW to SSE trending Sesfontein-Thrust and the NNW to SSE trending Puros Mylonite Zone represent the boundaries between these tectono-stratigraphic zones (Figure 1). The Eastern Kaoko Zone is characterized by low-grade metamorphic to unmetamorphosed sedimentary and carbonate rocks of the Otavi and Mulden groups. The Central Kaoko Zone includes Archean to Palaeoproterozoic orthogneiss and metasedimentary rocks. The metamorphic grade in the Central Zone increases from greenschist facies in the east to amphibolite facies in the west [27].

Mesoproterozoic migmatite and Palaeoproterozoic gneiss are dominant in the eastern part of the Western Kaoko Zone, whereas the western part is mostly composed of Neoproterozoic Pan-African granitoid and metagranitoid [31]. The metamorphic grade of the Western Kaoko Zone is characterized by a high-temperature/low-pressure Buchan-type metamorphism [27].

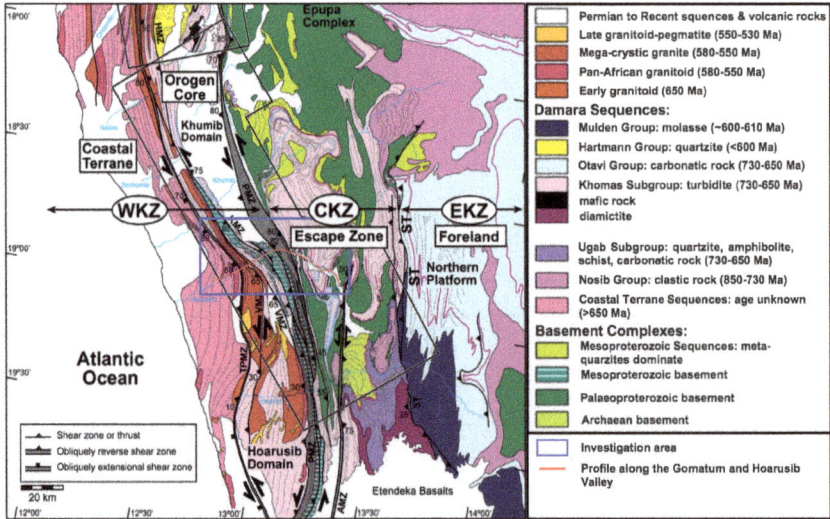

Figure 1. Geological and structural map of the Kaoko belt (modified after [32]) showing traces of the dominant foliation (fine lines) WKZ: Western Kaoko Zone, CKZ: Central Kaoko Zone, EKZ: Eastern Kaoko Zone, TPMZ: Three Palms Mylonite Zone, PMZ: Puros Mylonite Zone, HMZ: Hartmann Mylonite Zone, VMZ: Village Mylonite Zone, KMZ: Khumib Mylonite Zone, AMZ: Ahub Mylonite Zone, ST: Sesfontein Thrust.

The selected samples are mainly gneiss and mica schist of 500–750 °C metamorphic temperatures from a profile along the Gomatum and the Hoarusib valleys in the Central and Western Kaoko Zone (Figure 1). Different metamorphic zones with specific mineral associations can be described along this profile with increasing temperatures from east to west (Figure 2):

- Garnet zone: 500 ± 30 °C, 9 ± 1 kbar
- Staurolite zone: 580 ± 30 °C, 7–8 kbar
- Kyanite zone: 590 ± 30 °C, 6.5–8 kbar
- Kyanite-sillimanite-muscovite zone (ky-sill-mu zone): 650 ± 20 °C, 9 ± 1.5 kbar
- Sillimanite-K-feldspar zone (sill-ksp zone): 690 ± 40 °C, 4.5 ± 1 kbar
- Garnet-cordierite-sillimanite-K-feldspar zone (grt-cd-sill-ksp zone): 750 ± 30 °C, 4.0–5.5 kbar

The garnet, staurolite, kyanite, and kyanite-sillimanite-muscovite zones are located in the Central Kaoko Zone along the Gomatum Valley (Table 1, Figure 2). The samples from these metamorphic zones are mostly composed of biotite, muscovite, plagioclase, quartz, and some characteristic minerals from those specific zones such as garnet, staurolite, kyanite, and sillimanite.

The garnet-cordierite-sillimanite-K-feldspar zone is situated in the east of the Western Kaoko Zone and the rocks also consist of biotite, muscovite, plagioclase, and quartz, whereas the characteristic minerals are K-feldspar, garnet, sillimanite, and cordierite (Table 1, Figure 2).

Table 1. Analyzed samples with the respective metamorphic rock type and P-T conditions (data from [27]). WKZ: Western Kaoko Zone; CKZ: Central Kaoko Zone; Ky-sill-mu zone: Kyanite-sillimanite-muscovite zone; Grt-cd-sill-ksp zone: Garnet-cordierite-sillimanite-K-feldspar zone.

Sample	Metamorphic Zone	Rock Type	Temperature and Pressure Zone	Coordinates	Tectono-Stratigraphic Zone
GK 97-124	Garnet zone	Garnet mica schist	500 ± 30 °C, 9 ± 1 kbar	18°55,24 S 13°20,07 E	CKZ
GK 96-82	Garnet zone	Garnet mica schist	500 ± 30 °C, 9 ± 1 kbar	18°46,78 S 13°04,78 E	CKZ
GK 96-67	Staurolite zone	Garnet-staurolite mica schist	580 ± 30 °C, 7–8 kbar	18°48,17 S 13°07,48 E	CKZ
GK 97-47	Kyanite zone	Kyanite-staurolite mica schist	590 ± 30 °C, 6.5–8 kbar	18°48,06 S 13°07,35 E	CKZ
GK 97-48	Kyanite zone	Kyanite-staurolite mica schist	590 ± 30 °C, 6.5–8 kbar	18°48,06 S 13°07,35 E	CKZ
GK 97-127	Ky-sill-mu zone	Kyanite-sillimanite mica schist	650 ± 20 °C, 9 ± 1.5 kbar	18°47,88 S 13°04,94 E	CKZ
GK 97-06B	Grt-crd-sil-Kfs zone	Gneiss with garnet, cordierite and sillimanite	750 ± 30 °C, 4.0–5.5 kbar	18°52,76 S 12°51,14 E	WKZ
GK 97-14	Grt-crd-sil-Kfs zone	Migmatitic gneiss with garnet, cordierite and sillimanite	750 ± 30 °C, 4.0–5.5 kbar	18°52,76 S 12°51,14 E	WKZ
GK 96-116	Grt-crd-sil-Kfs zone	Migmatitic gneiss with garnet, cordierite and sillimanite	750 ± 30 °C, 4.0–5.5 kbar	18°52,77 S 12°50,63 E	WKZ
GK 96-110	Grt-crd-sil-Kfs zone	Garnet-sillimanite gneiss	750 ± 30 °C, 4.0–5.5 kbar	18°49,07 S 12°54,75 E	WKZ

3. Analytical Methods

Polished thin sections were prepared for microscopic and cathodoluminescence (CL) investigations from all samples listed in Table 1. Polarizing microscopy was carried out using a Zeiss Axio Imager A1m (ZEISS Microscopy, Jena, Germany) to document the mineral composition and micro-texture of the different rock types. Micrographs were obtained with a digital camera (MRc5) coupled with Axiovision software (ZEISS Microscopy, Jena, Germany).

CL microscopy and spectroscopy were performed on carbon-coated thin sections using a hot-cathode CL microscope HC1-LM (LUMIC, Bochum, Germany) [33]. The system was operated at 14 kV and 0.2 mA (current density ca. 10 $\mu A/mm^2$) with a defocused electron beam. Luminescence images were captured during CL operations using a peltier cooled digital video-camera (OLYMPUS DP72, OLYMPUS Deutschland GmbH, Hamburg, Germany). CL spectra in the wavelength range 370 to 920 nm were recorded with an Acton Research SP-2356 digital triple-grating spectrograph with a Princeton Spec-10 CCD detector (OLYMPUS Deutschland GmbH, Hamburg, Germany) that was attached to the CL microscope by a silica-glass fiber guide. CL spectra were measured under standardized conditions (wavelength calibration by an Hg-halogen lamp, spot width 30 μm, measuring time 5 s). Irradiation experiments were performed to document the behaviour of the quartz crystals under electron bombardment. Samples were irradiated 5 min under constant conditions (14 kV, 0.2 mA) and spectra were measured initially and after every 1 min. The evaluation of time-dependent spectral CL measurements provided further information about the stable or transient behavior under the electron beam and was indispensable for the identification of luminescence-active defect centres.

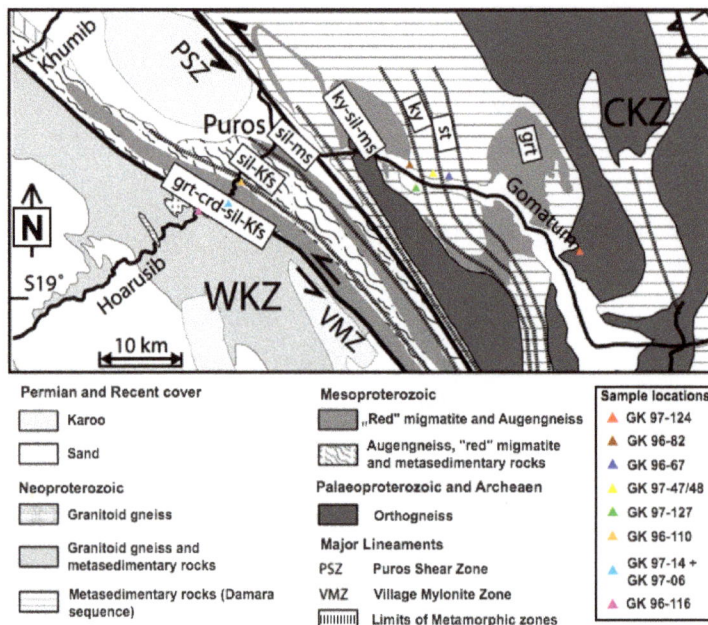

Figure 2. Schematic map of the metamorphic zones along the Gomatum and Hoarusib valleys with sample locations (modified after [34]); grt: garnet zone, st: staurolite zone, ky: kyanite zone, ky-sil-ms: kyanite-sillimanite-muscovite zone, sil-ms: sillimanite-muscovite zone, sil-Kfs: sillimanite-K-feldspar zone, grt-crd-sil-Kfs: garnet-cordierite-sillimanite-K-feldspar zone, WKZ: Western Kaoko Zone; CKZ: Central Kaoko Zone, VMZ: Village Mylonite Zone, PSZ: Puros Shear Zone.

4. Results

4.1. Garnet Zone

Sample GK-97-124, from the garnet zone in the eastern Gomatum Valley, is composed of quartz (10 vol % of the groundmass), biotite, muscovite, plagioclase, and distributed idiomorphic garnet porphyroblasts. Quartz is commonly xenomorphic, sometimes arranged in more or less monomineralic layers parallel to the foliation and frequently displays an undulatory extinction (Figure 3a,b). It has weak blue luminescence stable under the electron beam (Figure 3c) and the CL spectrum predominantly shows a weak emission band with a maximum around 450 nm and another band at about 650 nm (Figure 3d).

Sample GK 96-82, also from the garnet zone (in the western Gomatum Valley), which corresponds to the kyanite-sillimanite-muscovite zone (in the western part of the Central Kaoko Zone; Figure 2), is dominantly composed of quartz (40 vol %; Figure 3e), with similar properties to those of quartz described in GK 97-124. This quartz shows an intense short-lived blue-green luminescence, which is often heterogeneously distributed within the quartz crystals (Figure 3f) and decreases under electron irradiation turning into a weak purple CL (Figure 3g,h). The spectrum shows a broad emission band with a maximum around 490 nm and another weak emission band around 620–650 nm (Figure 3f).

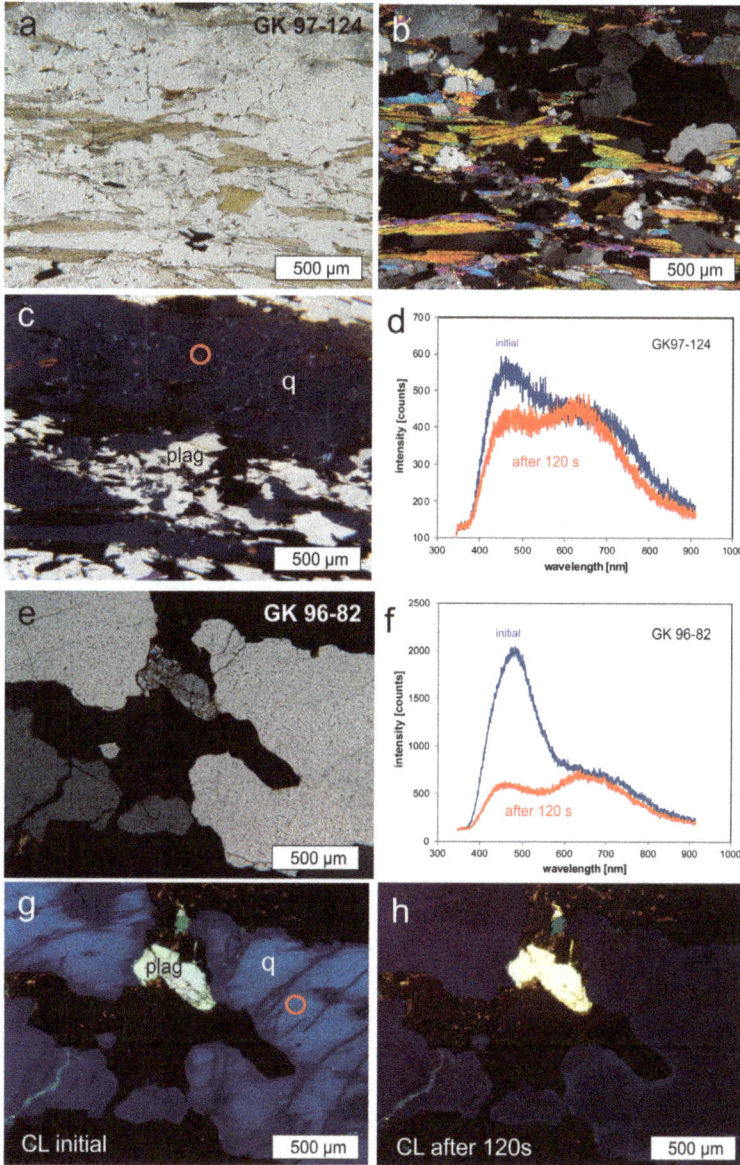

Figure 3. Polarized light and CL micrographs of two investigated metamorphic rocks from the garnet zone showing different luminescence characteristics; (**a,b**) transmitted light and polarized light micrographs of sample GK 97-124; (**c**) CL image of the same sample area showing dark blue CL of quartz (q) and bright CL of plagioclase (plag), mica is non-luminescent; the circle marks the spot for CL spectroscopy; (**d**) related CL emission spectra of sample GK 97-124; (**e**) micrograph in polarized light of a quartz-rich area in sample GK 96-82; (**f**) CL emission spectra of quartz in sample GK 96-82 showing a dominant initial emission band at ca. 490 nm and the spectrum after 120 s of electron irradiation; (**g,h**) CL images of sample GK 96-82 initially and after 120 s of electron irradiation; the short-lived blue-green CL of quartz (q) disappears and turns to a weak violet CL.

4.2. Staurolite Zone

GK 96-67 from the staurolite zone is composed of staurolite and garnet prophyroblasts in a quartz, biotite, plagioclase and muscovite groundmass. Quartz (up to 15 vol % of the rock) also forms layers parallel to the foliation direction and usually shows undulatory extinction. Its properties are similar to that of quartz characterized in GK 96-82, with an intense short-lived blue-green luminescence, a dominating 490 nm and a weak 620–650 nm emission band in the CL spectrum.

4.3. Kyanite Zone

Quartz is up to 15 vol % in rock samples from the kyanite zone (GK 97-47 and GK 97-48) and forms the groundmass together with biotite, muscovite and plagioclase. It appears in almost all monomineralic layers across the samples and shows undulatory extinction (Figure 4a). The is characterised by a strong short-lived blue-green luminescence, similar to that of sample GK 96-82 and GK 96-67 (Figure 4c,d). The CL spectrum shows a strong emission band around 490 nm and the additional weak 620–650 nm emission (Figure 4b), also similar to quartz in sample GK 96-82 (garnet zone) and GK 96-67 (staurolite zone).

4.4. Kyanite-Sillimanite-Muscovite Zone

The kyanite-sillimanite-muscovite zone (GK 97-127) is dominantly made up of quartz (30 vol %) together with kyanite, sillimanite, muscovite, biotite, and plagioclase. Quartz is anhedral with common undulatory extinction, and CL properties (an intense short-lived blue-green luminescence and emission band at ca. 490 nm) are similar to those of samples from the Central Kaoko zone.

The slight shoulder around 700–710 nm appearing in the CL spectra of GK 97-127 is probably not related to quartz but due to a CL signal from plagioclase, which is intimately intergrown with quartz (see bright areas in Figure 4c,d).

4.5. Garnet-Cordierite-Sillimanite-K-Feldspar Zone

Samples GK 97-06B, GK 97-14, GK 96-116 and GK 96-110 from the garnet-cordierite-sillimanite-K-feldspar zone show more or less similar mineralogical and spectroscopic properties. Rocks from this zone are made up of cordierite, sillimanite, K-feldspar, and garnet in the foliated matrix of biotite, plagioclase and quartz. Quartz (25 vol % of each sample) is mostly xenomorphic with undulatory extinction (Figure 4e). Its CL properties are characterized by an intense blue luminescence (Figure 4g,h). The spectrum shows an intense 450 nm emission band that slightly decreases during electron irradiation, whereas the intensity of the 650 nm emission increases (Figure 4f).

4.6. Summary of Results

The measured CL images for all studied quartz show exclusively bluish luminescence colors. Differences were detected in the intensity and homogeneity of the CL between different quartz grains and also within the same quartz crystals. In many samples, portions with short-lived bluish-green CL were observed, the intensity of which strongly decreased under the electron beam. The initial emission spectra of all studied quartz obtained by spectral measurements are made up of three main emission bands: a strong blue band at 450 nm, a second band at ca. 490 nm (a component band of a strong 500 nm emission overlapping with a weak 450 nm emission), and an emission band in the orange-red region at 620–650 nm. The presence or absence and relative intensities of these bands cause the visible CL colors.

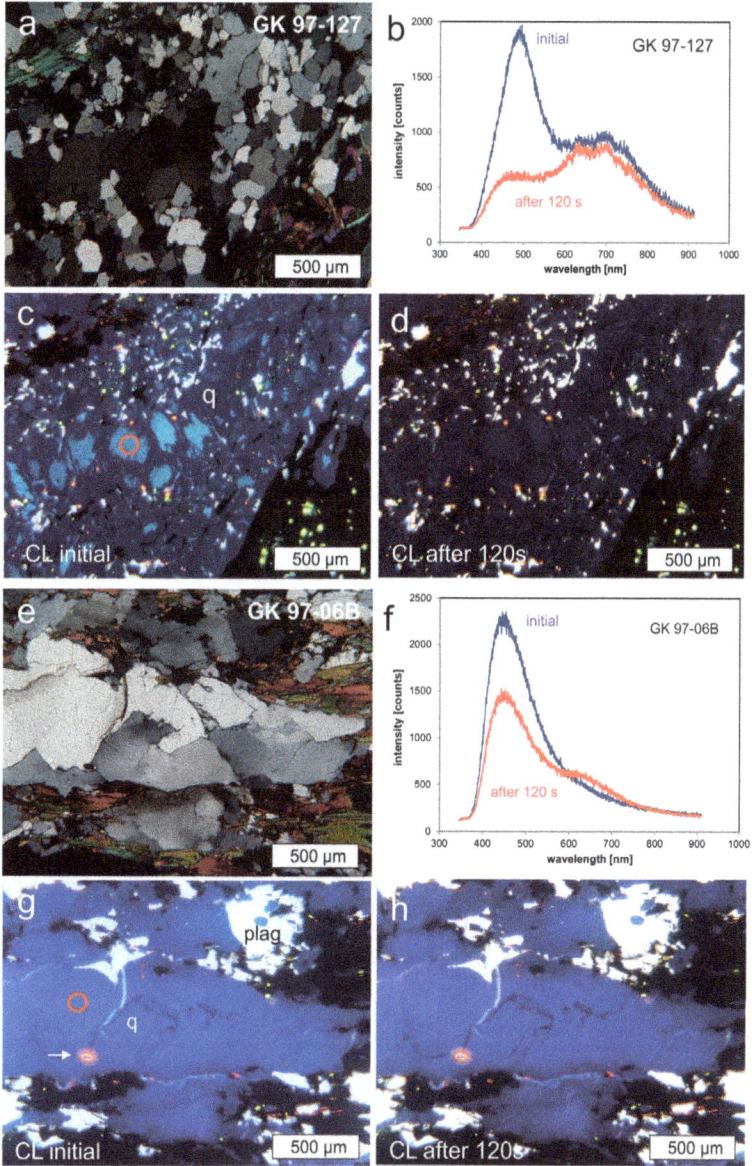

Figure 4. Polarized light and CL micrographs of two metamorphic rocks from the kyanite-sillimanite-muscovite zone (GK 97-127) and the garnet-cordierite-sillimanite-K-feldspar zone (GK 97-06B) showing different luminescence characteristics; (**a**) polarized light micrograph of sample GK 97-127; (**b**) CL emission spectra of quartz in sample GK 97-127 showing a dominant initial emission band at ca. 490 nm and the spectrum after 120 s of electron irradiation; (**c**,**d**) CL images of sample GK 96-82 initially and after 120 s of electron irradiation; the short-lived blue-green CL of quartz disappeared; (**e**) micrograph in polarized light of a quartz-rich area in sample GK 97-06B; (**f**) CL emission spectra of quartz in sample GK 97-06B showing a dominant initial emission band at ca. 450 nm and the spectrum after120 s of electron irradiation (**g**,**h**) CL images of sample GK 97-06B showing blue CL of quartz (q) and radiation halos around micro-inclusions (arrow); the circle marks the spot for CL spectroscopy.

5. Discussion

5.1. CL Colors and Emission Bands

The 450 nm emission is strongly polarized along the c-axis and is due to the recombination of the self-trapped exciton [6]. The polarization effect might influence the CL intensity due to varying crystallographic orientations of quartz grains in the cutting plane of the thin section. The self-trapped exciton involves an irradiation-induced oxygen Frenkel pair consisting of an oxygen vacancy and a peroxy linkage ($\equiv$Si–O–O–Si$\equiv$). The transient defect centre has a lifetime in the order of 1 ms. Stevens-Kalceff and Phillips [6] observed an increasing intensity of the ~2.7 eV CL emission with irradiation exposure due to the migration of competing radiative and non-radiative centres out of the interaction volume under the influence of the beam-induced electric field. In contrast, Luff and Townsend [35] reported a slight decrease of this emission band during CL experiments at room temperature, which is in accordance with the results of this study.

The bluish-green ca. 500 nm emission can be related to cation-compensated trace-element centres in the quartz structure [7,36]. Combined measurements on pegmatite quartz using electron paramagnetic resonance (EPR), CL and trace element analyses showed a complete lack of intrinsic lattice defects associated with O or Si vacancies, whereas some trace elements (Al, Ti, Ge, Li) were enriched and form paramagnetic centres [37]. The strong decrease of this blue-green luminescence during electron irradiation can be related to ionization enhanced diffusion of luminescence centres as it was shown by Ramseyer and Mullis [36] by electro-diffusion experiments. The sensibility of this luminescence emission is caused by the presence of charge-balancing cations (e.g., Li^+, Na^+, H^+) and their interaction with the electron beam.

The orange to red emission band at about 620–650 nm has been detected in almost all synthetic and natural quartz crystals. This emission is attributed to the recombination of electrons in the non-bridging oxygen band-gap state with holes in the valence band edge [13]. A number of different precursors of this non-bridging oxygen hole centre (NBOHC) have been proposed, such as hydrogen or sodium impurities ($\equiv$Si–O–H, $\equiv$Si–O–Na), peroxy linkages (oxygen-rich samples), or strained silicon-oxygen bonds [6]. Upon irradiation, precursor centres can be transformed into the NBOHC resulting in an initial increase of the CL emission at 1.9 eV during electron bombardment and subsequent stabilization.

5.2. CL Characteristics of the Metamorphic Quartz

Comparison of all investigated quartz samples revealed different groups concerning their observed luminescence characteristics (Table 2). Group 1 comprises quartz from the garnet-cordierite-sillimanite-K-feldspar zone from the Eastern part of the Western Kaoko zone. The samples come from a restricted area with similar geological history. Quartz shows a dominant 450 nm emission band with only slightly varying initial intensities (Figure 5). During electron irradiation the intensity of the 450 nm band decreases, whereas the intensity of the 650 nm emission increases. The CL colors and spectra are similar to those of quartz from igneous rocks (e.g., granite). The dominating blue CL with the characteristic emission band at 450 nm indicates recrystallization of quartz during metamorphism. The increase of the 650 nm band during electron irradiation can be related to the conversion of precursor defects into the NBOHC. Strained silicon-oxygen bonds that originate from the stress during metamorphic processes could be assumed to be one such preferred precursor.

Group 2 comprises samples of different metamorphic levels in the western part of the central zone. The material represents quartz from the garnet zone (GK 96-82), staurolite zone (GK 96-67), kyanite zone (GK 97-47) and kyanite-sillimanite-muscovite zone (GK 97-127), all situated east of the Puros mylonite zone. The CL of this quartz is characterized by a visible blue-green color that disappears under electron irradiation (Figure 6). The changes of the CL spectra during electron irradiation show that the broad emission band centered at 490 nm is composed of a transient 500 nm emission and a weak 450 nm emission band that is clearly visible after electron bombardment. The short-lived blue-green CL often shows a patchy texture indicating a heterogeneous distribution of relevant luminescence centers.

Assuming cation-compensated trace-element centers as responsible for the transient 500 nm emission, the observed features point to the heterogeneous incorporation of trace elements into the quartz structure. It can be assumed that the trace elements responsible for the CL variation were incorporated during recrystallization/neoformation of quartz and point to the participation of fluids in these metamorphic processes. Such metamorphogenic mobilisates would explain why quartz from these samples shows features that previously especially have been observed in quartz from pegmatites and of hydrothermal origin [4,5,7,21]. However, the relationship between trace elements and CL behavior in these samples should be analyzed in order to complete the interpretation of the present work.

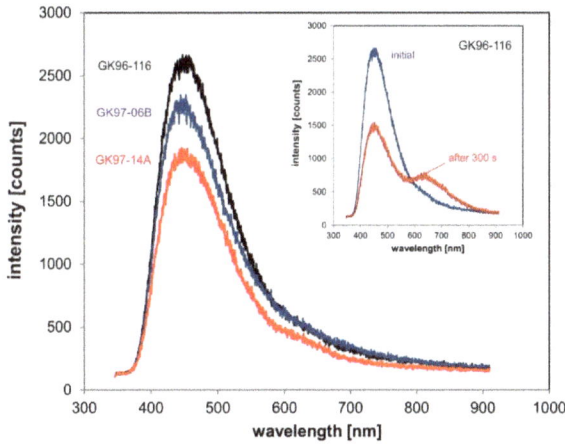

Figure 5. Initial CL emission spectra of quartz from the garnet-cordierite-sillimanite-K-feldspar zone; the dominating 450 nm band decreases during electron irradiation, whereas the 650 nm band intensity increases (see inset—the irradiation time of the electron beam was set to 300 s to see the changes of the spectra more clearly).

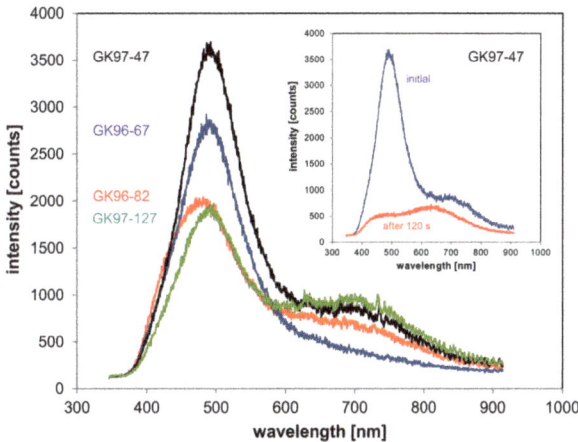

Figure 6. Initial CL emission spectra of quartz from the garnet zone (GK 96-82), staurolite zone (GK 96-67), kyanite zone (GK 97-47), and kyanite-sillimanite-muscovite zone (GK 97-127); the dominating 490 nm band strongly decreases during electron irradiation revealing the slight 450 nm band; the 650 nm emission intensity is more or less constant (see inset).

In GK 97-124 sampled from the garnet zone far away from the other sampling points, quartz is irregularly distributed along the foliation between muscovite and plagioclase and exhibits a weak blue CL. The spectrum shows two emission bands at 450 and 650 nm (Figure 3d). The intensity of the blue band at 450 nm slightly decreases during electron irradiation. The CL characteristics indicate that mineral recrystallization and fluid mobilization seem to have no remarkable influence on the quartz as it was shown for the samples of the first two groups. Neither a brownish CL, characterizing quartz in low-grade metamorphic rocks [3,21], nor CL behavior was observed for quartz in high-T metamorphic rocks [21,23].

In general, compared features for quartz from all studied samples show that the position of host rocks and the local geologic settings have more influence on quartz CL properties than the metamorphic grade. Recrystallization processes cause visible blue CL similar to that of quartz in magmatic rocks with a dominant 450 nm emission band [3–5]. The intensity of the 450 nm band probably increases from east to west. Mobilization of mineralizing fluids connected with the formation of quartz seems to result in quartz with short-lived blue-green CL mostly visible in samples from the central part of the metamorphic belt. A typical brown CL that is assumed as characteristic feature of low-T metamorphic rocks [3,4,21] was not observed in any of the investigated quartz since the metamorphic profile provided no indications for such conditions. Most of the samples derive from metamorphic conditions with temperatures above 500 °C, which was reported by Augustsson and Reker [21] to be the temperature range of quartz with blue CL.

Table 2. Different CL groups with metamorphic zone, p-T conditions (data from [27]), initial CL bands and CL colors. Grt: garnet zone, St: staurolite zone, Ky: kyanite zone, Ky-sil-ms: kyanite-sillimanite-muscovite zone; Grt-cd-sill-ksp: Garnet-cordierite-sillimanite-K-feldspar zone.

Group/Sample	Metamorphic Zone	Temperature Zone	Pressure Zone	Initial CL Color	Initial CL Bands
Group 1	Grt-crd-sil-Kfs	750 ± 30 °C	4.0–5.5 kbar	Intense blue	450 nm
Group 2	Grt, St, Ky, Ky-sil-ms	500 ± 30 to 650 ± 20 °C	9 ± 1.5 kbar	Short-lived Blue-green	450 nm, 500 nm, 650 nm
GK 97-124	Grt	500 ± 30 °C	9 ± 1 kbar	Weak blue	450 nm, 650 nm

6. Conclusions

The CL properties of quartz in metamorphic rocks from Kaoko belt (Namibia) are variable and mainly depend on the regional geological conditions than the metamorphic grade. Quartz deriving from rocks covering P-T conditions from the garnet zone (500 ± 30 °C, 9 ± 1 kbar) up to the garnet-cordierite-sillimanite-K-feldspar zone (750 ± 30 °C, 4.0–5.5 kbar) exclusively exhibits visible blue CL. However, CL spectra reveal differences in the luminescence characteristics, which enable to distinguish different quartz types.

Mineral recrystallization, fluid mobility and incorporation of trace elements during quartz neoformation resulted in CL properties similar to those of quartz from igneous rocks (dominant blue band at 450 nm) and from hydrothermal/pegmatitic origin (short-lived bluish-green CL emission at 500 nm). Often these quartz grains show heterogeneous internal textures, sometimes with patchy bluish-green CL indicating a heterogeneous distribution of related luminescence centres.

The presented results have also a significant relevance for the use and interpretation of quartz CL for provenance studies in sedimentary petrology. Up to now, in particular the occurrence of quartz with brown CL (low-T) and blue CL (high-T) was used as indicator for metamorphic source rocks of detrital quartz in sediments. Because of the fact that bright blue quartz commonly indicates a plutonic origin, only in the case that little plutonic material is expected in the potential source areas, blue quartz can be assumed to be derived from high-T metamorphic rocks. In addition, the detection of transient bluish-green CL in high-T metamorphic quartz similar to that of hydrothermal/pegmatitic quartz limits its application in provenance studies of quartz-rich sediments.

Minerals **2018**, *8*, 190

One limitation of the present investigation is that the sample material represents exclusively polymineralic rock samples. Therefore, the data have to be proved by further investigations of monomineralic metamorphic rocks consisting only of quartz.

Author Contributions: Jonathan Sittner and Jens Götze performed the experiments, analyzed the data, and wrote the paper.

Acknowledgments: We are grateful to Birgit Gruner for providing us with sample material for the present study. The reviews of Carita Augustsson and two anonymous reviewers improved the quality of the paper significantly.

Conflicts of Interest: The authors declare no conflict of interest.

References

1. Götze, J. Chemistry, textures and physical properties of quartz—geological interpretation and technical application. *Miner. Mag.* **2009**, *73*, 645–671. [CrossRef]
2. Götze, J.; Möckel, R. *Quartz: Deposits, Mineralogy and Analytics*; Springer Geology: Heidelberg, Germany; New York, NY, USA; Dordrecht, The Netherlands; London, UK, 2012; p. 360, ISBN 978-3-642-22160-6.
3. Zinkernagel, U. Cathodoluminescence of quartz and its application to sandstone petrology. *Contrib. Sedimentol.* **1978**, *8*, 1–69.
4. Ramseyer, K.; Baumann, J.; Matter, A.; Mullis, J. Cathodoluminescence colours of alpha-quartz. *Miner. Mag.* **1988**, *52*, 669–677. [CrossRef]
5. Götze, J.; Plötze, M.; Habermann, D. Cathodoluminescence (CL) of quartz: Origin, spectral characteristics and practical applications. *Miner. Petrol.* **2001**, *71*, 225–250. [CrossRef]
6. Stevens-Kalceff, M.; Phillips, M.R. Cathodoluminescence microcharacterization of the defect structure of quartz. *Phys. Rev. B* **1995**, *52*, 3122–3134. [CrossRef]
7. Götze, J.; Plötze, M.; Trautmann, T. Structure and luminescence characteristics of quartz from pegmatites. *Am. Miner.* **2005**, *90*, 13–21. [CrossRef]
8. Stevens-Kalceff, M.A. Cathodoluminescence microcharacterization of point defects in α-quartz. *Miner. Mag.* **2009**, *73*, 585–606. [CrossRef]
9. Götze, J. Application of cathodoluminescence (CL) microscopy and spectroscopy in geosciences. *Microsc. Microanal.* **2012**, *18*, 1270–1284. [CrossRef] [PubMed]
10. Götte, T.; Ramseyer, K. Trace element characteristics, luminescence properties and real structure of quartz. In *Quartz: Deposits, Mineralogy and Analytics*; Götze, J., Möckel, R., Eds.; Springer Geology: Heidelberg, Germany; New York, NY, USA; Dordrecht, The Netherlands; London, UK, 2012; pp. 265–285, ISBN 978-3-642-22160-6.
11. Götze, J.; Pan, Y.; Stevens-Kalceff, M.; Kempe, U.; Müller, A. Origin and significance of the yellow cathodoluminescence (CL) of quartz. *Am. Miner.* **2015**, *100*, 1469–1482. [CrossRef]
12. Gorton, N.T.; Walker, G.; Burley, S.D. Experimental analysis of the composite blue cathodoluminescence emission in quartz. *J. Lumin.* **1997**, *72–74*, 669–671. [CrossRef]
13. Siegel, G.H.; Marrone, M.J. Photoluminescence in as-drawn and irradiated silica optical fibers: An assessment of the role of nonbridging oxygen defect centers. *J. Non Cryst. Solids* **1981**, *45*, 235–247. [CrossRef]
14. Matter, A.; Ramseyer, K. Cathodoluminescence microscopy as a tool for provenance studies of sandstones. In *Provenance of Arenites*; Zuffa, G.G.D., Ed.; Reidel Publishing Company: Dordrecht, The Netherlands, 1985; pp. 191–211.
15. Owen, M.R. Application of cathodoluminescence to sandstone provenance. In *Luminescence Microscopy: Quantitative and Qualitative Aspects*; Barker, C.E., Kopp, O.C., Eds.; SEPM: Dallas, TX, USA, 1991; pp. 67–76.
16. Seyedolali, A.; Krinsley, D.H.; Boggs, S.; O'Hara, P.F.; Dypvik, H.; Goles, G.G. Provenance interpretation of quartz by scanning electron microscope-cathodoluminescence fabric analysis. *Geology* **1997**, *25*, 787–790. [CrossRef]
17. Götze, J.; Zimmerle, W. Quartz and silica as guide to provenance in sediments and sedimentary rocks. *Contrib. Sediment. Petrol.* **2000**, *21*, 1–91.
18. Boggs, S., Jr.; Kwon, Y.I.; Goles, G.G.; Rusk, B.G.; Krinsley, D.; Seyedolali, A. Is quartz cathodoluminescence color a reliable provenance tool? A quantitative examination. *J. Sediment. Res.* **2002**, *72*, 408–415. [CrossRef]

19. Richter, D.K.; Götte, T.; Götze, J.; Neuser, R.D. Progress in application of cathodoluminescence (CL) in sedimentary geology. *Miner. Petrol.* **2003**, *79*, 127–166. [CrossRef]

20. Götte, T.; Richter, D.K. Cathodoluminescence characterization of quartz particles in mature arenites. *Sedimentology* **2006**, *53*, 1347–1359. [CrossRef]

21. Augustsson, C.; Reker, A. Cathodolumenescence Spectra of Quartz as Provenance Indicators Revisited. *J. Sediment. Res. SEPM* **2012**, *82*, 559–570. [CrossRef]

22. Götze, J. Classification, mineralogy and industrial potential of SiO_2 minerals and rocks. In *Quartz: Deposits, Mineralogy and Analytics*; Götze, J., Möckel, R., Eds.; Springer Geology: Heidelberg, Germany; New York, NY, USA; Dordrecht, The Netherlands; London, UK, 2012; pp. 1–27, ISBN 978-3-642-22160-6.

23. Sprunt, E.S.; Dengler, L.A.; Sloan, D. Effects of metamorphism on quartz cathodoluminescence. *Geology* **1978**, *6*, 305–308. [CrossRef]

24. Owen, M.R. Sedimentary petrology and provenance of the upper Jackford sandstone (Morrowan), Ouachita Mountains, Arkansas. Ph.D. Thesis, University of Illinois, Urbana, St. Louis, MO, USA, 1984.

25. Van den Kerkhof, A.M.; Kronz, A.; Simon, K.; Scherer, T. Fluid-controlled recovery in granulite as revealed by cathodoluminescence and trace element analysis (Bamble sector, Norway). *Contrib. Miner. Petrol.* **2004**, *146*, 637–652.

26. Spear, F.S.; Wark, D.A. Cathodoluminescence imaging and titanium thermometry in metamorphic quartz. *J. Metamorph. Geol.* **2009**, *27*, 187–205. [CrossRef]

27. Gruner, B. Metamorphoseentwicklung im Kaokogürtel, NW-Namibia: Phasenpetrologische und Geothermobarometrische Untersuchungen Panafrikanischer Metapelite. Ph.D. Thesis, University Würzburg, Würzburg, Germany, 2000.

28. Porada, H. The Damara-Ribeira Orogen of the Pan-African-Brasiliano cycle in Namibia (South West Africa) and Brazil as interpreted in terms of continental collision. *Tectonophysics* **1979**, *57*, 237–265. [CrossRef]

29. Frimmel, H.E.; Klötzli, U.S.; Siegfried, P.R. New Pb-Pb single zircon age constraints on the timing of Neoproterozoic glaciation and continental break-up in Namibia. *J. Geol.* **1996**, *104*, 459–469. [CrossRef]

30. Miller, R.M. The Pan-African Damara Orogen of South West Africa/Namibia. In *Evolution of the Damara Orogen*; Special Publications of the Geological Society of South Africa: Johannesburg, South Africa, 1983; pp. 31–515.

31. Kröner, S.; Konopásek, J.; Kröner, A.; Poller, U.; Wingate, M.W.D.; Passchier, C.W.; Hofmann, K.H. U-Pb and Pb-Pb zircon ages for metamorphic rocks in the Kaoko Belt of NW Namibia: A Palaeo- to Mesoproterozoic basement reworked during the Pan-African orogeny. *S. Afr. J. Geol.* **2004**, *107*, 455–476. [CrossRef]

32. Goscombe, B. and Gray, D.R. The coastal terrane of the Kaoko Belt, Namibia: Outboard arc-terrane and tectonic significance. *Precambrian Res.* **2007**, *155*, 139–158. [CrossRef]

33. Neuser, R.D.; Bruhn, F.; Götze, J.; Habermann, D.; Richter, D.K. Kathodolumineszenz: Methodik und Anwendung. *Zent. Geol. Paläontologie Teil I* **1995**, *H 1*, 287–306.

34. Kröner, S.; Jung, S.; Kröner, A. A ~700 Ma Sm Nd garnet whole rock age from the granulite facies Central Kaoko Zone (Namibia): Evidence for a cryptic high-grade polymetamorphic history? *Lithos* **2007**, *97*, 247–270.

35. Luff, B.J.; Townsend, P.D. Cathodoluminescence of synthetic quartz. *J. Phys. Condens. Matter* **1990**, *2*, 8089–8097. [CrossRef]

36. Ramseyer, K.; Mullis, J. Factors influencing short-lived cathodoluminescence of alpha-quartz. *Am. Miner.* **1990**, *75*, 791–800.

37. Götze, J.; Plötze, M.; Graupner, T.; Hallbauer, D.K.; Bray, C. Trace element incorporation into quartz: A combined study by ICP-MS, electron spin resonance, cathodoluminescence, capillary ion analysis and gas chromatography. *Geochim. Cosmochim. Acta* **2004**, *68*, 3741–3759. [CrossRef]

minerals

MDPI

Review

Lunar and Martian Silica

Masahiro Kayama [1,2,*], **Hiroshi Nagaoka** [3,4] and **Takafumi Niihara** [5,6]

1   Creative Interdisciplinary Research Division, Frontier Research Institute for Interdisciplinary Sciences, Tohoku University, Sendai 980-8578, Japan
2   Department of Earth and Planetary Materials Science, Graduate School of Science, Tohoku University, Sendai 980-8578, Japan
3   Research Institute for Science and Engineering, Waseda University, Tokyo 169-8555, Japan; hiroshi-nagaoka@asagi.waseda.jp
4   Institute of Space and Astronautical Science, Japan Aerospace Exploration Agency, Sagamihara 252-5210, Japan
5   Department of Systems Innovation, University of Tokyo, Tokyo 113-8656, Japan; niihara@sys.t.u-tokyo.ac.jp
6   University Museum, University of Tokyo, Tokyo 113-0033, Japan
*   Correspondence: masahiro.kayama.a3@tohoku.ac.jp; Tel.: +81-22-795-6687

Received: 1 May 2018; Accepted: 14 June 2018; Published: 25 June 2018

Abstract: Silica polymorphs, such as quartz, tridymite, cristobalite, coesite, stishovite, seifertite, baddeleyite-type SiO_2, high-pressure silica glass, moganite, and opal, have been found in lunar and/or martian rocks by macro-microanalyses of the samples and remote-sensing observations on the celestial bodies. Because each silica polymorph is stable or metastable at different pressure and temperature conditions, its appearance is variable depending on the occurrence of the lunar and martian rocks. In other words, types of silica polymorphs provide valuable information on the igneous process (e.g., crystallization temperature and cooling rate), shock metamorphism (e.g., shock pressure and temperature), and hydrothermal fluid activity (e.g., pH and water content), implying their importance in planetary science. Therefore, this article focused on reviewing and summarizing the representative and important investigations of lunar and martian silica from the viewpoints of its discovery from lunar and martian materials, the formation processes, the implications for planetary science, and the future prospects in the field of "micro-mineralogy".

Keywords: silica; moon; Mars; lunar and martian meteorites; Apollo samples; remote-sensing observation; igneous process; shock metamorphism; hydrothermal fluid activity

---

1. Introduction

Silica is one of the constituent minerals in the Earth's crustal rocks, and is one of the minor minerals that is often distributed in extraterrestrial materials such as lunar and martian meteorites, returned samples (Apollo and Luna collections), and the parent bodies (e.g., the Moon and Mars). Over the past ten years, various types of silica polymorphs (e.g., quartz, tridymite, cristobalite, moganite, coesite, stishovite, seifertite, and baddeleyite-type SiO_2) and non-crystalline silica (high-pressure (HP) silica glass and opal) have been reported in lunar and martian rocks using microanalyses and studying the celestial bodies through remote-sensing observations (Table 1), although a lack of the diversity and wide distribution was believed before. For example, several previous investigations have supported the hypothesis that few high-pressure phases are included in lunar meteorites, despite the existence of many craters and regoliths, because a part of them (e.g., the high-pressure SiO_2 phases) may be volatilized during impact collisions under the high vacuum condition of the Moon [1,2]. However, recent studies have revealed that coesite, stishovite, and seifertite lie in several lunar meteorites and an Apollo collection [3–5]. Hydrothermal silica was believed to be absent from extraterrestrial materials, but moganite and opal-A, originating from the fluid activity of the parent bodies, have recently been discovered from lunar and martian meteorites [6,7]. Opaline silica deposits have also been observed in association with martian volcanic materials by the Mars rover "Spirit and Opportunity" [8]. In addition, silica nanoparticles were detected in Enceladus by the Cassini spacecraft [9,10]. Of course, for decades, many researchers have reported igneous silica such as quartz, tridymite, and cristobalite in various types of extraterrestrial materials, especially in lunar samples (lunar meteorites and Apollo and Luna collections) and martian meteorites [11,12]. From the latest viewpoint of planetary science, there seems to be a wide variety of silica polymorphs in the solar system.

According to the pressure–temperature phase diagram for SiO_2 (Figure 1) (e.g., [13–17]), quartz, tridymite, and cristobalite, as well as coesite and stishovite, are stable at different pressure and temperature regions. The high-temperature and high-pressure SiO_2 phases are quenchable down to ambient conditions, and thus exist in extraterrestrial materials via rapid cooling processes during magmatic eruption, decompression after impact events, or aqueous fluid activity. For example, seifertite is thermodynamically stable at more than ~100 GPa, which is not realistic for an impact event on the Moon and Mars. However, seifertite can be converted as a metastable phase from cristobalite at ~11 GPa, and can consequently be quenched down to an ambient condition [18]. As a result of this process, seifertite remains present in some of the lunar and martian meteorites that have experienced the host bodies [4,18,19]. Thus, the occurrence of these high-temperature and high-pressure SiO_2 phases allows us to set constraints for the crystallization temperature, cooling rate, and shock pressure and temperature, which provide valuable information on magma eruption (e.g., [12,20]) and the sizes of the crater and impactor that collided on the parent bodies (e.g., [4,21]). In addition, the metastable SiO_2 phase moganite has been found to be distributed in a lunar meteorite, which is thought to have precipitated from high-pressure alkaline fluid activity on the Moon [7]. Based on the moganite precipitation model, the bulk content of H_2O ice in the Moon's subsurface can be theoretically calculated. Thus, igneous, shock metamorphic, and hydrothermal events of various celestial bodies have been interpreted by a mineralogical description of such extraterrestrial silica.

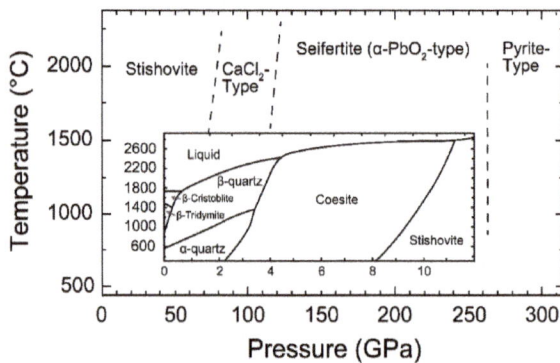

Figure 1. Phase diagram of SiO_2 modified after [15–17]. Metastable SiO_2 phases of moganite, HP-glass, and likely baddeleyite-type SiO_2 are not described here.

Table 1. Silica polymorphs discovered from lunar samples and martian meteorites.

Name	Crystal System	Space Group	Lattice Parameter (Å)				Density (g/cm³)	Occurrence	Reference
			a	B	c	β			
α-Quartz	Trigonal	$P3_121$ or $P3_221$	4.9137	4.9137	5.4047	90	2.6486	Lunar/martian samples	[22]
α-Tridymite	Monoclinic	Cc or C2/c	18.52	4.98	23.79	105.6	2.27	Lunar samples	[23]
	Triclinic (pseudo-orthorhombic)	F1	9.932	17.216	81.864	90	2.281	Terrestrial rocks *	[24]
α-Cristobalite	Tetragonal	$P4_12_12$	5.063	5.063	6.99	90	2.227	Lunar/martian samples	[4]
Coesite	Monoclinic	C2/c	7.14	12.45	7.17	120.02	2.89	Lunar/martian meteorites	[7]
Stishovite	Tetragonal	$P4_2/mnm$	4.204	4.204	2.678	90	4.216	Lunar/martian samples	[4]
Seifertite	Orthorhombic	Pbcn or Pnc2	4.097	5.0462	4.4946	90	4.2949	Lunar/martian meteorites	[25]
Baddeleyite-type	Monoclinic	P21/c	4.375	4.584	4.708	99.97	4.3	Martian meteorite	[26]
α-Moganite	Monoclinic	I2/a	8.77	4.90	10.77	90.38	2.59	Lunar meteorite	[7]
Opal-A	-	-	-	-	-	-	-	Martian meteorite	[6]

* Triclinic α-Tridymite has only been reported in terrestrial samples, but a monoclinic one has been invariably discovered from extraterrestrial materials.

Among the extraterrestrial materials, the Apollo and Luna collections, and lunar and martian meteorites, contain relatively abundant silica, reaching the bulk content of a few weight percentage at most. Some of the silica occurs as millimeter-sized to a few ten micrometer-sized euhedral or subhedral crystals, most of which have been identified as igneous tridymite and cristobalite (e.g., [27]). The coarse silica crystals affected by heavy impact events (high shock pressure-induced asteroidal collisions) frequently possess tweed or lamellae textures in their rim, where coesite, stishovite, and seifertite nanoparticles are present [4,5,19,26]. The other silica exists as subhedral to anhedral fine grains with a size of micrometers to submicrometers. In some cases, such fine silica grains appear as single crystals of igneous quartz, tridymite, and cristobalite, or consist of a mixture of numerous silica nanoparticles (quartz, cristobalite, coesite, stishovite, seifertite, and HP silica glass) because of impact events on the extraterrestrial materials. Recent investigations have also demonstrated the presence of microcrystalline aggregates of moganite with other silica polymorphs [7], as well as single nanoparticles of opal-A [6], in lunar and martian meteorites, respectively, both of which may have precipitated from hydrous fluid in the parent bodies. As described above, such micro-mineralogy of extraterrestrial silica is an important subject with regards to the implications for planetary science. However, some reports of extraterrestrial materials have only mentioned silica based on observations under optical microscope or scanning electron microscope (SEM), and did not determine the type of silica polymorphs, especially for the fine silica grains, using advanced microanalysis such as Raman microscopy, synchrotron angle dispersive X-ray diffraction (SR-XRD), transmission electron microscopy (TEM), and electron back scatter diffraction (EBSD).

Here, the purpose of this paper was to review and summarize representative and important micro-mineralogical studies of various silica polymorphs in the lunar and martian materials that have been reported up until now. Additionally, several case studies of silica that have been observed in celestial bodies by remote-sensing spacecraft have been included. In each category, named "Moon" and "Mars", we describe the types of SiO_2 phase, their occurrence, and the implications for planetary science, as well as the implied future prospects in the field of micro-mineralogy. Subsections within these categories address the origin of silica in the parent bodies.

2. Moon

2.1. Lunar Igneous Process

The Moon is a celestial body that has experienced differentiation from a lunar magma ocean (LMO) to its crust and mantle [28]. When we view the Moon from the Earth, the Moon is roughly divided into two regions, called terranes (Figure 2a,b). Bright and white-colored terranes are called "highland", and are thought to be a relic of an initial feldspathic crust solidified from a LMO. The feldspathic crust was formed by the flotation of light plagioclase due to the difference in densities between plagioclase and melt in the LMO. In contrast, heavy minerals, such as olivine and pyroxene, sank after crystallization and accumulated as a massive mafic mantle. Dark and black-colored terranes are known as maria, which are composed of basaltic magma originating from repeated partial melting and eruptions of the mafic mantle. Considering crater counts on the lunar surface (e.g., [29,30]), the basaltic igneous activities after the initial crust formation continued on the Moon for a long period of time (by 1–2 Ga). More concrete evidence of extended lunar igneous activity is evidenced by radiometric dating of lunar meteorites. The youngest sample of mare basalt dated so far, Northwest Africa 032, yielded a Rb–Sr age of 2947 ± 16 Ma [31]. Furthermore, the lunar surface can be geochemically divided into three major terranes (except for the maria) from the global standpoint of the Moon: the Feldspathic Highland Terrane (FHT), with low-Fe and low-Th contents; the Procellarum KREEP (high potassium, rare earth element, and phosphorus) Terrane (PKT), with medium-Fe and high-Th contents; and the South-Pole Aitken (SPA) terrane [32]. The FHT corresponds to the highland; the PKT is a terrane surrounding the Mare Imbrium and Oceanus Procellarum; and the SPA terrane is located on the farside of the moon, and is the largest impact basin on the Moon (Figure 2c,d). Details of each geochemical feature can be found in Jolliff et al. [32].

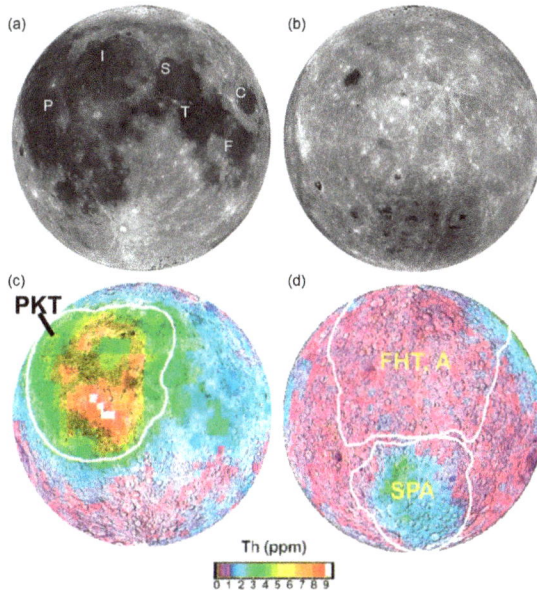

Figure 2. Albedo map of the Moon: (**a**) the nearside (face to earth), (**b**) the farside obtained by SELENE mission [33]. Label C points to the Mare Crisium, Label F points to the Mare Fecunditatis, Label T points to the Mare Tranquilitatis, Label S points to the Mare Serenitatis, Label I points to the Mare Imbrium, and Label P points to the Oceanus Procellarum. Th map of the Moon: (**c**) the nearside, (**d**) the farside [32]. PKT = Procellarum KREEP Terrane; FHT, A = Feldspathic Highland Terrane, Anorthositic; and SPA = South-Pole Aitken.

Silicate minerals are the most abundant minerals on the lunar surface, most of which belong to olivine ($(Mg,Fe)_2SiO_4$), pyroxene ($(Mg,Fe)_2Si_2O_6$–$(Ca,Mg,Fe)_2Si_2O_6$), and plagioclase ($CaAl_2Si_2O_8$–$NaAlSi_3O_8$) groups. Plagioclase has a low albite component, and K-feldspar ($KAlSi_3O_8$) is apparently a minor mineral on the Moon when compared with the Earth. Some locations with high abundances of specific minerals have been reported as endmembers of LMO: purest anorthosite (>98 vol % plagioclase) and olivine exposures with a dunite composition based on global remote-sensing data [34,35]. Oxide minerals are the second most abundant rock constituents. Among them, ilmenite ($FeTiO_3$) occurs most frequently on the Moon. It is commonly distributed in mare basalt, and its abundance varies largely from place to place [36,37]. Silica (SiO_2) is generally rare in lunar materials when compared with lunar silicate and ilmenite, as mentioned above. Phosphate minerals also occur as minor minerals in lunar rocks.

Silica found in lunar samples can be traced to three origins: igneous processes, shock metamorphism due to impact events, and hydrothermal fluid activity. The first origin leads to silica polymorphs of cristobalite, tridymite, and quartz; the second to coesite, stishovite, seifertite, baddeleyite-type SiO_2, and HP silica glass; and the third to moganite (Table 1). The mineralogy and petrology of shock-induced and aqueous silica polymorphs on the Moon are discussed in Sections 2.2 and 2.3, and lunar igneous silica is reviewed in light of recent studies of the Apollo collections, lunar meteorites, and remote-sensing observations.

2.1.1. Apollo Collections

The Apollo and Luna missions returned several kinds of samples from a total of nine locations distributed on the near side of the Moon from the 1960s to 1970s (Table 2). The total weight of all the returned samples was up to 382 kg. Six times, the Apollo astronauts collected rocks, regolith, soil, volcanic glass beads, and drill core samples. The Luna missions obtained regolith samples by drill core sampling automatically on three occasions. The locations and lithologies of the Apollo and Lunar collections are summarized as follows: the Apollo 11 and 12 and Luna 16 and 24 missions returned the mare basaltic samples. The Apollo 16 and Luna 20 missions obtained feldspathic samples from the highlands. The Apollo 15 and 17 missions set down and collected samples from the highland/mare boundary. The Apollo 14 mission landed in a geological and geochemical anomalous region, the PKT, of which the returned samples were highly enriched in incompatible elements such as rare earth elements (REE), potassium (K), and Th. Most of the Apollo-returned regolith samples were brecciated by impact events and were mixed with KREEP materials, which are enriched in elements such as potassium (K), rare earth elements (REE), and phosphorus (P) [2], because the Apollo landing sites were located around geochemically anomalous regions restricted in the central nearside of the Moon near the PKT. The geochemical anomaly of the Apollo 14 landing site was revealed later, where the Th abundances were highly enriched within PKT, as determined by a spacecraft mission by the Lunar prospector gamma-ray spectrometer [32].

Table 2. Summary of landing sites, returned sample weight, and returned date of the Apollo and Luna missions [38].

Mission	Landing Site	Sample Weight (kg)	Retuned Date
Apollo 11	Mare Tranquilitatis	21.6	24 July 1969
Apollo 12	Oceanus Procellarum	34.3	24 November 1969
Apollo 14	Mare Imbrium	42.3	9 February 1971
Apollo 15	Hadley Rille/Appenine Mts	77.3	7 August 1971
Apollo 16	Descartes Highlands	95.7	27 April 1972
Apollo 17	Mare Serenitatis	110.5	19 December 1972
Luna 16	Mare Fecunditatis	0.10	24 September 1970
Luna 20	Apollonius Highlands	0.03	25 February 1972
Luna 24	Mare Crisium	0.17	22 August 1976

Igneous silica was mostly found in the Apollo collections in the following rock types: mare basalt, quartz monzodiorite (QMD), granite, and felsite (fine-grained granite). Cristobalite is the most common silica in lunar basalts [2]. These lithological characteristics of lunar basalts are different from those of terrestrial basalts, which do not contain silica as a free phase. The Apollo sample, 15405 possesses several types of lithic fragments, such as coarse-grained granite, KREEP-rich QMD, and basalt with high-KREEP compositions (KREEP basalt), as described in [39]. Silica occurs in the granite as large, crushed, and discrete fragments, sometimes displaying grains with a characteristic fractured texture or fine-grained intergrowths with K-feldspar. This fractured textural pattern of the silica corresponds to a habit of cristobalite. QMD contains more ilmenite and phosphate, and less silica (5 or 10%), represented by assumed cristobalite, as in the case of the granite. Silica occurs in QMD in two different microtextural contexts (i.e., in the intergrowths with K-feldspar and/or as individual single grains), the same as in the granite [39]. Jolliff [11] described a SiO_2 phase in Apollo 14 QMD (14161,7069) occurring as single grains (up to 300 μm) being fractured into a mosaic pattern similar to the texture in 15405 granite [39]. However, this mineralogical description of possible cristobalite is only based on morphology and texture and no direct phase analysis [39].

Quartz is generally present in the granitic (or felsic) samples of the Apollo 12, Apollo 14, and Apollo 17 landing sites. These granitic rocks are rich in incompatible trace elements (ITE) and only comprise <0.03% of the mass of all Apollo samples [40]. Lunar granite is different from terrestrial ones because of the absence of mica and amphibole. Mineralogical and petrological investigations of silica in the Apollo samples up until a decade ago were reviewed by [2,27,37]. Therefore, this article focused on recent works on silica polymorphs in the Apollo samples.

A recent study of the Apollo 12 samples [41] reported a quartz-bearing granite fragment (12023,147-10), which possessed two different textures: (1) 80 vol % granophyric intergrowths of

K-feldspar and quartz, and (2) 15 vol % intergrowths (or mirmekitic texture) of plagioclase and quartz. Quartz within the former intergrowths was present as an inter-connected fretwork of elongate crystals typically 100 μm in length and 1–15 μm in width, or quartz laths within the K-feldspar grains. The latter was an anhedral, frequently curved quartz intergrown with K-feldspar or plagioclase. The remaining 5 vol % of this granite consisted of hedenbergitic pyroxene, fayalitic olivine, and ilmenite with other trace minerals such as zircon, yttrobetafite, thorite, apatite, Fe-metal, and monazite. The silica was identified as quartz by Raman spectroscopy. The crystallization age of this granite is 3.87 ± 0.03 Ga, using Th–U–Pb geochronology derived from an electron microprobe analyzer (EMPA) of thorite, which is relatively young among the lunar granites. The U/Pb method for zircon in lunar granophyre provides two different crystallization ages: (1) as old as 4.3 Ga, and (2) as young as 3.9 Ga. The older group was derived from the residue liquid of the LMO [42]. In contrast, the relatively young crystallization age of the Apollo 12 granite fragment (12023,147-10) suggested that the granite was related to the formation of the Imbrium impact basin where it was formed by heating and melting [41]. Seddio et al. [40] investigated other granitic fragments in the returned Apollo 12 samples: 12001,909-14, 12032,367-16, 12033,634-30, and 12033,634-34. Sample 12001,909-14 was a complex polymict granitic breccia, where the breccia phases were categorized in seven areas based on different textures and mineral compositions. Furthermore, the modal abundances of silica in the breccia phases were 4.5–37% among the seven areas. The other samples each included ~20 vol % silica. Quartz is the most common silica polymorph in lunar granite. A hackle fracture pattern of the quartz was caused by an inversion from the high-temperature and low-pressure silica polymorph of tridymite or cristobalite.

2.1.2. Lunar Meteorites

Lunar meteorites came to the Earth from the Moon by launching from the lunar surface via meteoroid or asteroid collisions. In other words, the Apollo and Luna missions only collected samples from a relatively small and geochemically anomalous region of the lunar nearside, but lunar meteorites come randomly from the entire Moon [43]. The launch processes of lunar meteorites provide global information on lunar geochemical and petrological features (e.g., [44–52]). The total estimated numbers show that 139 paired meteorites have been recovered on Earth [53]. Among them, most of the igneous silica polymorphs have been found in the basaltic or granitic clasts in lunar meteorites, similar to the Apollo returned samples.

Some of the silica-bearing igneous clasts have been recognized in the lunar meteorite, Northwest Africa (NWA) 773 clan, which is a series of paired lunar meteorites of NWA 773, NWA 2727, NWA 2977, NWA 6950, and others. Most are basaltic breccias containing igneous clasts of various lithologies including olivine gabbro, olivine phyric basalt, pyroxene phyric basalt, pyroxene gabbro, ferroan symplectite, and alkali-rich-phase ferroan rocks (ARFe) [54–56]. In fact, NWA 2977 and NWA 6950 consist entirely of olivine gabbro. Small angular fragments (<100–200 μm) of silica glass also occur in the breccia, which have two possible origins: (1) a transition of the crystalline phase to silica glass by shock metamorphism, or (2) an original amorphous phase [55]. Silica is included in ferroan symplectite, ARFe clasts, and clasts of a late-stage assemblage of silica–K–feldspar–plagioclase intergrowths plus troilite, baddeleyite, and REE-bearing-merrillite [54–56]. The symplectite clasts consist of fined-grained curved intergrowths of fayalite, hedenbergitic pyroxene, and silica [54,56]. It is difficult to determine whether the symplectic silica is crystalline or amorphous because of its fine-grained size [55]. Fagan et al. [56] suggested that the silica was formed by the breakdown of pyroxferroite on the basis of the mineral assemblage. The alkaline-phase-ferroan clasts are composed of fayalitic olivine, hedenbergitic pyroxene, silica, and Ca–phosphates. Silica occurs as elongate crystals or K-rich glass + silica. Their mineral assemblage and ferroan compositions are similar to those of the symplectite clasts, but their textural morphologies are different to those of the ARFe clasts [56]. Fagan et al. [56] implied a petrological connection between a basaltic magma system to form the olivine gabbro and a silicic magma system to generate the symplectite and ARFe clasts. The crystallization ages of the olivine gabbro lithology are much younger (3.0–3.1 Ga [31,57]) than those of the Apollo granophyric samples (3.9–4.3 Ga). On the Moon, complex igneous activities that produce silica-rich magma could have possibly continued longer than the Apollo granophyric samples suggest.

Cristobalite is a minor mineral, based on Raman spectroscopic measurements of a basaltic lunar meteorite, LaPaz Icefield (LAP) 02205, and is a low-Ti mare basalt found in Antarctica [58]. NWA 4734 is a low-Ti basalt, unbrecciated with a medium- to coarse-grained, subophitic texture. NWA 4734 contains 1.5% silica as subhedral to anhedral grains in the mesostasis, where fayalite, ilmenite, and Fe-sulfide coexist, but the crystallographic nature of the silica phase has not been described [59]. Some of the silica crystals in NWA 4734 were converted into high-pressure SiO_2 phases by impact events, as explained in the next section. Sayh al Uhaymir (SaU) 169, a lunar meteorite from Oman, has two different lithologies: the first is a polymict regolith breccia (RB), and the second is an impact-melt breccia (IMB). This lunar meteorite is the most enriched in Th among all lunar meteorites (8.44 ppm Th in RB and 32.7 ppm Th in IMB). A basaltic clast (Basalt 11) in the regolith consists of fayalite, ferroaugite, interstitial silica (tridymite), and ilmenite. The bulk composition indicates an origin from the breakdown of pyroxferroite [60], like in the case of symplectite in NWA 773 [56].

2.1.3. Remote-Sensing Observation

As shown in the above investigations of the returned samples and lunar meteorites, silica is one of the major minerals in the granitic samples. Where did the silicic volcanism that generated igneous silica take place on the Moon? Which magmatic systems formed those rocks? The first question can be answered by global remote-sensing observations. Remote-sensing data can provide us with global information on the Moon, and consequently, possible candidates of silicic volcanism on the lunar surface have already been reported, showing the following characteristics: the Christiansen Feature (CF) observed in the range of the mid-infrared wavelength, high albedo, high Th abundance, and dome-like structures (e.g., [61–65]). The Diviner Lunar Radiometer Experiment onboard the Lunar Reconnaissance Orbiter can detect the signal of silicic volcanism, mineralogically. The three spectral band-pass filters centered at 7.7, 8.25, and 8.55 µm used in the Diviner were designed to characterize silicate mineralogy and the bulk SiO_2 content by the CF. Silicic minerals and lithologies exhibiting short-wavelength positions were observed at several locations: Hansteen Alpha, Lassell, Gruithuisen, and Aristarhus. These are all located within the PKT, where Th is highly concentrated. Outside the PKT, Compton–Belkovich is also an explosive silicic volcanism, where Th abundance is enhanced (14–26 ppm Th) when compared with the surrounding highland [65]. The following three possibilities of magmatic mechanisms are discussed to explain the formation process of silicic magmas on the Moon: differentiation of a mafic magma (or KREEP basalt) [39], magma differentiation with silicate liquid immiscibility [66], and re-melting of the crust because of basaltic underplating [63].

2.2. Shock Metamorphism on the Moon

Since the Apollo era, maskelynite (naturally shock-induced plagioclase glass) and impact melt/glass have been reported in the Apollo collections and lunar meteorites as traces of impact events on the Moon. However, high-pressure minerals have lately been seemingly absent in the lunar samples, although they usually occur in impact craters of the Earth's surface [1,2]. Reasons for this have been discussed. First, high-pressure minerals, especially silica, might be eliminated by impact-induced volatilization under the high vacuum condition on the lunar surface [1,2]. Second, most high-pressure phases in terrestrial impactites, ordinary chondrites, and lunar and martian meteorites are found within and surrounding localized zones of shock melting as veins or pockets, or within amorphous glass (e.g., [17,67,68]). Thus, it seems to have been difficult to determine a structural phase of the nanoparticles using the technique of the time. However, recent investigations of lunar meteorites and an Apollo collection [3–5] have proven the existence of high-pressure minerals such as coesite, stishovite, seifertite, and baddeleyite-type SiO_2 (Table 1) by advanced microanalyses.

High-pressure silica polymorphs of lunar materials have been discovered for the first time from the gabbroic unbrecciated lunar meteorite Asuka-881757 using Raman, TEM, and EBSD analyses [3]. According to previous studies (e.g., [69,70]), Asuka-881757, found in Antarctica, consists of the constituting minerals of coarse-grained pyroxene (1–2 mm in diameter) and plagioclase (maskelynite) (1–5 mm in diameter), ilmenite with minor chromite, troilite, olivine, apatite, Fe–Ni metal, fayalite–hedenbergite–silica symplectite (at the boundaries between pyroxene and maskelynite), and several grains of silica (50–300 μm in diameter) (Figure 3a). The silica grains are entrained in the shock melt pockets that are composed of mixtures of partially melted, and then quenched, pyroxene and plagioclase glasses. There are many small granular inclusions with a size of 1–10 μm in the silica grains under optical microscope (Figure 3a). Raman spectroscopy demonstrated that most of the silica grains were amorphous because of missing Raman peaks, and the inclusions were coesite (522 cm^{-1}) (Figure 3a) and quartz (464 cm^{-1}). The coesite and quartz inclusions in the silica grains were also identified by EBSD measurements. The crystallographic orientation between the inclusions and the adjacent crystals of coesite is commonly based on the Kikuchi patterns obtained from EBSD data, implying that the inclusions are part of the same skeletal crystal. A similar orientation has been observed between the quartz inclusions and the adjacent quartz. TEM observations of the silica grains showed round-shaped coesite with a size of 300 nm and angular-shaped stishovite with a size of 100 nm.

As a result of these findings, Ohtani et al. [3] interpreted the genesis of the amorphous silica grains containing coesite, stishovite, and quartz to be a result of the transformation of a precursor cristobalite affected by shock metamorphism due to impact. According to a previous study by [1], the silica grains are considered to have originally been cristobalite that was crystallized during the final stage of the host basaltic magma. Because of shock compression during impact events, the cristobalite precursor may have been transformed to stishovite, which constrains the peak pressure to at least 8–30 GPa based on the SiO$_2$ phase diagram (Figure 1). Considering the angular-shaped morphology, stishovite formed via a solid-state reaction when the shock reached peak pressure. If crystallized from a melt, it should have shown acicular crystals or needles. Furthermore, the silica grains lay in contact with the quenched pyroxene–plagioclase glasses with a clear boundary between them, suggesting that they were not molten during impact events. The Raman spectrum of amorphous portions of the silica grains did not show the characteristic defect band of HP silica glass at about 602 cm^{-1}, where the intensity decreased by annealing at high temperatures, and therefore, the amorphous silica may have been back-transformed from the high-pressure silica polymorphs, such as stishovite, during shock decompression. The preferred crystallographic orientation of coesite and quartz crystals indicated that they were generated in topotactic relation with preexisting crystals such as stishovite. The quartz Raman band is located at 464 cm^{-1}, and did not shift to lower wavenumbers, as in the case of the shocked quartz at 456 cm^{-1} in terrestrial impact craters [71]. In the cases of the Apollo collection 14,163 and 15,271 soils, the main Raman peaks of the quartz shifted to a lower wavenumber and were broadened in width, which may have been caused by a distortion of the SiO$_2$ framework because of impact events [72]. Therefore, the quartz inclusions seem to have formed because of partial back-transformation from stishovite and/or coesite during decompression. The isotopic ages of Asuka-881757 defined the crystallization age of 3871 ± 51 Ma by ^{147}Sm-^{143}Nd dating and the impact age of 3798 ± 12 Ma by ^{39}Ar/^{40}Ar chronology [73], which belong to the putative heavy bombardment period on the lunar surface [74]. The cosmic-ray exposure age of Asuka-881757 implies that this meteorite was exposed in space after launching from the lunar surface for perhaps one million years [75]. According to Ohtani et al. [3], the impact age obtained from ^{39}Ar/^{40}Ar chronology [73] is thought to represent the time when the shock melt pocket, maskelynite, and the amorphous silica grains containing coesite, stishovite, and quartz were produced. In contrast, the high-pressure silica polymorphs are unlikely to be a product of the impact event when the meteorite was launched from the Moon, based on the space exposure age [75]. This means that the impact event at perhaps one million years ago seems to be energetically lower than that during the putative heavy bombardment period. In addition, their possible formation by impacts to the terrestrial surface can be excluded because of the cutting of a shock vein by a fusion crust (melting textures that formed because of aerodynamic frictional heating during the atmospheric entry). By combining descriptions of high-pressure silica polymorphs in lunar meteorites with various isotopic ages, we can thus obtain valuable information on the impact processes on the Moon during the putative heavy bombardment period.

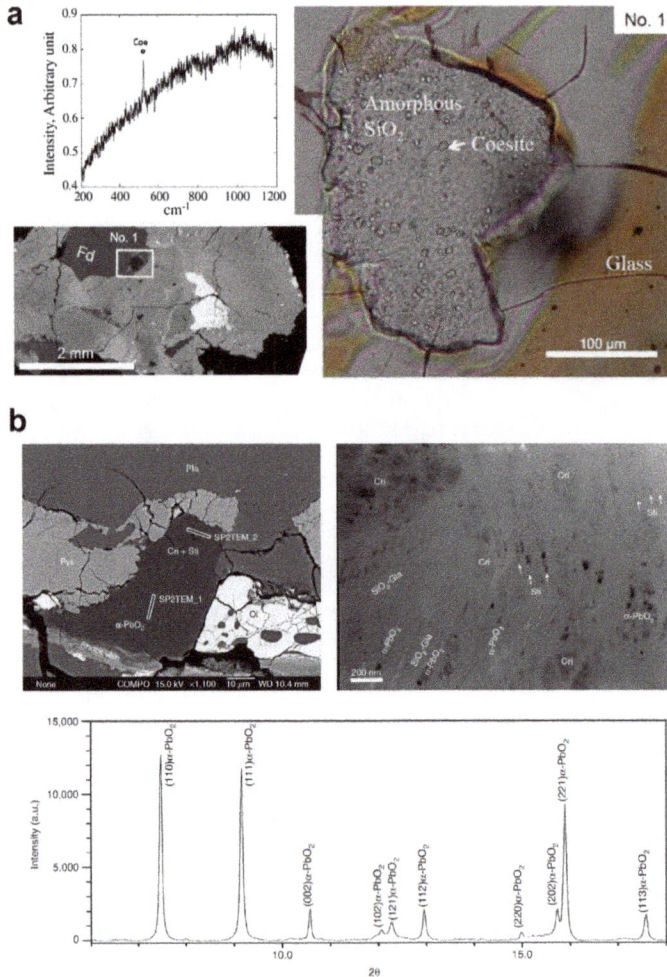

Figure 3. (**a**) Raman spectrum (**left upper**), BSE (backscattered electron) image (**left bottom**), and optical microscopic photograph (**right**) of silica grain in Asuka-881757. The amorphous silica grain (No. 1) is surrounded by glass with radiating cracks, indicating a volume increase upon pressure release. It contains many coesite (Coe) inclusions that are 1–10 µm in diameter under optical microscopy and is surrounded by feldspar (Fd) in the BSE image. A Raman spectrum of the inclusion shows the typical peak of coesite (521 cm^{-1}) (modified after [3]). (**b**) BSE image (**left upper**), transmission electron microscopy (TEM) image, and synchrotron angle dispersive X-ray diffraction (SR-XRD) pattern of the silica grain in NWA 4734. Seifertite (α-PbO$_2$)-, cristobalite (Cri)-, and stishovite (Sti)-bearing silica grain is enclosed by olivine (Olv), pyroxene (Pyx), and plagioclase (Plg), and lies adjacent to a shock vein in the BSE image. TEM observation reveals that nano-fragments of seifertite, cristobalite, and stishovite exist in the matrix of amorphous silica glass (SiO$_2$–Gla). The SR-XRD profile of the silica grain can be indexed into a seifertite structure (modified after [4]).

High-pressure SiO$_2$ phases that are stable at higher pressures and temperatures than stishovite are called "post-stishovite" (e.g., α-PbO$_2$-type named seifertite and unnamed baddeleyite-, CaCl$_2$-, and pyrite-type structures) (Figure 1). The series of post-stishovite phases are considered to be unquenchable when the static pressure releases during experimental decompression, and their existence have been confirmed with XRD analyses [76–78] only during in-situ diamond anvil cell high-pressure and temperature experiments. Until recently, the phase seifertite has been identified in some martian meteorites, as explained in the next chapter. In 2013, Miyahara et al. [4] proved the presence of seifertite in a lunar meteorite, NWA 4734, by field-emission SEM (FE-SEM), SR-XRD, TEM, Raman, and cathodoluminescence (CL) analyses. NWA 4734 is an unbrecciated lunar meteorite with a large amount of highly fractured pyroxene and lath-shaped plagioclase that is partly converted to maskelynite (Figure 3b). Relatively minor silica grains and intersertal silica–feldspar glass, and even smaller amounts of fayalite, ilmenite, baddeleyite, zirconolite, tranquilityite, pyrrhotite, and metal, were also distributed in NWA 4734 (Figure 3b). Shock veins (melting textures due to the impact event) were also seen in NWA 4734, which were either continuous or intersected by other shock veins (Figure 3b). Silica grains (~100 mm in size) were entrained in, close to, or far from the shock veins of NWA 4734. The TEM and SR-XRD analyses of excavated samples from the silica micrograins with a focused ion beam (FIB) system indicated that most of the silica grains in the shock veins were amorphous, but coesite was included in some of them close to the shock vein and occurred as nanometer-sized crystal assemblages (Figure 3b). NWA 4734 also contained tweed-like and lamellae-like textured silica grains, which lay close to and far from the shock veins, respectively. Silica grains with a lamellae-like texture become dominant as the distance from the shock vein increases. The XRD pattern of the tweed-like textured silica grain samples can be indexed to the seifertite structure (Figure 3b). Based on SR-XRD, Raman spectroscopy, and CL measurements, a small amount of stishovite coexists with seifertite. The TEM images and selected area electron diffraction (SAED) patterns of these samples demonstrated that rhomboid or spindle seifertite crystals with dimensions of 50–200 nm by 100–600 nm were surrounded by amorphous silica, and the seifertite crystals appeared to become coarser close to the shock vein. According to the SR-XRD and TEM analyses, the lamellae-like silica textured grains possess twinned α-cristobalite, platelet-shaped stishovite, and seifertite. The stishovite platelets were stacked in the twinned cristobalite under the TEM images. In some cases, amorphous silica was present between the cristobalite and stishovite subgrains.

These results indicate that α-cristobalite, originally crystallized via a rapid cooling process of the lunar magmatism, may have been converted into β-cristobalite by impact events, as indicated by the twinning and the stacking faults. Nucleation took place along the stacking faults under the shock-induced high-pressure and high-temperature conditions, followed by stishovite platelet growth. According to the pressure–temperature phase diagram of SiO$_2$ [79,80], seifertite is stable at similar pressures, but higher temperatures than stishovite. This is consistent with the conversion of cristobalite into seifertite closer to the shock veins and into stishovite farther away from the shock veins. Moreover, the seifertite in the NWA 4734 crystals became coarser as they were closer to the shock veins. Therefore, seifertite and stishovite may have been formed by the transition from cristobalite during the compression by impact events, where seifertite formed in the hotter regions when compared with stishovite. During decompression, seifertite or other high-pressure silica polymorphs would have been vitrified, resulting in a formation of abundant amorphous silica accompanied with seifertite in NWA 4734. Finally, coesite might have formed in the amorphous silica via rapid-growth processes. Considering the phase diagram [4,80] and impurity contents of aluminum and sodium in the silica grains of NWA 4734, the presence of seifertite would set a constraint for the peak-shock pressure of ~40 GPa or more. An expected temperature of 2573 K or more in the shock veins could also be deduced based on the melting temperature of the KLB-1 peridotite and Allende meteorite, which are chemically similar to NWA 4734. The duration of high-pressure waves required for stishovite formation in NWA 4734 could be estimated by equations related to the grain growth rate and thermal history, which would at least be ~0.1 s. By applying the estimated shock pressure, temperature, and duration to

Rankine-Hugoniot's relation, the impact velocity of the impactor that produced stishovite in NWA 4734 can be obtained [4]. The size of the impactor and the impact crater can be calculated based on Melosh's impact cratering law [81], as discussed in the next section. However, a recent investigation suggested that seifertite metastably appears at a much lower pressure than initially estimated [18].

Seifertite is believed to be unquenchable, that is, it cannot be recovered after an experiment from high-pressure and high-temperature conditions. In addition, this phase is thermodynamically stable at more than ~100 GPa (Figure 1), and heavy impact events that can generate such high pressure conditions are unlikely to occur on the parent bodies of meteorites in the solar system. However, seifertite has been recognized in some lunar and martian shock-metamorphosed meteorites. Recent high-pressure and high-temperature experiments, started from cristobalite with a multi-anvil apparatus and in situ SR-XRD measurements [18], promise new insight into the formation processes of seifertite as a solid-state reaction in meteorites [18]. Kubo et al. [18] revealed the pressure-, temperature-, and time-dependent appearance of seifertite based on the fact that this phase metastably formed during the compression of cristobalite at ~40 GPa and room temperature [80].

According to Kubo et al. [18], cold compressions were first performed for synthetic α-cristobalite and quartz up to ~30 GPa at room temperature, and then heated by ~1450 K with a step of 100 K, where the temperature was kept constant for ~10 to 50 min at each step. During the cold compressions, α-cristobalite was transformed into cristobalite-II and X-I. Subsequent heating experiments caused the transitions of these high-pressure cristobalite phases into metastable seifertite over a wide range of temperatures at pressures greater than ~11 GPa. Finally, a transition of metastable seifertite into the stable stishovite occurred because of further heating. Moreover, this synthetic seifertite could be quenchable down to ambient conditions via the decompression, of which the SR-XRD patterns and lattice parameters corresponded to those reported in meteorites (Table 1). In contrast, quartz was converted into stishovite at approximately 18 GPa and 800 K, and seifertite, did not appear in the high-pressure and high-temperature experiments of quartz up to ~25 GPa and 900 K.

The kinetics of the cristobalite X-I–seifertite and the seifertite–stishovite transitions were analyzed using data obtained from time-resolved SR-XRD measurements. Seifertite formation from cristobalite had very low activation energy (~10 kJ/mol), indicating fast kinetics even at low temperatures, as predicted by [82]. However, the activation energy for the stishovite formation from seifertite was relatively high (~110 kJ/mol). Therefore, seifertite can start to form as a solid-state reaction at rather low temperatures because of its low activation energy. The time–temperature–transformation curves obtained based on these kinetic parameters demonstrated that seifertite formation is time-sensitive, requiring a shock duration time of at least ~0.01 s to even start at a temperature of more than ~2000 K (not completion in the time scale of impact events). In contrast, stishovite formation is temperature-sensitive, requiring temperatures higher than ~1200 to 1500 K to start, and can complete at less than ~2000 K. Considering the solid state-reaction, the existence of seifertite in meteorites constrains the peak pressure of at least ~11 GPa and the duration time of at least ~0.01 s. The impactor size requiring the seifertite formation was inferred to be ~50 to 100 m based on the estimated impact velocities of ~5 to 10 km/s on the Moon [83] and Melosh's impact cratering law [81].

Recent FIB, SR-XRD, and TEM analyses have confirmed the existence of stishovite in Apollo collection sample 15299, returned by the Apollo 15 mission [5]. This is the first report of high-pressure polymorphs from returned lunar samples. Previous reports have indicated that Apollo 15299 is a regolith breccia, and was recovered near the Hadley Valley of the Moon in 1971 (e.g., [84,85]). Apollo 15299 contains relatively low abundances of lithic fragments (mare basalt, gabbro, anorthosite, and pre-existing breccia), mineral fragments (bytownite, clinopyroxene, orthopyroxene, olivine, spinel, and opaques), glass fragments, and glass spheres, with a large portion of glassy matrix. Traces of impact events, such as mafic impact melt breccia and shock veins, were also found in this sample. The mafic melt breccia was composed of fragments of olivine, pyroxene, plagioclase, silica and ilmenite, and glass. Vesicular melt veins of less than 200 μm width crosscut the breccia matrix and mineral fragments. Some silica grains (10–100 μm across) were found in the mafic melt breccia, a part of which was located in the shock veins. Raman spectroscopy determined that most of the silica grains were quartz, tridymite, or cristobalite based on peak positions in the spectra. One of the silica grains in a shock vein of the breccia matrix consisted of an assemblage of fine quartz and tridymite crystals (~30 μm). Most of the SR-XRD signals of the excavated blocks from this silica grain could be indexed to unit-cell parameters of stishovite and tridymite. A diffraction peak assigned to seifertite appeared in the SR-XRD spectra, although additional peaks would be required to confirm the existence of this phase in the samples. TEM observations indicated that stishovite occurred as needle-like in habit with an ~400 nm grain size, coexisting with poorly crystallized or amorphous silica.

Stishovite, in a shock vein of the breccia matrix, seems to have been transformed from quartz. However, the possibility that the stishovite formed from melt silica cannot be excluded as the needle-shaped habit is considered to be a product because of impact melting [4,86–89]. In any case, the presence of stishovite constrains the shock pressure experienced by Apollo 15299 to be >8 GPa, according to the pressure–temperature phase diagram [90,91] (Figure 1). The relatively high abundance of silica in Apollo 15299 appears to be consistent with KREEP-like basalt or impact melt derived from them. It is possible that Apollo 15299 originated from the Imbrium impact or subsequent local cratering events that occurred in the Procellarum KREEP Terrane (PKT) of the near side of the Moon as the KREEPy rocks were likely concentrated there.

As described above, various types of high-pressure silica polymorphs have been found in lunar meteorites and in the Apollo collection, and appear to be widely distributed in not only the lunar samples, but also materials around the Moon. Therefore, we emphasize that the re-examination of silica in other lunar samples from the viewpoint of a high-pressure phase is important for future micro-mineralogy.

2.3. Alkaline Fluid Activity of the Moon

Aqueous silica polymorphs, originating from fluid activity, are believed to be absent from lunar samples because the Moon is thought to be a water-depleted celestial body [92]. However, various water species (e.g., H_2O ice, OH bound to minerals, and hydrated phases) have been detected at various sites of the lunar surface by recent remote-sensing (e.g., [93,94]). Recent research has discovered the existence of moganite in the lunar meteorite NWA 2727 and indicated the likelihood that this mineral formed as a result of lunar fluid activity [7]. Moganite is a metastable phase of monoclinic SiO_2 in the *I2/a* space group (Table 1). This SiO_2 phase has been synthesized by hydrothermal experiments by high-pressure induced poly-condensation (ca. >100 MPa) and de-hydroxylation of colloidal silicic acid upon changes in pH from 9.5 to 12.0–13.0 at 373–418 K (e.g., [95,96]). On Earth, moganite has been shown to precipitate nano- to micro-crystalline SiO_2 from alkaline fluids only in sedimentary environments that produce high consolidation pressure [97–100]. It readily converts into quartz or dissolves during silica–water interaction at ambient pressure because of its thermochemical instability. Thus, moganite has a limited occurrence in unaltered sedimentary rocks. These findings, reported by previous studies, have become one of the most important pieces of evidence that moganite in NWA 2727 is indigenous to the Moon [7].

NWA 2727 is a gabbroic–basaltic breccia lunar meteorite and is paired with NWA 773, 2977, 3333, 6950, and so on, which together are named the "NWA 773 clan" (e.g., [31,54]). The NWA 773 clan is characterized by KREEP-like compositions with very low Ti, with almost the same crystallization (3.0–3.1 Ga), shock metamorphism (<2.67 ± 0.04 Ga), transition (1–30 Ma), and terrestrial ages (17 ± 1 ka) (e.g., [31,54,101,102]). NWA 2727 consists of an olivine–cumulate (OC) gabbro and pyroxene phyric basalt lithic clasts within a breccia matrix (Figure 4). The OC gabbroic clasts in NWA 2727 contain abundant euhedral olivine and clinopyroxene (0.1–0.5 mm in radius) with minor amounts of anhedral plagioclase [7]. The basaltic clasts are composed of clinopyroxene phenocrysts with a groundmass of clinopyroxene, plagioclase, and small fine-grained silica (10–20 μm in radius). The breccia matrix fills the interstices between these clasts and is composed of numerous fine to coarse grains of the OC gabbro and basalt lithic minerals with small amygdaloidal silica micrograins (Figure 4). Coarse grains of euhedral silica are also distributed in the felsic lithologies. Continuous shock veins, characterized by impact melt glasses of the constituent minerals with bubbles and flow textures, crosscut both the clasts and the breccia matrix of NWA 2727, indicating their formation after brecciation. Some amygdaloidal silica grains are entrained in the shock veins.

Raman spectroscopy, SR-XRD, and TEM were performed for the silica in the breccia matrix, basaltic clasts, and felsic lithologies (Figure 4) [7]. As a result of these microanalyses, aggregates of moganite, coesite, stishovite, and cristobalite nanoparticles (4.5 nm in average radius) were confirmed in the amygdaloidal silica micrograins of the breccia matrix (Figure 4 and Table 1). In contrast, there was only a quartz phase in the fine-grained silica of the basaltic clasts and only tridymite and cristobalite phases in the euhedral coarse silica grains of the felsic lithologies.

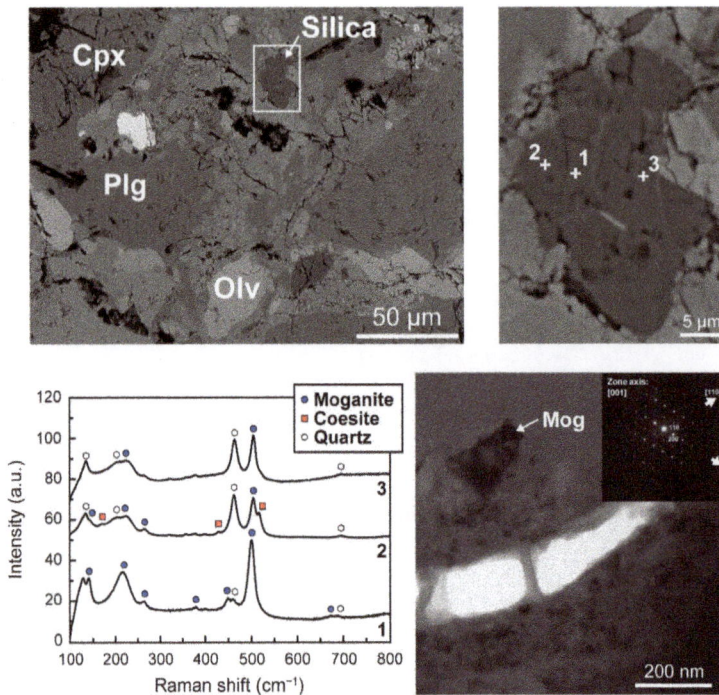

Figure 4. Low-magnification (**left upper**) and high-magnification (**right upper**) BSE images, Raman spectra (**left bottom**), and TEM analytical results of a silica micrograin in NWA 2727. BSE images of NWA 2727 demonstrate that amygdaloidal silica micrograins occur in the breccia matrix of NWA 2727 and coexist with olivine (Olv), clinopyroxene (Cpx), and plagioclase (Plg). Raman signatures of moganite, coesite, and quartz are seen in the silica micrograin. The TEM images and selected area electron diffraction (SAED) patterns of the focused ion beam (FIB)-sliced samples for the silica micrograin indicate the presence of microcrystalline aggregates of moganite nanoparticles with other silica polymorphs (modified after [7]).

Kayama et al. [7] suggested a formation process for lunar moganite based on the comparison of their results with previously reported findings (e.g., [31,54,101,102]). Host rock bodies of the OC gabbroic and basaltic clasts were formed by magmatic processes at the PKT at 2.993 ± 0.032 Ga. Subsequent carbonaceous chondrite collisions occurred there at $<2.67 \pm 0.04$ Ga, resulting in the formation of the breccia bodies on the impact basin. The alkaline water delivered by these collisions is highly likely to have been captured inside the breccia bodies via consolidation. On the sunlit surface (363–399 K), moganite could have formed via precipitation from the captured alkaline water in the breccia matrix. Simultaneously, the captured water was cold-trapped as H_2O ice in the subsurface of the brecciated bodies. The NWA 773 clan was launched from the surface of the brecciated bodies by the most recent impact event at 8–22 GPa and >673 K. These shock conditions could be constrained by the coexistence of moganite with coesite, stishovite, and cristobalite in the amygdaloidal silica micrograins of the breccia matrix in NWA 2727, according to the SiO_2 phase diagram (Figure 1). This impact event was possibly generated at ca. 1–30 Ma. Finally, the NWA 773 clan may have fallen to Earth at a terrestrial age of 17 ± 1 ka. Such H_2O ice cold-trapped in the subsurface is expected to still remain today as it can theoretically survive over billions of years [103]. Assuming the moganite precipitation model, an amount of at least 0.6 wt % of H_2O ice in the subsurface was calculated [7]. This value is in excellent agreement with concentrations of H_2O ice on the surface of the lunar poles estimated from spacecraft observations (e.g., [94,104]). Therefore, the subsurface H_2O ice is one of the most abundant water resources for future lunar explorations.

As introduced here, moganite has been overlooked for many years, but it is an important silica polymorph in extraterrestrial materials that can be used as a marker for the existence of H_2O ice in the Moon's subsurface. Further discoveries of such aqueous silica polymorphs, including moganite from other lunar samples and/or future sample return programs, might provide new insight into water on the Moon.

3. Mars

3.1. Martian Igneous Process

Martian meteorites are launched from Mars as a result of heavy impact and the ejected fragments are attracted toward Earth. In this decade, the number of martian meteorites has drastically increased because of collecting programs in hot desert areas, currently reaching over 200 meteorites [105]. Martian meteorites have mostly formed via igneous processes and can be classified into five groups: shergottite, nakhlite, chassignite, orthopyroxenite, and basaltic breccia. Their igneous ages vary from the 4.5 Ga of Allan Hills (ALH) 84001 to as young as 180 Ma shergottite. Therefore, they reveal the environmental history of Mars across various periods of time (e.g., [106,107]). Martian meteorites have experienced heavy impact events at least in the moment of ejection from Mars, as indicated by the transformation of plagioclase into maskelynite [108]. Igneous or secondary silica has been reported in most types of martian meteorites.

Only one meteorite has been classified as orthopyroxenite, Allan Hills (ALH) 84001. This martian meteorite (1931 g), found in Antarctica in 1984, has quite an old radiogenic age of 4.5–4.0 Ga and a cosmic ray exposure age of 15 Ma [106]. The silica in this meteorite is associated with various occurrences; igneous silica has been reported as the intergrowth with plagioclase, which has currently transformed into maskelynite by shock metamorphism [109]. This meteorite is unique because of the existence of carbonates, which are often mixed with silica. The amount of carbonates in the meteorite is estimated to be around 0.46% to 1.0% of the rock [110–112]. Silica glasses are thought to be contemporaneously formed with carbonates (Figure 5a) [109,113–120]. Furthermore, silica occurs as patchy or vein-like structures in the fractures of pyroxene grains [121–123]. Silica phases in the meteorite have been identified by mineralogical investigations. Cooney et al. [124] analyzed these silica grains using Raman spectroscopy, which showed a broad feature with a weak sharp peak at 462 cm^{-1}, probably less than 5% of quartz, and the peak at 400 cm^{-1} may come from trace amounts of tridymite. Formational and alteration processes are still under discussion. The existence of feldspathic and silica glasses constrains the relative timing of impact events and carbonate deposition in the meteorite [119]. Melwani Daswani et al. [120] computed their results using a one-dimensional transport thermochemical model and suggested that the alteration process produced carbonate assemblages, including amorphous silica. Therefore, silica materials in ALH84001 were formed in association with aqueous alteration and impact events.

Nakhlites are mostly composed of clinopyroxene formed at 1.3 Ga. Currently, 19 meteorites have been classified into this group. Some nakhlites that have experienced aqueous alteration contain carbonate veins and altered minerals such as iddingsite (e.g., [6,125]). Silica in Nakhla meteorites (Nakhlite) is also considered to be an alteration product [6]. Imae et al. [126] identified silica grains such as tridymite in Yamato 000593 using a Gandolfi camera, but did not discuss their origin. A representative report of hydrothermally precipitated silica was later introduced.

Shergottites are the largest group of martian meteorites. This group is sub-classified into three types: basaltic, olivinephyric, and lherzolitic, and can be further subdivided into three types by incompatible element features (enriched, depleted, and intermediate). These features indicate a wide variation in igneous process [106,127]. Furthermore, shergottites have a wide range of crystallization age of about 500–200 Ma [106] and consist of the constituent minerals of pyroxene, olivine, and plagioclase that were transformed into maskelynite by the impact event (igneous plagioclase with minor amorphization has been reported from two shergottites; [128]). Fine-grained silica, including silica glass, is part of the mesostasis of shergottites [129–137]. NWA 5298 shows relatively large grains of silica polymorphs up to 0.4 mm in size [138,139]. A shergottite named Los Angeles has large grains (up to 1 mm in size) of silica [140,141]. In Los Angeles, silica is an intergrowth with anorthoclase in late stage residual melts and transformed by secondary impact events (Figure 5b). Almost all identified silica consists of high-pressure polymorphs formed by shock metamorphism together with pre-existing low-pressure silica polymorphs [25,26,71,142,143], described in later section). Quartz and tridymite have been identified in Dhofar 378 using Raman spectroscopy [144]. Coesite has also been reported from NWA 8675 [145]. Late stage products of fayalite, hedenbergite, and pyroxferroite are present in the evolved shergottites of Zagami [146,147], Los Angeles [140,148–150], QUE 94201 [151], Ksar Ghilane 002 [152,153], NWA 2800 [154], and NWA 7320 [155]. Pyroxferroite is a metastable phase [156] that occurs marginally on pyroxene grains or in the mesostasis, and is currently symplectite (pyroxene–fayalite–silica) that might have been formed by a later thermal event. The formation event is still under debate, but was probably not affected by shock metamorphism [151].

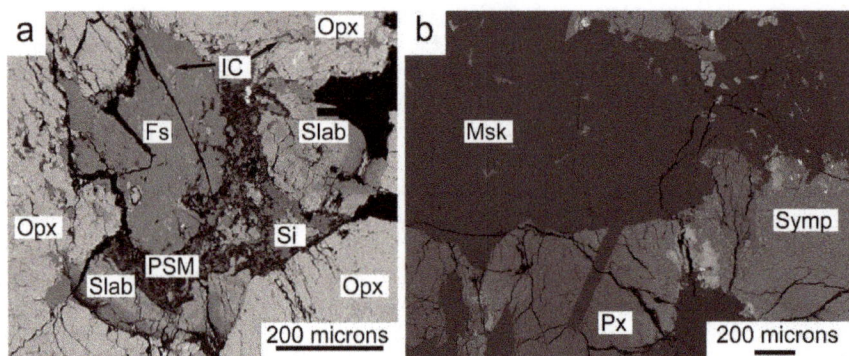

Figure 5. Occurrence of silica in martian meteorites. (**a**) Silica in association with carbonate materials in ALH84001 (image from [117]). (**b**) Pyroxferroite break down products (pyroxene–olivine–SiO$_2$) in the Los Angeles shergottite. Slab: Slab carbonates, Opx: Orthopyroxene, Fs: Feldspathic glass, IC; Interstitial carbonate, PSM: Post slab magnesite, Si: Silica glass, Msk: Maskelynite, Px: Pyroxene, Symp: Symplectite.

Only three meteorites (Chassigny, NWA 2737, and NWA 8695) have been classified in the Chassignite group. The dominant mineral in this group of meteorites is mostly olivine, with ~5% of pyroxene. Recently, new types of rocks (basaltic breccia and martian regolith) from Mars have been added to the martian meteorite collections [157,158]. These are known as "black beauty" and are 10 times richer in water, whereas silica has not been identified from these two types of martian meteorites.

Martian meteorites have been launched from several places on Mars according to the formation and ejection ages [106]. However, recent martian explorations have not found similar materials on Mars, indicating that martian meteorites are not representative of martian surface materials. Silica was observed by the martian exploration rovers Spirit and Opportunity and the Mars Science Laboratory's Curiosity. Spirit identified opaline silica deposit in the Gusev crater [8], which seems to form similar deposits as terrestrial silica, for example, at hot springs in Chile [159]. Curiosity found ~14 wt % of tridymite, ~2 wt % cristobalite, and silica-rich amorphous (opal-A and opal-CT) in the drill sample of the sedimentary rocks in the Gale crater according to an X-ray diffraction instrument [160].

Silica in martian materials formed via igneous processes, shock metamorphism, and hydrothermal alteration. Martian meteorites have been overprinted by shock metamorphism, at least during ejection from Mars. However, most of the silica has not been directly identified by mineralogical analysis, especially from the viewpoint of micro-mineralogy. Remote-sensing data show aqueous activity on Mars, as well as martian meteorites, thus, silica in martian meteorites is important for solving questions about past and present hydrous activity on Mars, as discussed later.

3.2. Shock Metamorphism on Mars

Shergotty, which fell in 1865 near Shergahti in Bihar State, India, is texturally and mineralogically similar to terrestrial diabases, and a martian meteorite family petrologically similar to it is called "shergottite" (e.g., Shergotty, Tissint, Zagami, NWA 480, 856, 1068, 2975, 4468, and SaU 005). Almost all shergottites record a trace of heavy impact events (maskelynite and melting textures) on Mars. Shergotty is mainly composed of prism-shaped pyroxene up to 1 cm size and lath-like and interstitial maskelynite. It also contains minor minerals of fayalite, Ti-magnetite, and large silica grains (>150 mm), most of which are wedge shaped [19]. The silica grains are either enclosed in clinopyroxene or exist among clinopyroxene, mesostasis, and maskelynite [26]. Many of the silica grains are surrounded by radiating cracks that were caused by a rapid increase of the volume with decompression. TEM observations and SAED analysis confirmed, for the first time, the existence of post-stishovite SiO_2 with an α-PbO_2 structure in the martian meteorite Shergotty [19]. Ten years later, this high-pressure SiO_2 phase has been named "seifertite", based on more accurate data [25].

In general, martian meteorites have been subject to higher shock pressure during impact events on the parent body when compared with other meteorites such as lunar meteorites and carbonaceous chondrites (e.g., [1,2,17,21]). Therefore, various types of high-pressure minerals, which have already been named, have been reported in martian meteorites (e.g., ringwoodite, akimotoite, and majorite), and new high-pressure minerals have been found in them (e.g., tissintite, liebermannite, and zagamiite) [17,161]. In the case of silica, Sharp et al. [19] and El Goresy et al. [25,26] reported the first discoveries of an α-PbO_2-type (named seifertite) and a baddeleyite-type (unnamed) high-pressure SiO_2 phase from martian meteorites.

Sharp et al. [19] investigated the morphology, crystallinity, and structure of the silica grains in Shergotty by using FE-SEM and TEM. Lamellae textures of the silica grains were observed using FE-SEM. The three zone-axes SAED patterns of crystalline domains in the silica grains were inconsistent with known SiO_2 polymorphs, including post-stishovite, at the time. Following this initial characterization of seifertite in Shergottite, El Goresy et al. [25] acquired SAED and XRD diffraction data of the silica grains in Shergotty that could be interpreted in terms of α-PbO_2-type structure (Table 1), and they named it "seifertite". The large seifertite-bearing silica grains have a typical pre-shock morphology and habit of tridymite or cristobalite, one of them seemed to be igneous SiO_2 phase in Shergotty before shock metamorphism [19,25]. Considering the kinetic data reported in [18], it is reasonable to suppose that cristobalite was a preexisting mineral before the impact, because it transforms into seifertite at relatively low pressures. Mineralogical and petrological descriptions of seifertite in Shergotty have also been reported in [132], where shock pressure and temperature were estimated based on mineral paragenesis of various high-pressure minerals including silica.

Post-stishovite with a baddeleyite-type structure has also been discovered from the silica grains in Shergotty [26]. Previous crystallographic experimental investigations indicated that α-PbO$_2$-like silica (space group *Pbcn* or *Pnc2*) is stable above 70 GPa and theoretically above 85 GPa, and is related to the baddeleyite-type SiO$_2$ structure by its seven-coordination of oxygen (e.g., [26,80,162]). Most silica in Shergotty occurs as wedge-shaped coarse crystals (150 to 900 μm), and are surrounded by pervasive radiating cracks [26]. BSE images of every silica grain obtained by FE-SEM showed mosaics of domains (10 to 60 μm) with orthogonal sets of lamellae.

XRD analysis of the silica grains showed characteristic signals of both a broad halo and reflections. The broad halo belonged to amorphous silica. Some of the XRD reflections were consistent with a stishovite structure. However, most of the XRD reflections were unsuitable for any known silica polymorphs, and could be indexed to a monoclinic unit cell with a cell parameter that has not been recognized in natural silica polymorphs [26]. This monoclinic phase had a calculated density higher than stishovite, that is, post-stishovite produced by impact events on Mars. The calculated intensities of the reflections for baddeleyite-structured SiO$_2$ were similar to the observed monoclinic phase (Table 1). The TEM observations revealed that the silica grains had intergrowths of crystalline and amorphous lamellae. The SAED patterns of the crystalline lamellae corresponded to a mixture of stishovite and cristobalite structures. However, baddeleyite-type SiO$_2$ was not seen in the SAED patterns of the silica grains, although it was identified in the XRD analysis. This may be caused by the high instability of baddeleyite-type SiO$_2$ during sample preparation for TEM. Moreover, this baddeleyite-type SiO$_2$ appeared to be metastable and readily changed to a more stable, low-pressure phase because of its sensitivity to X-ray and electron irradiation. Although uncertainties remain regarding the formation process of baddeleyite-type SiO$_2$ instead of the expected α-PbO$_2$-type phase (e.g., [76,163,164]), El Goresy et al. [26] speculated that small amounts of impurities (such as Na$_2$O and Al$_2$O$_3$) and heterogeneous stress distribution during compression could be factors preserving the baddeleyite-type SiO$_2$ phase from destruction. However, this baddeleyite-type SiO$_2$ has yet to be accepted as a new mineral as it has not been well-characterized, and for this reason it has not been given a name. Thus, further crystallographic research on this silica phase in martian meteorites is timely. Shock melt pockets in Shergotty and other martian meteorites also possess intergrowths of stishovite with a (Na,Ca)–hexaluminosilicate mineral (hexagonal barium ferrite-type) [(Ca$_x$Na$_{1-x}$)Al$_{3+x}$Si$_{3-x}$O$_{11}$], as described at the end of this chapter.

In martian meteorites, other high-pressure silica phases have also been reported. Stishovite was discovered in Zagami using TEM analysis [86–88]. Zagami (fell in 1962 in Nigeria) is a basaltic shergottite and is composed of several different lithologies. The basaltic lithology (called fine-grained lithology) is petrologically similar to Shergotty, and is characterized by a foliation of pyroxene and maskelynite that is crosscut by shock veins.

Langenhorst and Poirier [86,87] examined mineral assemblages in shock veins of the basaltic lithology of Zagami. TEM observation of the wider shock veins with a thickness of 100 μm indicated a tiny mineral phase embedded in a glassy matrix [86]. One of the mineral phases was needle-shaped silica. SAED patterns and TEM images consistently established this phase to be stishovite. Stishovite (10 to 100 nm in size) mostly occurred as isolated crystallites in the glassy matrix. Some stishovite coexisted with the hollandite-type KAlSi$_3$O$_8$ (liebermannite) as rounded polycrystalline aggregates. This assemblage is considered to have been crystallized from a melt at a high shock pressure. Amorphous grains of silicate perovskite were surrounded by stishovite and wüstite. On the basis of mineralogical description and thermodynamic calculation of the shock vein, most of the mineral phases, including silicate perovskite, directly crystallized from shock-induced melt at high pressures and very high temperatures, and then the silicate perovskite became amorphous during the decompression. Finally, silicate perovskite changed into the lower pressure assemblage of stishovite and wüstite because of solid-state diffusion at its grain boundaries. Additionally, CAS-phases intergrown with stishovite have been observed in the shock melt pockets of Zagami [88]. This CAS phase is now referred to as zagamiite (IMA No. 2015-022a, as described by [165]).

Tissint is a unique martian meteorite, because its fall was witnessed by many people (2011 in Morocco), so the rock is thus unweathered. Tissint is a heavily shocked olivine–phyric shergottite composed of subhedral to euhedral olivine (up to 2 mm in size) in a groundmass of smaller olivine, pyroxene, maskelynite, and minor minerals (Ti–chromite to ulvöspinel, ilmenite, merrillite). Shock melt pockets and shock veins contained various high-pressure minerals, for example, ringwoodite, tissintite, and stishovite [166,167].

Raman spectroscopy and SEM imaging suggested the existence of stishovite associated with olivine inside shock melt pockets [166]. However, EMPA analysis could not be performed for stishovite as the grains were too small (1 to 2 μm microcrystalline of stishovite + olivine). According to Walton et al. [167], within a shock vein of 120 μm in thickness, sparse anhedral crystals of clinopyroxene and ringwoodite and tiny needles of stishovite were embedded in a predominantly amorphous matrix and represented a metastable assemblage (stishovite + clinopyroxene + glass in shock veins) crystallized via rapid quenching during decompression.

In other martian meteorites, stishovite, seifertite, and HP silica glass have been observed using various microanalyses. Stishovite was first reported in Los Angeles by Walton and Spray [168] as small (10 to 20 μm size), needle-shape crystals that were quench-crystallized in shock melt pockets. Beck et al. [88] reported that a new mineral of the (Na,Ca)–hexaluminosilicate phase (hexagonal barium ferrite-type) $[(Ca_xNa_{1-x})Al_{3+x}Si_{3-x}O_{11}]$, referred to as the CAS-phase, was intimately intergrown with stishovite in the shock melt pockets of NWA 480, 856, 1068, and Sayh Al Uhaymir 005, similar to the cases of Shergotty and Zagami. Chennaoui Aoudjehane et al. [141] studied silica in shergottites NWA 480, NWA 856, Zagami, Shergotty, and Los Angeles using SEM and CL, and confirmed stishovite in all meteorites, with the exception of Los Angeles. According to the phase diagram of a basaltic composition at high pressures and temperatures, stishovite and CAS-phase appear as liquidus phases in the shock-induced melt. El Goresy et al. [143] found oblong polycrystalline stishovite with an orthogonal intergrowth of presumable post-stishovite (either residual seifertite, or monoclinic ZrO_2-type silica polymorph) inverted from stishovite in NWA 480. In the shock vein of NWA 856, they also encountered a coarse silica grain with a faulted orthogonal pattern of seifertite. He et al. [169] described irregularly shaped silica grains between plagioclase, clinopyroxene, and Fe–Ti oxides in NWA 2975, which according to Raman spectroscopy, were stishovite. Furthermore, such lamellae-textured intergrowths were present in silica grains that might be a mixture of seifertite and HP silica glass, based on the SEM observations. In addition to these investigations, Boonsue and Spray [170] reported stishovite in NWA 4468. As introduced in this section, high-pressure silica polymorphs are a valuable recorder of impact events on Mars. A mineral assemblage of high-pressure minerals, including silica, gives us hints to constrain the shock pressure and temperature of those impact events, and in deducing the impactor and crater sizes based on these factors. This information hints at the birthplace of each martian meteorite.

3.3. Acid Fluid Activity of Mars

Minerals precipitated from Mars's liquid water have been reported in some martian meteorites, and provide valuable information on the martian hydrosphere, including its temperature, elemental composition, pH, Eh (i.e., redox potential), and longevity (e.g., [6,171–175]). Recently, Lee et al. [6] confirmed the existence of opal-A nanoparticles in the martian meteorite, Nakhla. The fall of the Nakhla meteorites in Egypt in 1911 were recorded, and some of the fragments were collected by a researcher. Nakhla is pristine and free from terrestrial contamination because of its rapid recovery, therefore, alteration products described from Nakhla can be inferred to be martian in origin. Nakhla is a cumulate clinopyroxenite with small amounts of olivine, intercumulate mesostasis, and minor iddingsite (1 wt %), and is similar to other martian meteorites called nakhlites. A series of cumulate clinopyroxenitic martian meteorites is referred to as "nakhlite" (e.g., NWA 998, MIL 03346, and Yamato 000593), and most of them consist of clinopyroxene, olivine, and water-based minerals such as iddingsite. In Nakhla, this alteration product from olivine was a result of water–mineral interactions and is a mixture of hydrous Fe–Mg silicate (phyllosilicate), carbonate, sulfate, and halide. This alteration product seems to be indigenous to Mars, rather than terrestrial weathering as it is crosscut by the fusion crust and its radiometric formation age is 633 ± 23 Ma. Recent TEM analyses of the iddingsite from Nakhla have proven the presence of opal-A (~12 nm in diameter) in the Fe–Mg silicate, which is thought to have been formed by martian fluid activity.

Scanning TEM (STEM)-high angle annular dark-field (HAADF) imaging showed that the Fe–Mg silicate is comprised of closely packed particles that are circular to oval in shape. The cores ranged in diameter from 5 nm to 29 nm. As a result of STEM-electron energy loss spectroscopy (EELS) and energy dispersive X-ray (EDX) measurements, one of the largest cores corresponded to pure silica nanoparticles (Figure 6). Orange-colored discontinuous iddingsite veins were readily observed in a thin section of Nakhla using a plane polarized transmitted optical microscope, and were distributed within all olivine grains (300 to 1000 μm in size) of Nakhla [6]. Most of them were located—with a few ten micrometers in length and 1.9 to 7.0 μm—along the olivine–augite grain boundaries, and rarely within augite, and Fe–Mg silicate, siderite, and ferric (oxy)hydroxide were included in the iddingsite veins (Figure 6). The SAED patterns of the silica nanoparticles showed a broad diffuse ring assigned to amorphous SiO_2 (Figure 6). According to previous studies that assumed a 10 wt % of bulk water of Fe–Mg silicate of nakhlite [172,176–180], these silica nanoparticles were most probably opal-A (amorphous $SiO_2 \cdot nH_2O$) [6]. Narrow rings in the SAED pattern were consistent with two-line ferrihydrite ($Fe_{8.2}O_{8.5}(OH)_{7.4} + 3H_2O$), suggesting its coexistence with opal-A nanoparticles. In fact, the TEM observation revealed that each opal-A nanoparticle was enclosed within a ferrihydrite shell (Figure 6).

According to these micro-mineralogical descriptions [6], olivine changed into iddingsite in the host rocks of Nakhla during discrete events with the variation of pH, Eh, and the chemical composition of solutions, rather than from a single evolving fluid. During the discrete fluid activities, Fe–Mg silicate formed first, then siderite grew by replacement, and finally, ferric (oxy)hydroxide was produced by the oxidation of the carbonate; preexisting olivine, a precursor of Fe–Mg silicate, was dissolved during initial the water–mineral interaction, and simultaneously, opal-A was precipitated in the host rock of Nakhla, in agreement with the previous report [181]. The initial acidic (pH $\leq$ 4) and oxidizing fluid activities were considered to have occurred according to the previous chemical analyses of the Fe–Mg silicate and spacecraft observations (e.g., [8,175,182]). The fluids dissolved olivine and precipitated opal-A and ferrihydrite. The secondary fluid solutions had a high pH and low Eh, and introduced siderite into the iddingsite veins of Nakhla, which were formed by the replacement of olivine and the opal-A plus ferrihydrite. Findings of opal-A coexisting with ferrihydrite in the Nakhla iddingsite are in good agreement with the result obtained from theoretical [183] and laboratory experiments of silica under martian conditions [184,185]. The final fluid activity recorded by the iddingsite veins was the partial oxidation of siderite to a ferric (oxy)hydroxide, which probably took place on Mars.

Furthermore, spectroscopic spacecraft observations of the martian surface by orbiters and landers have also discovered extensive deposits of hydrous and amorphous opaline silica on the surface of Mars (e.g., [186–188]). Although opal-A has been recognized in only one of the nakhlite samples, amorphous and hydrous silica such as opal-A may be widespread on Mars. Therefore, Lee et al. [6] interpreted that Amazonian clinopyroxenite lava flow/sill seemed to be the martian host of opaline silica, and Syrtis Major (hosts a volcanic hydrothermal deposit of silica) was proposed as a source of the nakhlites [189]. As shown here and above, aqueous silica polymorphs in extraterrestrial materials, such as opal-A in the martian meteorite Nakhla and moganite in the lunar meteorite NWA 2727, are indicative of fluid activities on the parent bodies.

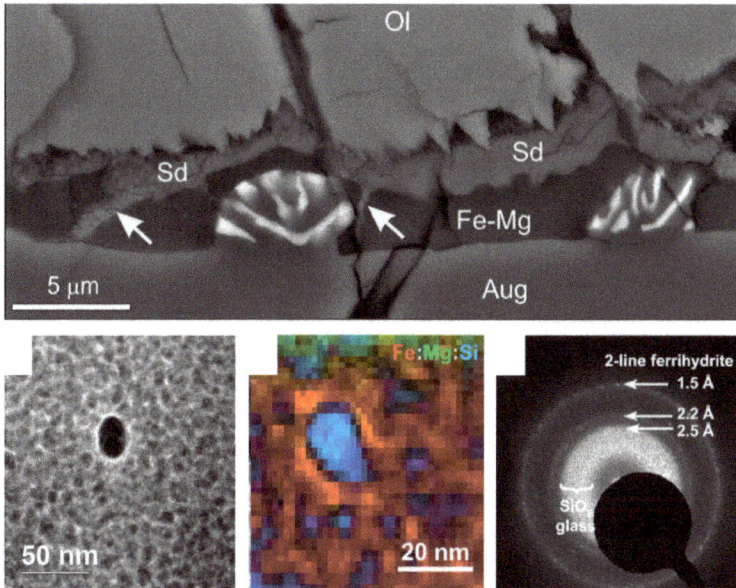

Figure 6. BSE images of iddingsite veins within grains of olivine (Ol) (**top**); the high angle annular dark-field (HAADF) scanning TEM (STEM) image of nanoparticles in the Fe–Mg silicate (**left bottom**); false-color elemental map of Fe, Mg, and Si (**middle bottom**) obtained from the EELS spectrum imaging of the nanoparticles; and the SAED pattern from a group of the nanoparticles (**right bottom**). The BSE image demonstrated that the iddingsite veins contained Fe–Mg silicate (Fe–Mg) and siderite (Sd), and lay at the interface between olivine and augite (Aug). The HAADF STEM observation of the Fe–Mg silicate revealed a large amount of nanoparticles. Pure silica nanoparticles (aqueous blue) were found in the Fe–Mg silicate, using the false-color silicate elemental map, and may be opal-A enclosed within ferrihydrite, based on the results of the SAED analysis (modified after [6]).

4. Conclusions

The micro-mineralogy of silica in lunar and martian materials provides new insight into geological events in the host celestial bodies. Of course, to achieve this purpose, it is essential to clarify the mineralogical properties of micro- to nanometer-sized silica crystals using Raman microscopy, SR-XRD, TEM, and EBSD. However, some works on extraterrestrial materials, including lunar and martian materials, have only reported silica based on conventional observations, and did not determine the type of silica polymorphs, especially for the fine silica grains, by using advanced microanalyses. Here, we introduced representative and important micro-mineralogical studies of silica polymorphs that were formed by igneous, shock metamorphic, and fluid processes on the Moon and Mars. Furthermore, we reviewed the details from the viewpoints of the discovery from lunar and martian materials, the formation processes, the implications for planetary science, and future prospects in the field of micro-mineralogy in each section:

(1) Lunar igneous process

Igneous silica polymorphs are known as cristobalite, tridymite, and quartz. On the basis of micro-mineralogical investigations of the Apollo collections and lunar meteorites, silica is one of the major minerals in late-stage products such as granitic rocks. A combination of micro-mineralogical and remote-sensing data can provide hints to determine the candidates of silicic volcanism that formed on the lunar surface.

(2)   Shock metamorphism on the Moon

Recently, high-pressure silica polymorphs such as coesite, stishovite, and seifertite have been discovered from a part of lunar meteorites and an Apollo collection, although they were previously believed to be absent from the lunar materials because of their elimination from impact-induced volatilization and the difficulty in finding fine-grained high-pressure silica using conventional methods. Of course, various types of high-pressure silica polymorphs are likely to be included in other parts of lunar meteorites and in the Apollo and Luna collections, therefore, further findings of high-pressure silica polymorphs in future micro-mineralogical studies are important for understanding impact events on the Moon.

(3)   Alkali fluid activity of the Moon

Aqueous silica polymorphs, formed by fluid activity, were believed to be absent from lunar samples. However, a recent study reported the presence of moganite in the lunar meteorite of NWA 2727 and demonstrated the likelihood that this silica was precipitated from lunar fluid activity. An amount of at least 0.6 wt % of H_2O liquid is theoretically necessary for moganite precipitation and would accumulate as ice in the subsurface. Therefore, the subsurface H_2O ice is one of the most abundant water resources for future lunar explorations.

(4)   Martian igneous process

Most and various types of martian meteorites contain silica polymorphs, however, micro-mineralogical investigation in this area seems to be limited. Therefore, types of some martian silica have yet to be identified. More detailed works on the micro-mineralogy of silica could assist in answering questions regarding igneous processes on Mars.

(5)   Shock metamorphism on Mars

Martian meteorites are known to have been affected by heavy impact events. For this reason, various types of high-pressure minerals, including silica, have been discovered from them for the first time, and one of the high-pressure silica phases has been named seifertite. A possible baddeleyite-type high-pressure SiO_2 phase has also been reported in a martian meteorite. The presence of high-pressure silica polymorphs sets a constraint for peak-shock pressure and shock duration time on the celestial bodies, and provides us with clues to deduce the impactor and crater sizes, and the birthplace of each martian meteorite, as well as the lunar meteorites.

(6)   Acid fluid activity of Mars

The existence of opal-A nanoparticles, precipitated from Mars's liquid water, has been confirmed in the martian meteorite, Nakhla. This finding provides valuable information on the martian hydrosphere. Moreover, spacecraft observations have also discovered extensive deposits of hydrous and amorphous opaline silica on the surface of Mars. Aqueous silica polymorphs in extraterrestrial materials such as opal-A in Nakhla and moganite in NWA 2727 are indicative of fluid activities on the parent bodies.

In the near future, the micro-mineralogy of igneous silica in lunar and martian materials (that have yet to be fully studied) will definitely give us new insight into the thermal histories of the Moon and Mars. Furthermore, new silica polymorphs (e.g., $CaCl_2$- and pyrite-type) and the unnamed baddeleyite-type will likely be discovered from lunar and martian materials that have experienced heavy impact events. Additionally, the martian aqueous silica polymorphs (e.g., moganite, opal-A, and maybe opal-CT) will make the fluid activities of the Moon and Mars clear. Other lunar and martian meteorites, the Apollo and Luna collections, and other types of meteorites such as eucrite and carbonaceous/ordinary chondrites should be investigated using advanced microanalysis for the micro-mineralogy of silica.

Author Contributions: M.K. outlined and designed this article, and also wrote the paper relevant to shock metamorphic and fluid activity martian igneous silica, respectively. All authors participated in discussing the manuscript and checking all content.

Funding: This work was partly supported by the Astrobiology Center of National Institutes of Natural Sciences (NINS) (Grant Number AB291023) and Kurita Water and Environment Foundation (Grant Number 17D006) to T.N.

Acknowledgments: We would like to thank Jens Götze (TU Bergakademie Freiberg, Institute of Mineralogy,) for inviting Masahiro Kayama to submit to the Special Issue "Mineralogy of Quartz and Silica Minerals" of "Minerals". We are deeply indebted to Loker He (Assistant Editor, Minerals) for coordinating our paper. This work was partly supported by the Astrobiology Center of National Institutes of Natural Sciences (NINS) (Grant Number AB291023) and Kurita Water and Environment Foundation (Grant Number 17D006) to T.N.

Conflicts of Interest: The authors declare no conflict of interest.

References

1. Papike, J.J. Comparative planetary mineralogy: Chemistry of melt-derived pyroxene, feldspar, and olivine. In *Planetary Materials; Reviews in Mineralogy & Geochemistry*; Papike, J.J., Ed.; Mineralogical Society of America: Chantilly, VA, USA, 1998; pp. 7-1–7-11.

2. Lucey, P.; Korotev, R.L.; Gillis, J.J.; Taylor, L.A.; Lawrence, D.; Campbell, B.A.; Elphic, R.; Feldman, B.; Hood, L.L.; Hunten, D.; et al. Understanding the lunar surface and space-moon interactions. In *New Views of the Moon, Reviews in Mineralogy & Geochemistry*; Jolliff, B.L., Wieczorek, M.A., Shearer, C.K., Neal, C.R., Eds.; Mineralogical Society of America: Chantilly, VA, USA, 2006; pp. 83–220.

3. Ohtani, E.; Ozawa, S.; Miyahara, M.; Ito, Y.; Mikouchi, T.; Kimura, M.; Arai, T.; Sato, K.; Hiraga, K. Coesite and stishovite in a shocked lunar meteorite, Asuka-881757, and impact events in lunar surface. *Proc. Natl. Acad. Sci. USA* **2011**, *108*, 463–466. [CrossRef] [PubMed]

4. Miyahara, M.; Kaneko, S.; Ohtani, E.; Sakai, T.; Nagase, T.; Kayama, M.; Nishido, H.; Hirao, N. Discovery of seifertite in a shocked lunar meteorite. *Nat. Commun.* **2013**, *4*, 1737. [CrossRef] [PubMed]

5. Kaneko, S.; Miyahara, M.; Ohtani, E.; Arai, T.; Hirao, N.; Sato, K. Discovery of stishovite in Apollo 15299 sample. *Am. Mineral.* **2015**, *100*, 1308–1311. [CrossRef]

6. Lee, M.R.; MacLaren, I.; Andersson, S.M.L.; Kovács, A.; Tomkinson, T.; Mark, D.F.; Smith, C.L. Opal-A in the Nakhla meteorite: A tracer of ephemeral liquid water in the Amazonian crust of Mars. *Meteorit. Planet. Sci.* **2015**, *50*, 1362–1377. [CrossRef]

7. Kayama, M.; Tomioka, N.; Ohtani, E.; Seto, Y.; Nagaoka, H.; Götze, J.; Miyake, A.; Ozawa, S.; Sekine, T.; Miyahara, M.; et al. Discovery of moganite in a lunar meteorite as a trace of H_2O ice in the Moon's regolith. *Sci. Adv.* **2018**, *4*, eaar4378. [CrossRef] [PubMed]

8. Squyres, S.W.; Arvidson, R.E.; Ruff, S.; Gellert, R.; Morris, R.V.; Ming, D.W.; Crumpler, L.; Farmer, J.D.; Des Marais, D.J.; Yen, A.; et al. Detection of silica-rich deposits on Mars. *Science* **2008**, *320*, 1063–1067. [CrossRef] [PubMed]

9. Hsu, H.-W.; Postberg, F.; Sekine, Y.; Shibuya, T.; Kempf, S.; Horányi, M.; Juhász, A.; Altobelli, N.; Suzuki, K.; Masaki, Y.; et al. Ongoing hydrothermal activities within Enceladus. *Nature* **2015**, *519*, 207–210. [CrossRef] [PubMed]

10. Sekine, Y.; Shibuya, T.; Postberg, F.; Hsu, H.-W.; Suzuki, K.; Masaki, Y.; Kuwatani, T.; Mori, M.; Hong, P.K.; Yoshizaki, M.; Tachibana, S.; Sirono, S. High-temperature water–rock interactions and hydrothermal environments in the chondrite-like core of Enceladus. *Nat. Commun.* **2015**, *6*, 1–8. [CrossRef] [PubMed]

11. Jolliff, B.L. Fragments of quartz monzodiorite and felsite in Apollo 14 soil particles. In Proceedings of the Twenty-First Lunar and Planetary Science Conference, Houston, TX, USA, 12–16 March 1991; pp. 101–118.

12. Leroux, H.; Cordier, P. Magmatic cristobalite and quartz in the NWA 856 Martian meteorite. *Meteorit. Planet. Sci.* **2006**, *41*, 913–923. [CrossRef]

13. Akimoto, S.; Yagi, T.; Inoue, K. High temperature-pressure phase boundaries in silicate systems using in situ X-ray diffraction. In *High-Pressure Research, Applications in Geophysics*; Academic: New York, NY, USA, 1977; pp. 595–602.

14. Presnall, D.C. Phase diagrams of Earth-forming minerals. In *Mineral Physics & Crystallography: A Handbook of Physical Constants*; Ahrens, T.J., Ed.; American Geophysical Union: Washington, DC, USA, 1995; pp. 248–268.

15. Swamy, V.; Saxena, S.K.; Sundman, B.; Zhang, J. A thermodynamic assessment of silica phase diagram. *J. Geophys. Res.* **1994**, *99*, 11787–11794. [CrossRef]

16. Kuwayama, Y. Ultrahigh pressure and high temperature experiments using a laser heated diamond anvil cell in multimegabar pressures region. *Rev. High Press. Sci. Technol.* **2008**, *18*, 3–10. [CrossRef]

17. Tomioka, N.; Miyahara, M. High-pressure minerals in shocked meteorites. *Meteorit. Planet. Sci.* **2017**, *52*, 2017–2039. [CrossRef]

18. Kubo, T.; Kato, T.; Higo, Y.; Funakoshi, K. Curious kinetic behavior in silica polymorphs solves seifertite puzzle in shocked meteorite. *Sci. Adv.* **2015**, *1*, e1500075. [CrossRef] [PubMed]

19. Sharp, T.G.; El Goresy, A.; Wopenka, B.; Chen, M. A post-stishovite SiO₂ polymorph in the meteorite Shergotty: Implications for impact events. *Science* **1999**, *284*, 1511–1513. [CrossRef] [PubMed]

20. Kring, D.A.; Gleason, J.D. Magmatic temperatures and compositions on early Mars as inferred from the orthopyroxene-silica assemblage in ALH84001. In Proceedings of the Annual Meteoritical Society Meeting, Maui, HI, USA, 21–25 July 1997.

21. Pang, R.-L.; Zhang, A.-C.; Wang, S.-Z.; Wang, R.-C.; Yurimoto, H. High-pressure minerals in eucrite suggest a small source crater on Vesta. *Sci. Rep.* **2016**, *6*, 26063. [CrossRef] [PubMed]

22. Kihara, K. An X-ray study of the temperature dependence of the quartz structure. *Eur. J. Mineral.* **1990**, *2*, 63–78. [CrossRef]

23. Dollase, W.A.; Cliff, R.A.; Wetherill, G.W. Note on tridymite in rock 10021. *Proc. Lunar Sci. Conf.* **1971**, *1*, 141–142.

24. Konnert, J.H.; Appleman, D.E. The crystal structure of low tridymite. *Acta Cryst.* **1978**, *34*, 391–403. [CrossRef]

25. El Goresy, A.; Dera, P.; Sharp, T.G.; Prewitt, C.T.; Chen, M.; Dubrovinsky, L.; Wopenka, B.; Boctor, N.Z.; Hemley, R.J. Seifertite, a dense orthorhombic polymorph of silica from the Martian meteorites Shergotty and Zagami. *Eur. J. Mineral.* **2008**, *20*, 523–528. [CrossRef]

26. El Goresy, A.; Dubrovinsky, L.; Sharp, T.G.; Saxena, S.K.; Chen, M. A monoclinic post-stishovite polymorph of silica in the Shergotty meteorite. *Science* **2000**, *288*, 1632–1634. [CrossRef] [PubMed]

27. Papike, J.J.; Taylor, L.; Simon, S. Lunar minerals. In *Lunar Sourcebook*; Heiken, G.H., Vaniman, D.T., French, B.M., Eds.; Cambridge University Press: New York, NY, USA, 1991; pp. 121–181.

28. Warren, P.H. The magma ocean concept and lunar evolution. *Annu. Rev. Earth Planet. Sci.* **1985**, *13*, 201–240. [CrossRef]

29. Hiesinger, H.; Head, J.W.; Wolf, U.; Jaumann, R.; Neukum, G. Ages and stratigraphy of mare basalts in Oceanus Procellarum, Mare Nubium, Mare Cognitum, and Mare Insularum. *J. Gephys Res.* **2003**, *108*, E001985. [CrossRef]

30. Morota, T.; Haruyama, J.; Ohtake, M.; Matsunaga, T.; Honda, C.; Yokota, Y.; Kimura, J.; Ogawa, Y.; Demura, H.; Iwasaki, A.; et al. Timing and characteristics of the latest mare eruption on the Moon. *Earth Planet. Sci. Lett.* **2011**, *302*, 255–266. [CrossRef]

31. Borg, L.E.; Gaffney, A.M.; Shearer, C.K.; DePaolo, D.J.; Hutcheon, I.D.; Owens, T.L.; Ramon, E.; Brennecka, G. Mechanisms for incompatible-element enrichment on the Moon deduced from the lunar basaltic meteorite Northwest Africa 032. *Geochim. Cosmochim. Acta* **2009**, *73*, 3963–3980. [CrossRef]

32. Jolliff, B.L.; Gillis, J.J.; Haskin, L.A.; Korotev, R.L.; Wieczorek, M.A. Major lunar crustal terranes: Surface expressions and crust-mantle origins. *J. Geophys. Res.* **2000**, *105*, 4197–4216. [CrossRef]

33. Ohtake, M.; Haruyama, J.; Matsunaga, T.; Yokota, Y.; Morota, T.; Honda, C.; LISM Team. Performance and scientific objectives of the SELENE (KAGUYA) Multiband Imager. *Earth Planets Space* **2008**, *60*, 257–264. [CrossRef]

34. Ohtake, M.; Matsunaga, T.; Haruyama, J.; Yokota, Y.; Morota, T.; Honda, C.; Ogawa, Y.; Torii, M.; Miyamoto, H.; Arai, T.; et al. The global distribution of pure anorthosite on the Moon. *Nature* **2009**, *461*, 236–240. [CrossRef] [PubMed]

35. Yamamoto, S.; Nakamura, R.; Matsunaga, T.; Ogawa, Y.; Ishihara, Y.; Morota, T.; Hirata, N.; Ohtake, M.; Hiroi, T.; Yokota, Y.; et al. Possible mantle origin of olivine around lunar impact basins detected by SELENE. *Nat. Geosci.* **2010**, *3*, 533–536. [CrossRef]

36. Neal, C.R.; Taylor, T.A. Petrogenesis of mare basalts: A record of lunar volcanism. *Geochim. Cosmochim. Acta* **1992**, *56*, 2177–2211. [CrossRef]

37. Taylor, G.J.; Warren, P.H.; Ryder, G.; Delano, J.; Pieters, C.; Lofgren, G. Lunar rocks. In *Lunar Sourcebook*; Heiken, G.H., Vaniman, D.T., French, B.M., Eds.; Cambridge University Press: New York, NY, USA, 1991; pp. 183–284.

38. Neal, C.R. The Moon 35 years after Apollo: What's left to learn? *Chem. Erde-Geohcem.* **2008**. [CrossRef]

39. Ryder, G. Lunar sample 15405: Remnant of a KREEP basalt-granite differentiated pluton. *Earth Planet. Sci. Lett.* **1976**, *29*, 255–268. [CrossRef]

40. Seddio, S.M.; Korotev, R.L.; Jolliff, B.L.; Wang, A. Silica polymorphs in lunar granite: Implications for granite petrogenesis on the Moon. *Am. Mineral.* **2015**, *100*, 1533–1543. [CrossRef]

41. Seddio, S.M.; Jolliff, B.L.; Korotev, R.L.; Carpenter, P.K. Thorite in an Apollo 12 granite fragment and age determination using the electron microprobe. *Geochim. Cosmochim. Acta* **2014**, *135*, 307–320. [CrossRef]

42. Meyer, C.; Williams, I.S.; Compston, W. Uranium-lead ages for lunar zircons: Evidence for a prolonged period of granophyre formation from 4.32 to 3.88 Ga. *Meteorit. Planet. Sci.* **1996**, *31*, 370–387. [CrossRef]

43. Warren, P.H.; Kallemeyn, G.W. Geochemical investigations of five lunar meteorites: Implications for the composition, origin and evolution of the lunar crust. *Antarct. Meteor. Res.* **1991**, *4*, 91–117.

44. Arai, T.; Takeda, H.; Yamaguchi, A.; Ohtake, M. A new model of lunar crust: Asymmetry in crustal composition and evolution. *Earth Planet. Space* **2008**, *60*, 433–444. [CrossRef]

45. Gross, J.; Treiman, A.H.; Mercer, C.N. Lunar feldspathic meteorites: Constraints on the geology of the lunar highlands, and the origin of the lunar crust. *Earth Planet. Sci. Lett.* **2014**, *388*, 318–328. [CrossRef]

46. Korotev, R.L. Lunar geochemistry as told by lunar meteorites. *Chem. Erde-Geochem.* **2005**, *65*, 297–346. [CrossRef]

47. Korotev, R.L.; Jolliff, B.L.; Zeigler, R.A.; Gillis, J.J.; Haskin, L.A. Feldspathic lunar meteorites and their implications for compositional remote sensing of the lunar surface and the composition of the lunar crust. *Geochim. Cosmochim. Acta* **2003**, *67*, 4895–4923. [CrossRef]

48. Nagaoka, H.; Takeda, H.; Karouji, Y.; Ohtake, M.; Yamaguchi, A.; Yoneda, S.; Hasebe, N. Implications for the origins of pure anorthosites found in the feldspathic lunar meteorites, Dhofar 489 group. *Earth Planets Space* **2014**, *66*, 115. [CrossRef]

49. Takeda, H.; Yamaguchi, A.; Bogard, D.D.; Karouji, Y.; Ebihara, M.; Ohtake, M.; Saiki, K.; Arai, T. Magnesian anorthosites and a deep crustal rock from the farside crust of the moon. *Earth Planet. Sci. Lett.* **2006**, *247*, 171–184. [CrossRef]

50. Treiman, A.H.; Maloy, A.K.; Shearer, C.K., Jr.; Gross, J. Magnesian anorthositic granulites in lunar meteorites Allan Hills A81005 and Dhofar 309: Geochemistry and global significance. *Meteorit. Planet. Sci.* **2010**, *45*, 163–180. [CrossRef]

51. Warren, P.H.; Ulff-Møller, F.; Kallemeyn, G.W. 'New' lunar meteorites: Impact melt and regolith breccias and large-scale heterogeneities of the upper lunar crust. *Meteorit. Planet. Sci.* **2005**, *40*, 989–1014. [CrossRef]

52. Yamaguchi, A.; Karouji, Y.; Takeda, H.; Nyquist, L.E.; Bogard, D.D.; Ebihara, M.; Shih, C.-Y.; Reese, Y.D.; Garrison, D.; Park, J.; et al. The variety of lithologies in the Yamato-86032 lunar meteorite: Implications for formation processes of the lunar crust. *Geochim. Cosmochim. Acta* **2010**, *74*, 4507–4530. [CrossRef]

53. List of Lunar Meteorites. Available online: http://meteorites.wustl.edu/moon_meteorites_list_alumina.htm (accessed on 9 April 2018).

54. Jolliff, B.L.; Korotev, R.L.; Zeigler, R.A.; Floss, C. Northwest Africa 773: Lunar mare breccia with a shallow-formed olivine-cumulate component, inferred very-low-Ti (VLT) heritage, and a KREEP connection. *Geochim. Cosmochim. Acta* **2003**, *67*, 4857–4879. [CrossRef]

55. Fagan, T.J.; Taylor, G.J.; Keil, K.; Hicks, T.L.; Killgore, M.; Bunch, T.E.; Wittke, J.H.; Mittlefehldt, D.W.; Clayton, R.N.; Mayeda, T.K.; et al. Northwest Africa 773: Lunar origin and iron-enrichment trend. *Meteorit. Planet. Sci.* **2003**, *38*, 529–554. [CrossRef]

56. Fagan, T.J.; Kashima, D.; Wakabayashi, Y.; Suginohara, A. Case study of magmatic differentiation trends on the Moon based on lunar meteorite Northwest Africa 773 and comparison with Apollo 15 quartz monzodiorite. *Geochim. Cosmochim. Acta* **2014**, *133*, 97–127. [CrossRef]

57. Zhang, A.-C.; Hsu, W.-B.; Floss, C.; Li, X.-H.; Li, Q.-L.; Liu, Y.; Taylor, L.A. Petrogenesis of lunar meteorite Northwest Africa 2977: Constraints from in situ microprobe results. *Meteorit. Planet. Sci.* **2011**, *45*, 1929–1947. [CrossRef]

58. Collins, S.J.; Righter, K.; Brandon, A. Mineralogy, petrology and oxygen fugacity of the LaPaz Ice Field lunar basaltic meteorites and the origin and evolution of evolved lunar basalts. In Proceedings of the Thirty-Sixth Lunar and Planetary Science Conference, League City, TX, USA, 14–18 March 2005.

59. Wang, Y.; Hsu, W.; Guan, Y.; Li, X.; Li, Q.; Liu, Y.; Tang, G. Petrogenesis of the Northwest Africa 4734 basaltic lunar meteorite. *Geochim. Cosmochim. Acta* **2012**, *92*, 329–344. [CrossRef]

60. Al-Kathiri, A.; Gnos, E.; Hofmann, B.A. The regolith portion of the lunar meteorite Sayh al Uhaymir 169. *Meteorit. Planet. Sci.* **2007**, *42*, 2137–2152. [CrossRef]

61. Ashley, J.M.; Robinson, M.S.; Stopar, J.D.; Glotch, T.D.; Hawke, B.R.; van der Bogert, C.H.; Hiesinger, H.; Lawrence, S.J.; Jolliff, B.L.; Greenhagen, B.T.; et al. The Lassell massif-A silicic lunar volcano. *Icarus* **2016**, *273*, 248–261. [CrossRef]

62. Glotch, T.D.; Lucey, P.G.; Bandfield, J.L.; Greenhagen, B.T.; Thomas, I.R.; Elphic, R.C.; Bowles, N.; Wyatt, M.B.; Allen, C.C.; Hanna, K.D.; et al. Highly silicic compositions on the Moon. *Science* **2010**, *329*, 1510–1513. [CrossRef] [PubMed]

63. Hagerty, J.J.; Lawrence, D.J.; Hawke, B.R.; Vaniman, D.T.; Elphic, R.C.; Feldman, W.C. Refined thorium abundances for lunar red spots: Implications for evolved, nonmare volcanism on the Moon. *J. Geophys. Res.* **2006**, *111*, E06002. [CrossRef]

64. Jolliff, B.L.; Wiseman, S.A.; Lawrence, S.J.; Tran, T.N.; Robinson, M.S.; Sato, H.; Hawke, B.R.; Scholten, F.; Oberst, J.; Hiesinger, H.; et al. Non-mare silicic volcanism on the lunar farside at Compton-Belkovich. *Nat. Geosci.* **2011**, *4*, 566–571. [CrossRef]

65. Wilson, J.T.; Eke, V.R.; Massey, R.J.; Elphic, R.C.; Jolliff, B.L.; Lawrence, D.L.; Llewellin, E.W.; McElwaine, J.N.; Teodoro, L.F.A. Evidence for explosive silicic volcanism on the Moon from the extended distribution of thorium near the Compton-Belkovich Volcanic Complex. *J. Geophys. Res.* **2015**, *120*, 92–108. [CrossRef]

66. Roedder, E.; Weiblen, P.W. Lunar petrology of silicate melt inclusions, Apollo 11 rocks. *Geochim. Cosmochim. Acta Suppl.* **1970**, *1*, 801–837.

67. Sharp, T.G.; DeCarli, P.S. Shock effects in meteorites. In *Meteorites and the Early Solar System*; Lauretta, D.S., McSween, H.Y., Eds.; Arizona University Press: Tucson, AZ, USA, 2006; pp. 653–678.

68. Biren, M.B.; Spray, J.G. Shock veins in the central uplift of the Manicouagan impact structure: Context and genesis. *Earth Planet. Sci. Lett.* **2011**, *303*, 310–322. [CrossRef]

69. Koeberl, C.; Kurat, G.; Brandstätter, F. Gabbroic lunar mare meteorites Asuka-881757 (Asuka-31) and Yamato 793169: Geochemical and mineralogical study. *Antarct. Meteor. Res.* **1993**, *6*, 14–34.

70. Arai, T.; Takeda, H.; Warren, P.H. Four lunar meteorites: Crystallization trends of pyroxenes and spinels. *Meteorit. Planet. Sci.* **1996**, *31*, 877–892. [CrossRef]

71. El Goresy, A.; Gillet, P.; Chen, M.; Stähle, V.; Graup, G. In situ finding of the fabric settings of shock-induced quartz/coesite phase transition in crystalline clasts in suevite of the Ries crater, Germany. In Proceedings of the 64th Annual Meteoritical Society Meeting, Vatican City, Italy, 10–14 September 2001.

72. Ling, Z.C.; Wang, A.; Jolliff, B.L. Mineralogy and geochemistry of four lunar soils by laser-Raman study. *Icarus* **2011**, *211*, 101–113. [CrossRef]

73. Misawa, K.; Tatsumoto, M.; Dalrymple, G.B.; Yanai, K. An extremely low U/Pb source in the Moon: U-Th-Pb, Sm-Nd, Rb-Sr, and ^{40}Ar/^{39}Ar isotopic systematics and age of lunar meteorite Asuka 881757. *Geochim. Cosmochim. Acta* **1993**, *57*, 4687–4702. [CrossRef]

74. Hiesinger, H.; Head, J.W., III. New Views of Lunar Geoscience: An Introduction and Overview. In *New Views of the Moon, Reviews in Mineralogy & Geochemistry*; Jolliff, B.L., Wieczorek, M.A., Shearer, C.K., Neal, C.R., Eds.; Mineralogical Society of America: Chantilly, VA, USA, 2006; Volume 60, pp. 1–81.

75. Nishiizumi, K.; Arnold, J.R.; Caffee, M.W.; Finkel, R.C.; Southon, J. Cosmic ray exposure histories of lunar meteorites Asuka 881757, Yamato 793169, and Calcalong Creek. In Proceedings of the 17th Symposium on Antarctic Meteorites, Tokyo, Japan, 19–21 August 1992; pp. 129–132.

76. Hemley, R.J.; Prewitt, C.T.; Kingma, K.J. High-pressure behavior of silica. *Rev. Mineral.* **1994**, *29*, 41–81.

77. Murakami, M.; Hirose, K.; Ono, S.; Ohishi, Y. Stability of CaCl$_2$-type and α-PbO$_2$ at high pressure and temperature determined by in-situ X-ray measurements. *Geophys. Res. Lett.* **2003**, *30*, 1207. [CrossRef]

78. Kuwayama, Y.; Hirose, K.; Sata, N.; Ohishi, Y. The pyrite-type high-pressure form of silica. *Science* **2005**, *309*, 923–925. [CrossRef] [PubMed]

79. Tsuchida, Y.; Yagi, T. New pressure-induced transformations of silica at room temperature. *Nature* **1990**, *347*, 267–269. [CrossRef]

80. Dubrovinsky, L.S.; Saxena, S.K.; Lazor, P.; Ahuja, R.; Eriksson, O.; Wills, J.M.; Johansson, B. Pressure-induced transformations of cristobalite. *Chem. Phys. Lett.* **2001**, *333*, 264–270. [CrossRef]

81. Melosh, H.J. A Geologic Process. In *Impact Cratering*; Oxford University Press: New York, NY, USA, 1989.

82. Blaß, U.W. Shock-induced formation mechanism of seifertite in shergottites. *Phys. Chem. Miner.* **2013**, *40*, 425–437. [CrossRef]

83. Marchi, S.; Bottke, W.F.; Cohen, B.A.; Wuennemann, K.; Kring, D.A.; McSween, H.Y.; De Sanctis, M.C.; O'Brien, D.P.; Schenk, P.; Raymond, C.A.; et al. High-velocity collisions from the lunar cataclysm recorded in asteroidal meteorites. *Nat. Geosci.* **2013**, *6*, 303–307. [CrossRef]

84. Taylor, S.R.; Gorton, M.P.; Muir, P.; Nance, W.; Rudowski, R.; Ware, N. Lunar highlands composition: Apennine Front. *Geochim. Cosmochim. Acta* **1973**, *2*, 1445–1459.

85. McKay, D.S.; Bogard, D.D.; Morris, R.V.; Korotev, R.L.; Wentworth, S.J.; Johnson, P. Apollo 15 regolith breccias: Window to a KREEP regolith. In Proceedings of the 19th Lunar and Planetary Science Conference, Houston, TX, USA, 14–19 March 1989; pp. 19–41.

86. Langenhorst, F.; Poirier, J.P. "Eclogitic" minerals in a shocked basaltic meteorite. *Earth Planet. Sci. Lett.* **2000**, *176*, 259–265. [CrossRef]

87. Langenhorst, F.; Poirier, J.P. Anatomy of black veins in Zagami: Clues to the formation of high-pressure phases. *Earth Planet. Sci. Lett.* **2000**, *184*, 37–55. [CrossRef]

88. Beck, P.; Gillet, P.; Gautron, L.; Daniel, I.; El Goresy, A. A new natural high-pressure (Na, Ca)-hexaluminosilicate [$(Ca_xNa_{1-x})Al_{3+x}Si_{3-x}O_{11}$] in shocked Martian meteorites. *Earth Planet. Sci. Lett.* **2004**, *219*, 1–12. [CrossRef]

89. Miyahara, M.; Ohtani, E.; Yamaguchi, A.; Ozawa, S.; Sakai, T.; Hirao, N. Discovery of coesite and stishovite in eucrite. *Proc. Natl. Acad. Sci. USA* **2014**, *111*, 10939–10942. [CrossRef] [PubMed]

90. Akaogi, M.; Navrotsky, A. The quartz–coesite–stishovite transformations: New calorimetric measurements and calculation of phase diagrams. *Phys. Earth Planet. Inter.* **1984**, *36*, 124–134. [CrossRef]

91. Zhang, J.; Liebermann, R.C.; Gasparik, T.; Herzberg, C.T.; Fei, Y. Melting and subsolidus relations of SiO_2 at 9–14 GPa. *J. Geophys. Res.* **1993**, *98*, 19785–19793. [CrossRef]

92. Duke, M.B.; Gaddis, L.R.; Taylor, G.J.; Schmitt, H.H. Earth-Moon System, Planetary Science, and Lessons Learned. In *New Views of the Moon, Reviews in Mineralogy & Geochemistry*; Jolliff, B.L., Wieczorek, M.A., Shearer, C.K., Neal, C.R., Eds.; Mineralogical Society of America: Chantilly, VA, USA, 2006; pp. 657–704.

93. Sunshine, J.M.; Farnham, T.L.; Feaga, L.M.; Groussin, O.; Merlin, F.; Milliken, R.E.; A'Hearn, M.F. Temporal and spatial variability of lunar hydration as observed by the deep impact spacecraft. *Science* **2009**, *326*, 565–568. [CrossRef] [PubMed]

94. Colaprete, A.; Schultz, P.; Heldmann, J.; Wooden, D.; Shirley, M.; Ennico, K.; Hermalyn, B.; Marshall, W.; Ricco, A.; Elphic, R.C.; et al. Detection of water in the LCROSS ejecta plume. *Science* **2010**, *330*, 463–468. [CrossRef] [PubMed]

95. Schäf, O.; Ghobarkar, H.; Garnier, A.; Vagner, C.; Lindner, J.K.N.; Hanss, J.; Reller, A. Synthesis of nanocrystalline low temperature silica polymorphs. *Solid State Sci.* **2006**, *8*, 625–633. [CrossRef]

96. Kyono, A.; Yokooji, M.; Chiba, T.; Tamura, T.; Tuji, A. Pressure-induced crystallization of biogenic hydrous amorphous silica. *J. Mineral. Petrol. Sci.* **2017**, *112*, 324–335. [CrossRef]

97. Heaney, P.J.; Post, J.E. The widespread distribution of a novel silica polymorph in microcrystalline quartz varieties. *Science* **1992**, *255*, 441–443. [CrossRef] [PubMed]

98. Heaney, P.J. Moganite as an indicator for vanished evaporites: A testament reborn? *J. Sediment. Res.* **1995**, *65*, 633–638.

99. Petrovic, I.; Heaney, P.J.; Navrotsky, A. Thermochemistry of the new silica polymorph moganite. *Phys. Chem. Miner.* **1996**, *23*, 119–126. [CrossRef]

100. Götze, J.; Nasdala, L.; Kleeberg, R.; Wenzel, M. Occurrence and distribution of "moganite" in agate/chalcedony: A combined micro-Raman, Rietveld, and cathodoluminescence study. *Contrib. Mineral. Petrol.* **1998**, *133*, 96–105. [CrossRef]

101. Fernandes, V.A.; Burgess, R.; Turner, G. ^{40}Ar-^{39}Ar chronology of lunar meteorites Northwest Africa 032 and 773. *Meteorit. Planet. Sci.* **2003**, *38*, 555–564. [CrossRef]

102. Nishiizumi, K.; Hillegonds, D.J.; McHargue, L.R.; Jull, A.J.T. Exposure and terrestrial histories of new lunar and martian meteorites. In Proceedings of the 35th Lunar and Planetary Science Conference, League City, TX, USA, 15–19 March 2004.

103. Schorghofer, N.; Taylor, G.J. Subsurface migration of H_2O at lunar cold traps. *J. Geophys. Res.* **2007**, *112*, E02010. [CrossRef]

104. Zuber, M.T.; Head, J.W.; Smith, D.E.; Neumann, G.A.; Mazarico, E.; Torrence, M.H.; Aharonson, O.; Tye, A.R.; Fassett, C.I.; Rosenburg, M.A.; et al. Constraints on the volatile distribution within Shackleton crater at the lunar south pole. *Nature* **2012**, *486*, 378–381. [CrossRef] [PubMed]

105. Meteoritical Bulletin Database. Available online: https://www.lpi.usra.edu/meteor/ (accessed on 15 June 2018).

106. Nyquist, L.E.; Bogard, D.D.; Shih, C.-Y.; Greshake, A.; Stoffler, D.; Eugster, O. Ages and geologic histories of Martian meteorites. *Chronol. Evol. Mars* **2001**, *96*, 105–164.

107. Niihara, T. Uranium-lead age of baddeleyite in shergottite Roberts Massif 04261: Implications for magmatic activity on Mars. *J. Geophys. Res.* **2011**, *116*, E12008. [CrossRef]

108. Fritz, J.; Artemieva, N.; Greshake, A. Ejection of Martian meteorites. *Meteorit. Planet. Sci.* **2005**, *40*, 1393–1411. [CrossRef]

109. Mittlefehldt, D.W. ALH84001, a cumulate orthopyroxenite member of the Martian meteorite clan. *Meteoritics* **1994**, *29*, 214–221. [CrossRef]

110. Romanek, C.S.; Grady, M.M.; Wright, I.P.; Mittlefehldt, D.W.; Socki, R.A.; Pillinger, C.T.; Gibson, E.K. Record of fluid-rock interactions on Mars from the meteorite ALH84001. *Nature* **1994**, *372*, 655–657. [CrossRef] [PubMed]

111. Romanek, C.S.; Thomas, K.L.; Gibson, E.K.; McKay, D.S.; Socki, R.A. Carbon and sulfur-bearing minerals in the Martian meteorite Allan Hills 84001. *Meteoritics* **1995**, *30*, 567–568.

112. Dreibus, G.; Burghele, A.; Jochum, K.P.; Spettel, B.; Wlotzka, F.; Wänke, H. Chemical and mineral composition of ALH84001: A Martian orthopyroxenite. *Meteoritics* **1994**, *29*, 461.

113. Harvey, R.P.; McSween, H.Y. A possible high-temperature origin for the carbonates in Martian meteorite ALH84001. *Nature* **1996**, *382*, 49–51. [CrossRef] [PubMed]

114. McKay, D.S.; Gibson, E.K.; Thomas-Keprta, K.L.; Vali, H.; Romanek, C.S.; Clemett, S.J.; Chillier, X.D.F.; Maechling, C.R.; Zare, R.N. Search for life on Mars: Possible relic biogenic activity in Martian meteorite ALH84001. *Science* **1996**, *273*, 924–930. [CrossRef] [PubMed]

115. McKay, D.S.; Thomas-Keprta, K.L.; Romanek, C.S.; Gibson, E.K.; Vali, H. Evaluating the evidence for past life on Mars: Response. *Science* **1996**, *274*, 2123–2124.

116. Corrigan, C.M.; Harvey, R.P.; Bradley, J. Sodium-bearing pyroxene in ALH 84001. In Proceedings of the Thirty-First Lunar and Planetary Science Conference, Houston, TX, USA, 13–17 March 2000; p. 1762.

117. Corrigan, C.M.; Harvey, R.P. Multi-generational carbonate assemblages in martian meteorite Allan Hills 84001: Implications for nucleation, growth and alteration. *Meteorit. Planet. Sci.* **2004**, *39*, 17–30. [CrossRef]

118. Brearley, A.J. Hydrous phases in ALH84001: Further evidence for preterrestrial alteration and a shock-induced thermal overprint. In Proceedings of the Thirty-First Lunar and Planetary Science Conference, Houston, TX, USA, 13–17 March 2000.

119. Greenwood, J.O.; McSween, H.Y., Jr. Petrogenesis of Allan Hills 84001: Constraints from impact-melted feldspathic and silica glasses. *Meteorit. Planet. Sci.* **2001**, *36*, 43–61. [CrossRef]

120. Melwani Daswani, M.; Schwenzer, S.P.; Reed, M.H.; Wright, I.P.; Grady, M.M. Alteration minerals, fluids, and gases on early Mars: Predictions from 1-D flow geochemical modeling of mineral assemblages in meteorite ALH 84001. *Meteorit. Planet. Sci.* **2016**, *51*, 2154–2174. [CrossRef]

121. Scott, E.R.D.; Yamaguchi, A.; Krot, A.N. Petrological evidence for shock melting of carbonates in the Martian meteorite ALH84001. *Nature* **1997**, *387*, 377–379. [CrossRef] [PubMed]

122. Turner, G.; Knott, S.F.; Ash, R.D.; Gilmour, J.D. Ar-Ar chronology of the Martian meteorite ALH84001: Evidence for the timing of the early bombardment of Mars. *Geochim. Cosmochim. Acta* **1997**, *61*, 3835–3850. [CrossRef]

123. Valley, J.W.; Eiler, J.M.; Graham, C.M.; Gibson, E.K.; Romanek, C.S.; Stolper, E.M. Low-temperature carbonate concretions in the Martian meteorite ALH84001: Evidence from stable isotopes and mineralogy. *Science* **1997**, *275*, 1633–1637. [CrossRef] [PubMed]

124. Cooney, T.F.; Scott, E.R.D.; Krot, A.N.; Sharma, S.K.; Yamaguchi, A. Vibrational spectroscopic study of minerals in the Martian meteorite ALH84001. *Am. Mineral.* **1999**, *84*, 1569–1576. [CrossRef]

125. White, L.M.; Gibson, E.K.; Thomas-Keprta, K.L.; Clemett, S.J.; McKay, D.S. Putative indigenous carbon-bearing-alteration features in Martian meteorite Yamato 000593. *Astrobiology* **2014**, *14*, 170–181. [CrossRef] [PubMed]

126. Imae, N.; Ikeda, Y.; Shinoda, K.; Kojima, H.; Iwata, N. Yamato Nakhlites: Petrography and mineralogy. *Antarct. Meteor. Res.* **2003**, *16*, 13–33.

127. Symes, S.J.; Borg, L.E.; Shearer, C.K.; Irving, A.J. The age of the Martian meteorite Northwest Africa 1195 and the differentiation history of the Shergottites. *Geochim. Cosmochim. Acta* **2008**, *72*, 1696–1710. [CrossRef]

128. Herd, C.D.K.; Walton, E.L.; Agee, C.B.; Muttik, N.; Ziegler, K.; Shearer, C.K.; Bell, A.S.; Santos, A.R.; Burger, P.V.; Simon, J.I.; et al. The Northwest Africa 8159 martian meteorite: Expanding the martian meteorite suite to the Amazonian. *Geochim. Cosmochim. Acta* **2017**, *218*, 1–26. [CrossRef]

129. Smith, J.V.; Hervig, R.L. Shergotty meteorite: Mineralogy, petrography, and minor elements. *Meteoritics* **1979**, *14*, 121–142. [CrossRef]

130. McSween, H.Y.; Taylor, L.A.; Stolper, E.M. Allan Hills 77005: A new meteorite type found in Antarctica. *Science* **1979**, *204*, 1201–1203. [CrossRef] [PubMed]

131. Ikeda, Y. Petrology of magmatic silicate inclusions in the Allan Hills 77005 Lherzolitic Shergottite. *Meteorit. Planet. Sci.* **1998**, *33*, 803–812. [CrossRef]

132. Malavergne, V.; Guyot, F.; Benzerara, K.; Martinez, I. Description of new shock-induced phases in the Shergotty, Zagami, Nakhla and Chassigny meteorites. *Meteorit. Planet. Sci.* **2001**, *36*, 1297–1305. [CrossRef]

133. Barrat, J.A.; Gillet, P.; Sautter, V.; Jambon, A.; Javoy, M.; Göpel, C.; Lesourd, M.; Keller, F.; Petit, E. Petrology and geochemistry of the basaltic Shergottite Northwest Africa 480. *Meteorit. Planet. Sci.* **2002**, *37*, 487–499. [CrossRef]

134. Jambon, A.; Barrat, J.-A.; Sautter, V.; Gillet, P.; Göpel, C.; Javoy, M.; Joron, J.-L.; Lesourd, M. The basaltic Shergottite Northwest Africa 856: Petrology and chemistry. *Meteorit. Planet. Sci.* **2002**, *37*, 1147–1164. [CrossRef]

135. Taylor, L.A.; Nazarov, M.A.; Shearer, C.K.; McSween, H.Y., Jr.; Cahill, J.; Neal, C.R.; Icanova, M.A.; Barsukova, L.D.; Lentz, R.C.; Clayton, R.N.; et al. Martian meteorite Dhofar 019: A new Shergottite. *Meteorit. Planet. Sci.* **2002**, *37*, 1107–1128. [CrossRef]

136. Mikouchi, T.; Barrat, J.A. NWA 5029 Basaltic Shergottite: A Clone of NWA 480/1460? In Proceedings of the 72nd Annual Meeting of the Meteoritical, Nancy, France, 13–18 July 2009.

137. Howarth, G.H.; Udry, A.; Day, J.M.D. Petrogenesis of basaltic shergottite Northwest Africa 8657: Implications for fO$_2$ correlations and element redistribution during shock melting in shergottites. *Meteorit. Planet. Sci.* **2018**, *53*, 249–267. [CrossRef]

138. Irving, A.J.; Kuehner, S.M. Northwest Africa 5298: A strongly shocked basaltic Shergottite equilibrated at QFM and high temperature. *Meteorit. Planet. Sci.* **2008**, *43*, A63.

139. Hui, H.; Peslier, A.; Lapan, T.; Schafer, J.; Brandon, A.; Irving, A.J. Petrogenesis of basaltic Shergottite Northwest Africa 5298: Closed system crystallization of an oxidized mafic melt. *Meteorit. Planet. Sci.* **2011**, *46*, 1313–1328. [CrossRef]

140. Mikouchi, T. Mineralogical similarities and differences between the Los Angeles basaltic shergottite and the Asuka-881757 lunar mare meteorite. *Antarct. Meteor. Res.* **2001**, *14*, 1–20.

141. Chennaoui Aoudjehane, H.; Jambon, A.; Reynard, B.; Blanc, P. Silica as a shock index in Shergottites: A catholuminescence study. *Meteorit. Planet. Sci.* **2005**, *40*, 1–14.

142. Stöffler, D.; Ostertag, R.; Jammes, C.; Pfannschmidt, G.; Sen Gupta, P.R.; Simon, S.B.; Papike, J.J.; Beauchamp, R.H. Shock metamorphism and petrography of the Shergotty achondrite. *Geochim. Cosmochim. Acta* **1986**, *50*, 889–913. [CrossRef]

143. El Goresy, A.; Gillet, P.; Miyahara, M.; Ohtani, E.; Ozawa, S.; Beck, P.; Montagnac, G. Shock-induced deformation of Shergottites: Shock–pressures and perturbations of magmatic ages on Mars. *Geochim. Cosmochim. Acta* **2013**, *101*, 233–262. [CrossRef]

144. Ikeda, Y.; Kimura, M.; Takeda, H.; Shimoda, G.; Kita, N.T.; Morishita, Y.; Suzuki, A.; Jagoutz, E.; Dreibus, G. Petrology of a new basaltic Shergottite: Dhofar 378. *Antarct. Meteor. Res.* **2006**, *19*, 20–44.

145. Hu, S.; Lin, Y.T.; Zhang, T.; Gu, L.X.; Tang, X. Discovery of first coesite in the martian meteorite Northwest Africa 8675. In Proceedings of the 80th Annual Meeting of the Meteoritical Society, Santa Fe, NM, USA, 23–28 July 2017.

146. McCoy, T.J.; Lofgren, G.E. Crystallization of the Zagami Shergottite: An experimental study. *Earth Planet. Sci. Lett.* **1999**, *173*, 397–411. [CrossRef]

147. Niihara, T.; Misawa, K.; Mikouchi, T.; Nyquist, L.E.; Park, J.; Yamashita, H.; Hirata, D. Complex formation history of highly evolved basaltic shergottite, Zagami. In Proceedings of the 75th Annual Meeting of the Meteoritical, Cairns, Australia, 12–17 August 2012.

148. Rubin, A.E.; Warren, P.H.; Greenwood, J.P.; Verish, R.S.; Leshin, L.A.; Hervig, R.L.; Clayton, R.N.; Mayeda, T.K. Los Angeles: The most differentiated basaltic Martian meteorite. *Geology* **2000**, *28*, 1011–1014. [CrossRef]

149. Warren, P.H.; Greenwood, J.P.; Rubin, A.E. Los Angeles: A tale of two stones. *Meteorit. Planet. Sci.* **2004**, *39*, 137–156. [CrossRef]

150. Rost, D.; Stephan, T.; Geshake, A.; Fritz, J.; Jesseberger, E.K.; Weber, I.; Stoffler, D. A combined TOF-SIMS, EMP/SEM of a three-phase symplectite in the Los Angeles basaltic Shergottite. *Meteorit. Planet. Sci.* **2009**, *44*, 1225–1237. [CrossRef]

151. Aramovich, C.J.; Herd, C.D.K.; Papike, J.J. Symplectites derived from metastable phases in Martian basaltic meteorites. *Am. Mineral.* **2002**, *87*, 1351–1359. [CrossRef]

152. Roszjar, J.; Bishoff, A.; Llorca, J.; Pack, A. Ksar Gilane 002 (KG002)—A new Shergottite: Discovery, mineralogy, chemistry and oxygen isotopes. In Proceedings of the 43nd Lunar and Planetary Science Conference, The Woodlands, TX, USA, 19–23 March 2012.

153. Llorca, J.; Roszjar, J.; Cartwright, J.A.; Bischoff, A.; Ott, U.; Pack, A.; Merchel, S.; Rugel, G.; Fimiani, L.; Ludwig, P.; et al. The Ksar Ghilane 002 shergottite—The 100th registered Martian meteorite fragment. *Meteorit. Planet. Sci.* **2013**, *48*, 493–513. [CrossRef]

154. Bunch, T.E.; Irving, A.J.; Wittke, J.H.; Kuehner, S.M. Highly evolved basaltic Shergottite Northwest Africa 2800: A clone of Los Angeles. In Proceedings of the Thirty-Ninth Lunar and Planetary Science Conference, League City, TX, USA, 10–14 March 2008.

155. Udry, A.; Howarth, G.H.; Lapen, T.J.; Righter, M. Petrogenesis of the NWA 7320 enriched martian gabbroic shergottite: Insight into the martian crust. *Geochim. Cosmochim. Acta* **2017**, *204*, 1–18. [CrossRef]

156. Lindsley, D.H. Pyroxene thermometry. *Am. Mineral.* **1983**, *68*, 477–493.

157. Agee, C.B.; Wilson, N.V.; McCubbin, F.M.; Ziegler, K.; Polyak, V.J.; Sharp, Z.D.; Asmerom, Y.; Nunn, M.H.; Shaheen, R.; Thiemens, M.H.; et al. Unique meteorite from early Amazonian Mars: Water-rich basaltic breccia Northwest Africa 7034. *Science* **2013**, *339*, 780–785. [CrossRef] [PubMed]

158. Beck, P.; Pommerol, A.; Zanda, B.; Remusat, L.; Lorand, J.P.; Göpel, C.; Hewins, R.; Pont, S.; Lewin, E.; Quirico, E.; et al. A Noachian source region for the "Black Beauty" meteorite, and a source lithology for Mars surface hydrated dust? *Earth Planet. Sci. Lett.* **2015**, *427*, 104–111. [CrossRef]

159. Ruff, S.W.; Farmer, J.D. Silica deposits in Mars with features resembling hot spring biosignatures at El Tatio in Chile. *Nat. Commun.* **2016**, *7*, 13554. [CrossRef] [PubMed]

160. Morris, R.V.; Vaniman, D.T.; Blake, D.F.; Gellert, R.; Chipera, S.J.; Rampe, E.B.; Ming, D.W.; Morrison, S.M.; Downs, R.T.; Treimann, A.H.; et al. Silicic volcanism on Mars evidenced by tridymite in high-SiO$_2$ sedimentary rock at Gale crater. *Proc. Natl. Acad. Sci. USA* **2016**, *113*, 7071–7076. [CrossRef] [PubMed]

161. Ma, C.; Tschauner, O.; Becett, J.R.; Liu, Y.; Rossman, G.R.; Zhuravlev, K.; Prakapenka, V.; Dera, P.; Taylor, L.A. Tissintite, $(Ca, Na, \Box)AlSi_2O_6$, a highly-defective, shock-induced, high-pressure clinopyroxene in the Tissint martian meteorite. *Earth Planet. Sci. Lett.* **2015**, *422*, 194–205. [CrossRef]

162. Teter, D.M.; Hemley, R.J.; Kresse, G.; Hafner, J. High pressure polymorphism in silica. *Phys. Rev. Lett.* **1998**, *80*, 2145–2148. [CrossRef]

163. Kingma, K.; Cohen, R.E.; Hemley, R.J.; Mao, H.-K. Transformation of stishovite to a denser phase at lower-mantle pressures. *Nature* **1995**, *374*, 243–245. [CrossRef]

164. Dubrovinsky, L.S.; Saxena, S.K.; Lazor, P.; Ahuja, R.; Eriksson, O.; Wills, J.M.; Johansson, B. Experimental and theoretical identification of a new high-pressure phase of silica. *Nature* **1997**, *388*, 362–365. [CrossRef]

165. Ma, C.; Tschauner, O. A new high-pressure calcium aluminosilicate $(CaAl_2Si_{3.5}O_{11})$ in martian meteorites: Another after-life for plagioclase and connections to the CAS phase. In Proceedings of the 48th Lunar and Planetary Science Conference, The Woodlands, TX, USA, 20–24 March 2017; p. 1128.

166. Baziotis, I.P.; Yang, L.; Paul, S.; DeCarli, H.; Melosh, J.; McSween, H.Y.; Bodnar, R.J.; Taylor, L.A. The Tissint Martian meteorite as evidence for the largest impact excavation. *Nat. Commun.* **2013**, *4*, 1404. [CrossRef] [PubMed]

167. Walton, E.L.; Sharp, T.G.; Hu, J.; Filiberto, J. Heterogeneous mineral assemblages in Martian meteorite Tissint as a result of a recent small impact event on Mars. *Geochim. Cosmochim. Acta* **2014**, *140*, 334–348. [CrossRef]

168. Walton, E.L.; Spray, J.G. Mineralogy, microtexture, and composition of shock-induced melt pockets in the Los Angeles basaltic shergottite. *Meteorit. Planet. Sci.* **2003**, *38*, 1865–1875. [CrossRef]

169. He, Q.; Xiao, L.; Balta, J.B.; Baziotis, I.P.; Hsu, W.; Guan, Y. Petrography and geochemistry of the enriched basaltic shergottite Northwest Africa 2975. *Meteorit. Planet. Sci.* **2015**, *50*, 2024–2044. [CrossRef]

170. Boonsue, S.; Spray, J. Shock-induced phase transformations in melt pockets within Martian meteorite NWA 4468. *Spectrosc. Lett.* **2012**, *45*, 127–134. [CrossRef]

171. Ashworth, J.R.; Hutchison, R. Water in noncarbonaceous stony meteorites. *Nature* **1975**, *256*, 714–715. [CrossRef]

172. Treiman, A.H.; Barrett, R.A.; Gooding, J.L. Preterrestrial aqueous alteration of the Lafayette (SNC) meteorite. *Meteoritics* **1993**, *28*, 86–97. [CrossRef]

173. Treiman, A.H.; Lindstrom, D.J. Trace element geochemistry of Martian iddingsite in the Lafayette meteorite. *J. Geophys. Res.* **1997**, *102*, 9153–9163. [CrossRef]

174. Bridges, J.C.; Schwenzer, S.P. The nakhlite hydrothermal brine on Mars. *Earth Planet. Sci. Lett.* **2012**, *359–360*, 117–123. [CrossRef]

175. Tomkinson, T.; Lee, M.R.; Mark, D.F.; Smith, C.L. Sequestration of Martian CO_2 by mineral carbonation. *Nat. Commun.* **2013**, *4*, 2662. [CrossRef] [PubMed]

176. Gooding, J.L.; Wentworth, S.J.; Zolensky, M.E. Aqueous alteration of the Nakhla meteorite. *Meteoritics* **1991**, *26*, 135–143. [CrossRef]

177. Gillet, P.; Barrat, J.A.; Deloule, E.; Wadhwa, M.; Jambon, A.; Sautter, V.; Devouard, B.; Neuville, D.; Benzerara, K.; Lesourd, M. Aqueous alteration in the Northwest Africa 817 (NWA 817) Martian meteorite. *Earth Planet. Sci. Lett.* **2002**, *203*, 431–444. [CrossRef]

178. Day, J.M.D.; Taylor, L.A.; Floss, C.; McSween, H.Y. Petrology and chemistry of MIL 03346 and its significance in understanding the petrogenesis of nakhlites on Mars. *Meteorit. Planet. Sci.* **2006**, *41*, 581–606. [CrossRef]

179. Treiman, A.H.; Irving, A.J. Petrology of Martian meteorite Northwest Africa 998. *Meteorit. Planet. Sci.* **2008**, *43*, 829–854. [CrossRef]

180. Noguchi, T.; Nakamura, T.; Misawa, K.; Imae, N.; Aoki, T.; Toh, S. Laihunite and jarosite in the Yamato 00 nakhlites: Alteration products on Mars? *J. Geophys. Res.* **2009**, *114*, E10004. [CrossRef]

181. Thomas-Keprta, K.L.; Wentworth, S.J.; McKay, D.S.; Gibson, E.K. Field emission gun scanning electron (FEGSEM) and transmison electron (TEM) microscopy of phyllosilicates in Martian meteorites ALH 84001, Nakhla, and Shergotty. In Proceedings of the 31st Lunar and Planetary Science Conference, Houston, TX, USA, 13–17 March 2000.

182. Morris, R.V.; Klingelhofer, G.; Schroder, C.; Rodionov, D.S.; Yen, A.; Ming, D.W.; de Souza, P.A.; Fleischer, I.; Wdowiak, T.; Gellert, R.; et al. Mössbauer mineralogy of rock, soil, and dust at Gusev crater, Mars: Spirit's journey through weakly altered olivine basalt on the plains and pervasively altered basalt in the Columbia Hills. *J. Geophys. Res.* **2006**, *111*, E02S13. [CrossRef]

183. McLennan, S.M. Sedimentary silica on Mars. *Geology* **2003**, *31*, 315–318. [CrossRef]

184. Baker, L.L.; Agenbroad, D.J.; Wood, S.A. Experimental hydrothermal alteration of a Martian analog basalt: Implications for Martian meteorites. *Meteorit. Planet. Sci.* **2000**, *35*, 31–38. [CrossRef]

185. Tosca, N.J.; McLennan, S.M.; Lindsley, D.H.; Schoonen, M.A.A. Acid-sulfate weathering of synthetic Martian basalt: The acid fog model revisited. *J. Geophys. Res.* **2004**, *109*, E050003. [CrossRef]

186. Bandfield, J.L. High-silica deposits of an aqueous origin in western Hellas Basin, Mars. *Geophys. Res. Lett.* **2008**, *35*, 142–147. [CrossRef]

187. Mustard, J.F.; Murchie, S.L.; Pelkey, S.M.; Ehlmann, B.L.; Milliken, R.E.; Grant, J.A.; Bibring, J.-P.; Poulet, F.; Bishop, J.; Dobrea, E.N.; et al. Hydrated silicate minerals on mars observed by the Mars reconnaissance orbiter CRISM instrument. *Nature* **2008**, *454*, 305–309. [CrossRef] [PubMed]

188. Ehlmann, B.L.; Mustard, J.F.; Swayze, G.A.; Clark, R.N.; Bishop, J.L.; Poulet, F.; Des Marais, D.J.; Roach, L.H.; Milliken, R.E.; Wray, J.J.; et al. Identification of hydrated silicate minerals on Mars using MRO-CRISM: Geologic context near Nili Fossae and implications for aqueous alteration. *J. Geophys. Res.* **2009**, *114*, E00D08. [CrossRef]

189. Harvey, R.P.; Hamilton, V.E. Syrtis Major as the source region of the nakhlite/chassignite group of Martian meteorites: Implications for the geological history of Mars. In Proceedings of the Thirty-Sixth Lunar and Planetary Science Conference, League City, TX, USA, 14–18 March 2005.

MDPI

St. Alban-Anlage 66

4052 Basel

Switzerland

Tel. +41 61 683 77 34

Fax +41 61 302 89 18

www.mdpi.com

Minerals Editorial Office

E-mail: minerals@mdpi.com

www.mdpi.com/journal/minerals

* 9 7 8 3 0 3 8 9 7 3 4 8 5 *